Industrial and Municipal Wastewater Treatment with a Focus on Water-Reuse

Industrial and Municipal Wastewater Treatment with a Focus on Water-Reuse

Editors

Martin Wagner
Sonja Bauer

MDPI • Basel • Beijing • Wuhan • Barcelona • Belgrade • Manchester • Tokyo • Cluj • Tianjin

Editors

Martin Wagner
Department of Wastewater
Technology
Technische Universität
Darmstadt
Germany

Sonja Bauer
Department of Electrical
Engineering, Media and
Computer Science
Ostbayerische Technische
Hochschule Amberg-Weiden
Amberg
Germany

Editorial Office
MDPI
St. Alban-Anlage 66
4052 Basel, Switzerland

This is a reprint of articles from the Special Issue published online in the open access journal *Water* (ISSN 2073-4441) (available at: www.mdpi.com/journal/water/special_issues/industrial_municipal).

For citation purposes, cite each article independently as indicated on the article page online and as indicated below:

LastName, A.A.; LastName, B.B.; LastName, C.C. Article Title. *Journal Name* **Year**, *Volume Number*, Page Range.

ISBN 978-3-0365-6256-8 (Hbk)
ISBN 978-3-0365-6255-1 (PDF)

© 2023 by the authors. Articles in this book are Open Access and distributed under the Creative Commons Attribution (CC BY) license, which allows users to download, copy and build upon published articles, as long as the author and publisher are properly credited, which ensures maximum dissemination and a wider impact of our publications.

The book as a whole is distributed by MDPI under the terms and conditions of the Creative Commons license CC BY-NC-ND.

Contents

About the Editors vii

Preface to "Industrial and Municipal Wastewater Treatment with a Focus on Water-Reuse" . . ix

Sonja Bauer and Martin Wagner
Possibilities and Challenges of Wastewater Reuse—Planning Aspects and Realized Examples
Reprinted from: *Water* 2022, 14, 1619, doi:10.3390/w14101619 1

Martin Zimmermann and Felix Neu
Social–Ecological Impact Assessment and Success Factors of a Water Reuse System for Irrigation
Purposes in Central Northern Namibia
Reprinted from: *Water* 2022, 14, 2381, doi:10.3390/w14152381 13

José de Anda and Harvey Shear
Sustainable Wastewater Management to Reduce Freshwater Contamination and Water
Depletion in Mexico
Reprinted from: *Water* 2021, 13, 2307, doi:10.3390/w13162307 35

Hana D. Dawoud, Haleema Saleem, Nasser Abdullah Alnuaimi and Syed Javaid Zaidi
Characterization and Treatment Technologies Applied for Produced Water in Qatar
Reprinted from: *Water* 2021, 13, 3573, doi:10.3390/w13243573 55

Ammar A. Albalasmeh, Ma'in Z. Alghzawi, Mamoun A. Gharaibeh and Osama Mohawesh
Assessment of the Effect of Irrigation with Treated Wastewater on Soil Properties and on the
Performance of Infiltration Models
Reprinted from: *Water* 2022, 14, 1520, doi:10.3390/w14091520 95

Arabel Amann, Nikolaus Weber, Jörg Krampe, Helmut Rechberger, Ottavia Zoboli and
Matthias Zessner
Operation and Performance of Austrian Wastewater and Sewage Sludge Treatment as a Basis
for Resource Optimization
Reprinted from: *Water* 2021, 13, 2998, doi:10.3390/w13212998 107

Weeraya Intaraburt, Jatuwat Sangsanont, Tawan Limpiyakorn, Piyatida Ruangrassamee,
Pongsak Suttinon and Benjaporn Boonchayaanant Suwannasilp
Feasibility Study of Water Reclamation Projects in Industrial Parks Incorporating
Environmental Benefits: A Case Study in Chonburi, Thailand
Reprinted from: *Water* 2022, 14, 1172, doi:10.3390/w14071172 123

Lam T. Phan, Heidemarie Schaar, Daniela Reif, Sascha Weilguni, Ernis Saracevic and Jörg
Krampe et al.
Long-Term Toxicological Monitoring of a Multibarrier Advanced Wastewater Treatment Plant
Comprising Ozonation and Granular Activated Carbon with In Vitro Bioassays
Reprinted from: *Water* 2021, 13, 3245, doi:10.3390/w13223245 143

Hiroshi Noguchi, Qiang Yin, Su Chin Lee, Tao Xia, Terutake Niwa and Winson Lay et al.
Performance of Newly Developed Intermittent Aerator for Flat-Sheet Ceramic Membrane in
Industrial MBR System
Reprinted from: *Water* 2022, 14, 2286, doi:10.3390/w14152286 159

Cláudia Pinto, Annabel Fernandes, Ana Lopes, Maria João Nunes, Ana Baía and Lurdes
Ciríaco et al.
Reuse of Textile Dyeing Wastewater Treated by Electrooxidation
Reprinted from: *Water* 2022, 14, 1084, doi:10.3390/w14071084 169

Mohamed H. EL-Saeid, Modhi O. Alotaibi, Mashael Alshabanat, Khadiga Alharbi, Abeer S. Altowyan and Murefah Al-Anazy
Photo-Catalytic Remediation of Pesticides in Wastewater Using UV/TiO$_2$
Reprinted from: *Water* **2021**, *13*, 3080, doi:10.3390/w13213080 . . . . **185**

Ilil Levakov, Yuval Shahar and Giora Rytwo
Carbamazepine Removal by Clay-Based Materials Using Adsorption and Photodegradation
Reprinted from: *Water* **2022**, *14*, 2047, doi:10.3390/w14132047 . . . . **197**

Mongi ben Mosbah, Abdulmohsen Khalaf Dhahi Alsukaibi, Lassaad Mechi, Fathi Alimi and Younes Moussaoui
Ecological Synthesis of CuO Nanoparticles Using *Punica granatum* L. Peel Extract for the Retention of Methyl Green
Reprinted from: *Water* **2022**, *14*, 1509, doi:10.3390/w14091509 . . . . **215**

Isabel Schestak, Jan Spriet, David Styles and A. Prysor Williams
Introducing a Calculator for the Environmental and Financial Potential of Drain Water Heat Recovery in Commercial Kitchens
Reprinted from: *Water* **2021**, *13*, 3486, doi:10.3390/w13243486 . . . . **231**

About the Editors

Martin Wagner

Prof. Dr.-Ing. habil. Martin Wagner graduated from the Technical University Darmstadt, Germany, in 1983 and holds a doctoral degree as Civil Engineer. Since 1996, he is General Manager of the Institute IWAR, Technical University Darmstadt. In 2006, he was nominated as Professor Honorarium at the Qingdao Technological University, China. Since 2009, he is Head of the Section China of German Water Partnership. In 2014, he received the certificate for Honorary Professorship from the Tongji University Shanghai. His expertise focuses on waste water treatment, aeration and gas transfer, energy in wastewater treatment plants and water infrastructure solutions for fast growing urban areas.

Prof. Wagner was honored with the "White Magnolia" of the Government of Shanghai, the top award for foreigners. Another award is the Qilu-laurate, the highest award for foreigners of Shangdong Province (China).

Sonja Bauer

Prof. Dr.-Ing. Sonja Bauer studied spatial planning at the Technische Universität Dortmund. This was followed by activities in a private planning office for urban development as well as with a municipality in the field of urban planning and urban development. From 2012 to 2019, Ms. Bauer worked as a research assistant at the Department of Land Management at the Technische Universität Darmstadt, where she received her PhD in 2016. From September 2019 until the end of August 2022, she was a professor for Land Management at Hochschule für Technik Stuttgart. As of September 01, 2022, Ms. Bauer was appointed to the Ostbayerische Technische Hochschule Amberg-Weiden to represent the professorship of Land Management. Her research focus is on water reuse for sustainable spatial development.

Preface to "Industrial and Municipal Wastewater Treatment with a Focus on Water-Reuse"

Population growth, climate change, but also rising prosperity are leading to increasing global water scarcity. Water shortages are thus hindering rural, urban and industrial development. Nowadays, approximately half of the world's population is affected temporarily by water scarcity. To enable a secure water supply, alternative water sources must be generated to tackle the challenge of water scarcity since ground and surface water sources are often overexploited. An important alternative resource is the reuse of treated wastewater. These days, water reuse processes are rarely considered and implemented. In contrast to the storage and use of rainwater, treated wastewater is a valuable resource, as it is available daily and in calculable quantities. Certain wastewater treatment processes within wastewater treatment plants are required to produce the new resource "reused water". The treatment processes depend on the quality of the wastewater inflow to the treatment plant since industrial and municipal wastewater flows are characterized, for example, by different concentrations. Moreover, water reuse methods must be developed in order to use the treated wastewater as efficiently as possible. Ideally, the reused water can be provided according to the "fit for purpose" principle and applied directly in areas such as irrigation, street cleaning, toilet flushing or make-up water for cooling systems.

The Special Issue "Industrial and Municipal Wastewater Treatment with a Focus on Water-Reuse" brings together new wastewater treatment technologies and emerging water reuse concepts to tackle the challenges of climate change with the aim of bringing the new resource "reused water" according to the "fit for purpose" principle to the subsequent user. This issue aims to draw on global experiences, approaches and solutions in the field of wastewater treatment and water reuse in rural, urban and industrial (park) areas. This allows a specific transferability to regions which have been subject to the long-term effects of climate change.

This book is focusing on the sections:

1. General challenges in the field of water scarcity, wastewater treatment and water reuse.

2. General approaches, processes, management and technologies in the field of municipal and industrial wastewater treatment and reuse.

3. Specific technologies for municipal and industrial wastewater treatment.

Regarding the first thematic part of the book the contribution of the Guest Editors Wagner and Bauer makes the prelude, in which different possibilities and challenges of the wastewater treatment and water reuse according to the status quo are presented. In addition, future needs and required innovations are also pointed out. The second paper by Zimmermann and Neu further addresses the challenges of water scarcity by providing a socio-ecological impact assessment and success factors of a water reuse system for irrigation purposes in central-northern Namibia. In order to achieve the best possible water quality from wastewater treatment plants, on the one hand to minimize the impact on surface waters, but also to prepare the water in the best possible way for water reuse, sustainable wastewater management is required. The article by de Anda and Shear presents therefore a sustainable wastewater management to reduce the freshwater contamination and water depletion in Mexico.

Concerning general approaches, processes, management and technologies in the field of municipal and industrial wastewater treatment and reuse the paper by Dawoud et al. deals with the optimization of wastewater treatment technologies in Qatar in order to reuse the wastewater, which is produced in high quantities in the industry, for irrigation purposes, among other things. Here, the analysis of the resulting sewage sludge also plays an important role. In the context of reusing treated wastewater, Albalasmeh et al. are focusing on the assessment of the effect of irrigation with treated wastewater on soil properties and on the performance of infiltration models. The contribution of Amann et al. extends the topic of wastewater treatment technologies by focusing not only on the removal of pollutants, but also on the recovery of resources from wastewater. In particular, the paper deals with operation and performance of Austrian wastewater and sewage sludge treatment as a basis for resource optimization. Beside all environmental aspects the financial feasibility is usually a concern in water reuse projects. Here, the paper by Intaraburt et al. investigated the influence of environmental benefits on the feasibility of water reclamation projects in industrial parks and focused on the case study in Chonburi, Thailand.

The third part of the book is focusing on specific technologies for municipal and also for industrial wastewater treatment. This section starts with Phan et al. which deals with Long-Term Toxicological Monitoring of a Multibarrier Advanced Wastewater Treatment Plant Comprising Ozonation and Granular Activated Carbon with In Vitro Bioassays. Nogushi et al. follow with Performance of Newly Developed Intermittent Aerator for Flat-Sheet Ceramic Membrane in Industrial MBR Systems. Reuse of Textile Dyeing Wastewater Treated by Electrooxidation is described by Pinto et al., followed by EL-Saeid et al. with the topic of Photo-Catalytic Remediation of Pesticides in Wastewater Using UV/TiO_2. The removal of Carbamazepine by Clay-Based Materials Using Adsorption and Photodegradation is discussed by Levakov et al. Mosbah et al. are dealing with Ecological Synthesis of CuO Nanoparticles Using Punica granatum L. Peel Extract for the Retention of Methyl Green. Finally, Schestak et al. introduce a Calculator for the Environmental and Financial Potential of Drain Water Heat Recovery in Commercial Kitchens.

Martin Wagner and Sonja Bauer
Editors

Perspective

Possibilities and Challenges of Wastewater Reuse—Planning Aspects and Realized Examples

Sonja Bauer [1,*] and Martin Wagner [2]

[1] Faculty of Geomatics, Computer Science and Mathematics, Hochschule für Technik Stuttgart, 70174 Stuttgart, Germany
[2] Department of Wastewater Technology, Technische Universität Darmstadt, 64287 Darmstadt, Germany; m.wagner@iwar.tu-darmstadt.de
* Correspondence: sonja.bauer@hft-stuttgart.de

Abstract: Population growth and climate change has a huge impact on water availability. To ensure a secure water supply, water-reuse concepts and its implementation are gaining more and more importance. Additionally, water saving potentials to optimize the drinking and water reuse availability have to be considered. However, limited spatial planning opportunities and missing regulation to provide treated wastewater according to the "fit-for-purpose" principle are often hindering its application. Some countries, such as the USA or Singapore, have been leading the way for decades in implementing water-reuse concepts and in treating wastewater for potable and non-potable reuse. The wastewater treatment technologies are currently providing solutions for an adequate provision of reclaimed water. Consequently, the opportunities for water reuse are given, but the challenge is largely in the implementation, which becomes necessary in water-scarce regions. This perspective is thus presenting the current possibilities and challenges of wastewater reuse with respect to existing examples of implementations but also shows the need for action in the future. The relevance of this topic is also underlined in particular by the Sustainable Development Goals (SDG), especially Goal 6 which is related to "Ensure availability and sustainable management of water and sanitation for all".

Keywords: wastewater treatment; spatial planning; water-reuse concepts

1. Introduction

1.1. The Impact of Population Growth and Climate Change on the Limited Resource Water

The growing population has a huge impact on the global water demand. The limited resources of water and land are increasingly being used, for example, for increased food production. Regarding this, it is assumed that global water demand will increase by about 1% per year until 2050 [1]. Nowadays, about half of the population worldwide is already temporarily affected by water scarcity [2]. Climate change further aggravates the situation since periods of drought are increasing. The total water demand is rising, especially for irrigating field crops. Therefore, water from natural resources such as ground- and surface water is often (over) used. Besides the higher food demand, both urban and industrial development is required to meet the challenges of the increasing population.

A well-known example for water scarcity is the U.S. State of California. It has to deal with the challenge of increasingly long drought periods [3]. A further example is China where huge regions particular in the north and the west have a high water-stress level due to the uneven distribution of natural water resources and the high levels of pollution of water bodies [4]. However, nowadays, countries, such as Germany, which are not well known for water shortages, are also increasingly dealing with this challenge. In particular, in 2018 and 2019, there were pronounced conflicts over the use of water as a resource in many regions of Germany. Here, conflicts have arisen over the use of water as a resource. Ground- and surface water were increasingly used for agricultural irrigation, but also for cooling and process water in the energy and manufacturing sectors, for public drinking water supplies and for shipping [5]. For example, the conflicts of use, caused by the low water level of the river Rhine, led to a reduction in production at the well-known worldwide operating chemical company BASF [6].

1.2. The Necessity to Ensure a Susainable Water Supply

The situations worldwide show that, especially in times of climate change, it is increasingly important to ensure a secure water supply. Only this way will it be possible to enable sustainable spatial development to meet the challenge of population growth and to secure human needs. For enabling a secure water supply, alternative water sources must be generated to tackle the challenge of increased water scarcity since ground- and surface water sources are frequently insufficient in many regions of the world. Currently, as in the past, new developments often take into account approaches to the storage of rainwater in order to use it for various purposes such as the irrigation of green spaces. However, this water source is uncertain because the quantity of required rainwater at a specific time is not available. Furthermore, storage tanks need a lot of space; however, in cities this space is limited and stored rainwater is definitely not enough to compensate water shortages.

Another alternative and sustainable resource is the reuse of treated wastewater. In contrast to the storage and use of rainwater, treated wastewater is a valuable resource, as it is available daily and in calculable quantities. Nowadays, concepts for water reuse are still rarely implemented [7]. In general, a distinction must be made between different fields of application for reused water: urban and industrial reuse, agricultural reuse, impoundments, environmental reuse, groundwater recharge as well as non-potable and potable reuse [8]. The focus in this paper lies on the type of municipal and industrial (non-potable) wastewater reuse relating to applications according to the "fit-for-purpose" principle with respect to urban, industrial and agricultural issues. Fit for purpose refers to wastewater treatment for the specific water reuse application. This means, for example, that reused water for toilet flushing requires a different and higher quality (disinfected) than water for the irrigation of green spaces.

Generally, the U.S. is leading the way [9], but countries such as China [10] and the EU [11], especially Spain [12], are also already practicing water reuse on a large scale and Spain in particular was ranked with the highest rates in the European Union in 2018 [1]. In the urban context, water reuse is practiced, for instance, for the irrigation of green spaces or for toilet flushing [1,9]. Regarding the industrial reuse, it mostly takes place within production processes [13] or cooling systems [14]. However, the reuse of recycled water for further purposes outside of production facilities, e.g., within an industrial park, is a new and efficient approach [15]. Applying reused water in agriculture is practiced worldwide and this sector is the major consumer of wastewater globally [16].

1.3. Possibilities for an Efficent Water Usage to Save Drinking and Reuse Water

Many ways exist to conserve water, but they are often neglected. For all areas of application, such as municipalities and households, industry and agriculture, there are possibilities to save water and to use the available, scarce water efficiently. Because consequently, more drinking water can be saved as well as the reuse water resource.

Municipalities and especially private households can save water by using water saving faucets and toilets. In addition, water can be saved by reducing existing leakage in the distribution network. Behavioral changes can also lead to water conservation, such as showering instead of bathing. Single-family house owners in particular can collect rainwater and use it for various purposes, such as watering the garden or even washing clothes. In addition, private gardens as well as public green spaces in cities can be planted with resilient greens that require only a small amount of water.

Water-saving measures can also be implemented in industrial parks and industrial sites. This includes, among other things, checking and if possible integrating water-saving production processes and/or using recycled water. Cooling towers are among the largest consumers of water in industry. However, the choice of cooling system affects the amount of water needed. Huge volumes of water can be saved by choosing a closed-circuit cooling system instead of a once-through cooling system. Here, only the evaporated water must be replenished.

Agriculture, the world's largest consumer of water, also has many opportunities to save water if irrigation becomes necessary. In particular, the best irrigation technology must be selected for the respective crop. For example, the efficient method of drip irrigation is often suitable for wine, vegetable and fruit cultivation. Optimal timing can also reduce evaporation losses, such as watering in the evening or during the night. Additionally, the way rainwater is stored also plays an important role in reducing evaporation. Cultivation of plants and crops in greenhouses can also reduce water requirements. In addition, it may be necessary to rethink the allocation of agricultural land in order to best implement irrigation techniques in relation to land use.

2. Methodology and Objectives

This perspective is intended to help explain the causes of water scarcity from a global perspective and to highlight the importance of water reuse for the future. The paper points out the different aspects for the planning, implementation and operation of water-reuse approaches and mentions the challenges of implementing water-reuse concepts due to different planning systems and regulations that exist worldwide. Therefore, different applications for water reuse are mentioned and planning concepts presented. The objective is to further show the enormous variety of different water-reuse solutions. Since water quality standards are essential for all water-reuse approaches, the paper gives an overview of the regulations of water reuse in selected countries as well as the state of the art in wastewater treatment for water reuse. Best-practice examples round off the topic to show the different possibilities that already exist in terms of implementation. Finally, the aim of this perspective is to show that, despite the many successfully implemented water-reuse approaches, transferability is not always possible due to the different legal requirements, and therefore there is still an essential need for action in order to realize water reuse.

3. Aspects for Planning, Implementing and Operation of Water-Reuse Approaches

Regarding the baselines for planning, implementation and operation, water reuse must take into account that water can be recovered from a variety of sources. Consequently, water reuse can provide an alternative to existing water supply systems and ensure water availability. Water reuse can be planned or unplanned. In unplanned water reuse, the water source consists largely of previously used water. An example of this are communities that draw their water from rivers into which previously treated wastewater from upstream communities has been discharged. Planned water reuse involves systems designed with the goal of putting treated wastewater to beneficial use [17].

Sources that are related to reuse include municipal wastewater, industrial processes and cooling water, stormwater, agricultural runoff and return flows and production water from natural resources. However, industrial wastewater that is particularly salty or toxic is not suitable for reuse and must be incinerated.

These water sources are treated as "fit for purpose" in a treatment plant. The use applications can vary widely. Common practices include irrigation for agriculture and landscaping such as parks, rights of way and golf courses. Further applications are the municipal water supply, process water for power plants, refineries, mills and factories, indoor uses such as toilet flushing, dust control or surface cleaning of roads, construction sites and other trafficked areas, concrete mixing and other construction processes, supplying artificial lakes and inland or coastal aquifers and environmental restoration [17].

An important question, whether in new developments or in already developed areas, is how to integrate water-reuse concepts into the environment and how ultimately reused water comes to the user. Various options are conceivable, such as the distribution via filling stations, which allows the reuse water to be transported by cars or trucks, or a separate network of pipes, which brings the product reuse water directly to the user or to the place of use. For the best possible use of the reuse water resource in both cases, implementations in planning strategies and instruments are necessary. Furthermore, the transportation infrastructure has to be designed appropriately. Consequently, for the provision of water reuse without a pipe network, specific issues have to be considered in integrated planning instruments on a city scale. Furthermore, approval issues for the installation of on-site advanced treatment units such as an additional chlorination or other disinfection systems for complying the quality requirements for reusing the water have to be considered. More complexity exists when the water is to be brought to the user via pipes. In this case, a holistic dual pipe network has to be implemented in the respective spatial situation. It is apparent that most use cases (see Section 5) of water reuse take place in cities with large water consumers such as universities, golf courses or large city parks. Many examples exist worldwide as to how reuse water is brought to the application through a second pipe system from the water resource recovery facility. Here, the city of Long Beach and Irvine in California with its dual pipe system are appropriate and prospective examples (see Section 5).

A long period of master planning enabled at least the city-wide provision of non-potable reuse water for irrigating large, landscaped areas such as parks, golf courses, community greenbelts and roadway medians. A separate system of pipes can deliver (drinking) water to homes and businesses [18]. In such cases, integrated planning is essential to drive water reuse in urban areas on a holistic scale.

Different use cases show that the implementation of water- reuse varies and therefore the planning steps also vary. An important question in this context is whether the reusable water is provided for private or public application, or whether it is used in a closed unit as it is the case within industrial parks as they have their own wastewater treatment plant on-site. Depending on this, water reuse must be integrated into the respective levels and (integrated) instruments of urban planning. Since nations around the world have different planning policies, it is necessary to analyze, for each system, at which planning levels water reuse can be implemented directly and indirectly. Integration at the regional and city level is particularly important in order to implement holistic approaches and thus integrate water-reuse sources, water scarcity and possible areas of application. In most countries, such approaches are missing so far.

For driving the efficient use of water, including reuse water, holistic water management approaches are required, and these have to be integrated in planning issues. In this context, water demands, availability, quality, regeneration and protections such as groundwater recharge have to be analyzed and included. Water as a sectoral specialist planning has to be considered in different planning strategies, guidelines or instruments. For instance, drought management plans, river basin management plans, drinking water protection areas, land-use planning instruments, irrigation plans as well as water supply and sanitation plans have to be taken into account. Further relevant plans can be also considered, such as rural development plans and infrastructure plans for utilities. These instruments are providing the basis for an integration of water reuse. This integrated approach shows the importance of cross-disciplinary collaboration among a wide range of stakeholders, such as wastewater treatment engineers, plant operators, urban planners and sociologists. The current challenge is that there is limited information available on how water reuse can be planned and integrated into planning instruments on different scales [19]. Since planning systems vary around the world, integrating water reuse into planning instruments is a major challenge for the future. Only through the implementation on different levels can an efficient water management be optimally realized. Furthermore, the issue of acceptance regarding the use of reuse water is very important especially in countries and regions where it has not been practiced so far. Consequently, the quality aspect is of great importance in the operation (see Section 4). To enable the use of reuse water (see Section 5), special quality requirements must be considered so that the use is safe and harmless to humans. Furthermore, the provision of reuse water according to the "fit-for-purpose" principle has to consider different quality requirements so that it can be directly applied for the subsequent use.

4. Regulations, State of the Art and Future Challenges for Municipal Wastewater Treatment and Water Reuse

Many regulations exist worldwide to meet the respective water quality requirements, e.g., discharging treated wastewater into water bodies or for groundwater recharge. Standards and thresholds vary depending on the respective country. Nevertheless, directives for discharging treated wastewater into waters is often the basis for the later developed reuse guidelines or regulations [1,20]. Although these guidelines are constantly evolving, the focus today is largely on developing new standards for the use of reuse water. In this context, water reuse is separated into potable and non-potable applications. Potable reuse generally refers to potable water supply augmentation, while non-potable reuse refers to all other (indoor and outdoor) uses such as toilet flushing or irrigation [9].

In the following considerations, the main focus is on quality requirements and treatment processes for non-potable water applications, since potable water reuse only accounts for a smaller portion of the total. In addition, industrial wastewater reuse, for instance, water reuse within the production process, will be excluded in the following, since there are different and specific quality requirements depending on the production plants and production process.

With regard to standards and guidelines, this section will only deal with the USA, the EU and China, and with global standards by way of example. This is due to the fact that the USA was one of the first countries to develop water-reuse standards at a very early stage. The EU is considered on the basis of the various different countries and situations and also in terms of how various sets of rules are implemented in national law. In China, there were different guidelines for specific reuse applications at an early stage. Well-known countries that implement water reuse such as Singapore, Australia, Israel are not considered in the following. Singapore is not included in the presentation of regulations and guidelines for water reuse, because it introduced recycled water very late in 2003, whereas the USA, among others, has been producing recycled water for several decades. Nevertheless, Singapore is mentioned in Section 5 because it is considered as one of the most innovative and well-known examples in the field of water reuse worldwide.

4.1. Regulations for Treated Wastewater into Water Bodies and for Reusing Water

4.1.1. Regulations for Wastewater Discharge into Waters

Many different standards and regulations for the discharge of pollutants into waters exist worldwide. In the United States, the Clean Water Act (CWA) is the basis for regulating discharges of pollutants into waters and was enacted in 1948 as the Federal Water Pollution Control Act. In 1972, it became the common name [21]. EPA has compiled water quality standards (WQS) from states, territories and authorized tribes. This will be updated as EPA approves new or revised WQS. The standards may contain additional provisions that fall outside the scope of the Clean Water Act. The EPA provides a tool to identify water quality criteria from different perspectives. Accordingly, the state specific WQS are provided online by state, territory or authorized tribe as well as by parameter [22].

Much later, in the European Union, the Council Directive 91/271/EEC was issued as early as 1991. The directive distinguishes effluent discharge requirements with respect to municipal and industrial wastewater into two categories: sensitive areas and less-sensitive areas. Accordingly, the requirements for wastewater treatment deviate with respect to the category. For sensitive areas, limit values for nitrogen and phosphorus concentrations are set in addition to the limit values for the parameters BOD_5, COD and suspended solids, which are set in less-sensitive areas. To achieve the required effluent quality, mechanical and biological as well as chemical treatment steps are necessary and used. The Council Directive from 1991 even integrates in Article 12 that treated wastewater shall be reused whenever appropriate. Additionally, it is said that by reusing the treated water the impact on the environment must be reduced to a minimum. Hence, a link was already made in 1991 in Europe to water reuse without having standards and treatment requirements for reuse water. Therefore, the definition of effluent discharge standards is often the baseline for additional treatment steps to receive the nowadays required standards for producing water reuse.

Furthermore, China started very early on with the regulation of water pollutant discharge. The country issued its first environmental protection standard, the Industrial Wastes Discharge Standards (GBJ 4-73), back in the 1970s due to rapid urbanization and industrialization. In the early 1980s, China enacted the first water quality standards: The Environmental Quality Standard for Surface Water (GB 3838-83) and Environmental Water Quality Standard for Sea Water (GB 3097-82). Afterwards, in the year 1988, the general water pollutant discharges standard, the Integrated Wastewater Discharge Standard (GB 8978-88) was issued. Further adjustments were made in the 1990s performed in the new Integrated Wastewater Discharge Standard (GB 8973-1996). Here, limits of ammonia and phosphate were integrated to reduce the eutrophication of surface water. In the 21st century, water pollution in China's major river basins has been devastating. Hence, further acting was required. In this context, the Discharge Standard of Pollutants for Municipal Wastewater Treatment Plant (GB 18918-2002) was issued for setting stricter requirements for wastewater discharge from wastewater treatment plants [23].

Even though the Chinese government has made considerable efforts to improve water quality for return to water bodies, the pollution level is much higher compared to the USA or the EU. For example, the wastewater recycling rate for industrial wastewater in China is also much lower than in the aforementioned industrialized countries [23].

4.1.2. Regulations for the Application of Non-Potable Water Reuse

Quality requirements are a prerequisite for the safe application of reused water. Hence, specific and advanced wastewater treatment processes are essential before reusing the water. It must be taken into account that standards or regulations may vary from country to country. Some countries have standards that specify only one water quality for water reuse, while other countries have very detailed standards for the specific application. In the following section, the examples are listed chronologically.

The State of California enacted a regulation for water reuse in agriculture as early as 1918, making it the first state in the U.S., and indeed in the world, to enact such a regulation [8]. In the U.S., specific regulations are given by the states, there are no federal regulations specifically governing water reuse [9,24]. In general, even if no guidelines or regulations exist, producing reuse water is permissible in the US by meeting the requirements of the Safe Drinking Water Act (SDWA), the Clean Water Act (CWA) and state requirements [24]. Regarding the regulation in California, it distinguishes two different types of recycled water: non-potable applications and potable applications. Non-potable regulations are separated into four different levels of treatment referring to the respective application. The California Code of Regulations (CCR) specifies in Section 60304 until 60307 the following types of use of recycled water: for irrigation, for impoundments and for cooling for other purposes. Potable recycled water-use applications are managed by the California Water Code (Section 13561) and are separated into "direct potable reuse", "indirect potable reuse for groundwater recharge" and "reservoir water augmentation" [25].

Beside regulations and standards of specific nations, in 1973, the World Health Organization (WHO) was the first international organization that issued a guideline for water reuse in agriculture followed by the FAO and the World Bank [8]. Nevertheless, these global regulations were issued much later on than those in the United States.

Another country that has issued very detailed water-reuse standards is China. To support the development of water reuse, the Chinese government has enacted a series of regulations for a variety of reuse applications for about 20 years. These are the Standard for Environment Reuse (GBT 18921-2002), the Standard for Miscellaneous Urban Reuse (GBT 18920-2002), the Standard for Industrial Reuse (GBT 19923-2005), the Standard for Farmland

Irrigation Reuse (GB20922-2007) and the Standard for Green Space Irrigation Reuse (GB/T 25499-2010) [10]. Hence, all wastewater treatment facilities producing reused water have specific treatment steps to meet the respective requirements for the respective application.

Compared to other countries, the European Union (EU) issued a guideline for water-reuse very late on, whereas individual countries, such as Spain and Portugal, enacted their own regulations much earlier [8]. In 2020 the EU published the regulation on minimum requirements for water reuse. The guideline must be transposed in national law until 26 June 2023 [11,26].

4.2. State of the Art and Future Challenges for Wastewater Treatment and Reuse

4.2.1. Advanced Treatment Steps for Wastewater Treatment

A current challenge is to eliminate micropollutants and microplastics from the wastewater, such as pharmaceutical residues, cosmetics, household and industrial chemicals, antibiotic-resistant pathogens or specific chemicals. These have a potential negative impact on ecosystems. In addition, such substances could have an impact on drinking water hygiene. Consequently, new or complementary processes for wastewater treatment are required to avoid endangering human health and to further minimize water pollution. Thus, advanced treatment steps within wastewater treatment plants are required. These already exist and are operational. For example, the further reduction in micropollutants will improve conditions for aquatic life. Thus, residual risks arising from the consumption of fish are further mitigated.

For eliminating micropollutants, a basic treatment step today is adsorption using activated carbon (powdered activated carbon or granulated activated carbon). Oxidation processes such as ozone or advanced oxidation process (AOP) are a further or complementary option. The separation by means of membranes is a third possibility.

For the elimination of antibiotic-resistant pathogens, the aforementioned separation as a treatment step using a membrane is especially suitable, if it is used simultaneously for the elimination of micropollutants. If it is required that the wastewater treatment plant effluent is additionally disinfected, then disinfection processes are essential. UV or chlorine can be principally used here. However, chlorine is not used in many countries, although it is common in the USA [27]. This often depends on the specific regulations. Often the use of chlorine is avoided due to its negative environmental impacts. For instance, the chlorination of municipal wastewater as well as drinking water can produce toxic chemical by-products [28,29].

4.2.2. Wastewater Treatment Steps to Produce Reuse Water "Fit-for-Purpose"

To use treated wastewater for simple reuse purposes (for example irrigation of city parks and street cleaning), a conventional wastewater treatment plant (with mechanical and biological treatment steps) as described above is usually sufficient if no further disinfection is required due to low bacterial counts. This includes applications such as irrigation of green areas and street cleaning. However, if drip irrigation is used for irrigation purposes, a conventional wastewater treatment plant with an additional sand filter to remove suspended solids is required.

To achieve a good water quality for reuse purposes, such as toilet flushing, an advanced treatment step, such as adsorption (activated carbon) or oxidation (ozone), is required to eliminate micropollutants (see above). Desalination processes up to reverse osmosis (RO) are required to achieve higher water quality up to drinking water quality.

5. Existing and Innovative "Fit-for-Purpose" Water-Reuse Approaches

5.1. Selection of the Best-Practice Examples

In the following, some successful water-reuse implementations where the water is provided fit for purpose are presented.

Here, a comprehensive approach is presented that shows the recycled water system where the reused water is provided via purple pipes (internationally agreed color for pipes transporting reused water) and distributed by several filling stations. Another example shows a holistic water-reuse approach on a city scale. Therefore, examples from the U.S. are taken into account based on the respective water-reuse scale and dual pipe systems for providing recycled water. The dual pipe system is related to a separate pipe system for transporting reused water beside the pipe system for providing potable, drinking water. The first one relates to Long Beach and the second one to Irvine, both located in California. Examples from the U.S., particularly in California, show that treated wastewater is used intensively, in some cases even through a citywide water-reuse network. In comparison, examples in Europe tend to be on a smaller scale and will be not considered here.

The third example relates to Singapore with regard to the production of "fit-for-purpose" recycled water (potable and non-potable). Singapore should be mentioned here as an example, as it is one of the best known and most innovative examples worldwide.

5.2. Long Beach Recycled Water Expansion System

The Long Beach recycled water system is a suitable example for the provision of reusable water via filling stations. The reused wastewater is treated in the Long Beach Water Reclamation Plant, located on the east side of the city and treats approx. 68 million liters per day. The disinfected recycled water can be used for irrigating parks, golf courses, cemeteries and athletic fields. Furthermore, it can be used to recharge the groundwater basins or for sweeping streets. The program is primarily aimed at connecting the recycled water system to new customers and increasing the reliability of the distribution system. To this end, the construction of recycled water pipelines, new pumping stations, enlargement of the water storage system and completion of new service connections are included. When completed, the expansion program is aimed to meet 15 percent of the city's total water demand. Among others, customers with large irrigation systems such as California State University Long Beach and the Long Beach Unified School District, as well as large parks, golf courses, cemeteries and sports fields, will be connected to the system. The city also uses recycled water instead of potable water to clean streets, saving millions of liters of water annually [30]. The water can be collected at the distributed filling stations. This avoids, for instance, long transport routes for tank trucks. The pipe network provides the reclaimed water by filling stations for the respective water consumer such as parks or schools. By recycled meter systems, water consumptions and requirements can be analyzed [31,32].

5.3. Irvine Ranch in California

The example of Irvine Ranch in California was chosen because it shows very impressively how water reuse was already implemented on a large scale for an entire city in the 1960s. A special feature is that Irvine was planned as a satellite city at the time [33]. A municipal master planning enabled a holistic planning, so that, e.g., plants and trees can be irrigated completely with recycled water. The baseline is the integrated dual pipe system so that drinking water and recycled water is provided separately [18]. The holistic network of the purple pipe distribution with a length of round about 450 miles provides reusable water [34].

The non-potable water source is reclaimed water and is used for several non-potable purposes. Most of the water is used for landscape irrigation, such as road medians, golf courses, parks, playgrounds, residential areas and schools. Additionally, the reclaimed water is used for toilet flushing mostly in commercial and residential buildings as well as for cooling towers [35]. The system delivers up to 106 million liters per day. In total, 91% of the water which is required for landscape irrigation is recycled water. Furthermore, due to technological advances in indoor plumbing, new homes are 50% more water efficient than older ones. The water is also used in toilets and urinals [18].

5.4. Water Reuse in Singapore

In Singapore, the so-called NEWater is part of a comprehensive water resource policy. In general, the reuse of water covers up to 40 percent of the water demand. By 2060, this percentage is expected to increase to 55 percent. Singapore uses treated wastewater for potable and non-potable applications instead of discharging it into the ocean after treatment. This way, the water cycle has been completed, bringing significant economic, social and environmental benefits [36].

Singapore is home to about 5.5 million people and the population density is 7485 people per km^2 [37]. Despite an average annual precipitation of about 2340 mm, it is not possible to store the rainwater because of the limited area available for reservoirs and the lack of aquifers. As a result, Singapore relies on imported water from Malaysia, as well as the production of recycled and desalinated water [36]. Singapore's strategy to secure water supply began in 1965 after independence due to the scarcity of water resources. For both non-conventional water sources, the production of reused water from municipal sources (known as NEWater) and desalination, significant investments have been made since the 1970s in research and development to advance technological developments [36].

The current NEWater Technology consists of three stages. Stage 1 of the NEWater production process is known as microfiltration (MF) or ultrafiltration (UF). Microscopic particles and bacteria are filtered out by using membranes. Stage 2 is known as the process of reverse osmosis (RO). Therefore, a semi-permeable membrane is used, which has very small pores to allow only very small molecules such as water molecules to pass through. This prevents unwanted contaminants, including viruses, from passing through the membrane. The process of ultraviolet or UV disinfection constitutes the third stage of treatment. This process is able to eliminate bacteria and viruses and serves as an additional safety measure to ensure the purity of NEWater [38].

NEWater extends water resources to various non-domestic users. Water fabrication plants are the largest user. However, industrial and commercial buildings are also supplied with NEWater [36]. In dry periods, NEWater is also used for indirect drinking water purposes to conserve surface water. It is mixed with raw water and treated conventionally before it is distributed [36,39].

Singapore introduced recycled water in 2003, while Orange County in California, for example, has been producing recycled water for several decades. Singapore created its own system. Thanks to comprehensive education and communication strategies, it was able to achieve large-scale industrial implementation and broad public acceptance of indirect drinking water use [36,40]. Water reuse under the circular economy concept focuses on implementing a closed system where treated wastewater is not discharged into the ocean but is further treated to produce NEWater [36].

6. Conclusions

The increasing water scarcity worldwide highlights the urgency of implementing water-reuse concepts. Countries that have been under high water stress for decades, such as the USA or China, have already reacted at an early stage and introduced appropriate regulations for water reuse to ensure sufficient water quality. Hence, to achieve Goal 6 of the Sustainable Development Goals (SDG), it is essential to exploit all opportunities with respect to reuse water.

In general, the reuse of wastewater is not the only solution to improve the situation in water-stressed regions. It is especially important to save water in order to produce less reuse water. Since its production is expensive, this can generally save costs.

For municipalities, industry and agriculture, there are still far-reaching implementation possibilities to use water more efficiently. In addition to the reuse of treated wastewater, efficient water-use methods make a major contribution to reducing water consumption, including the use of reuse water. Furthermore, there is still a great need for research to establish various correlations between the disciplines. One approach is seen in agriculture, for example, to reconcile the use of reuse water, irrigation techniques and optimal distribution of land with specific land uses.

Generally, regulations for the discharge of treated wastewater into water bodies exist in most countries, but further regulations based on them to produce reuse water according to the "fit-for-purpose" principle are partly missing. In Europe, the regulations came relatively late. The 1991 Council Directive recommended the reuse of treated wastewater, but at that time there were no standards or quality requirements for reuse. It is not until 2023 that the new Regulation 2020/741 of the European Parliament and the Council has to be adapted into national law. Consequently, advanced requirements and treatment steps as well as their implementation are essential for wastewater treatment, as for instance evidenced by Singapore, so that the treated water has a quality corresponding to the "fit-for-purpose" principle.

However, a current challenge is the implementation of water-reuse concepts. Only a few countries have realized holistic-integrated water management concepts that also include water reuse. In Germany, for instance, as of yet there are no strategic spatial planning concepts on a city scale to drive water reuse. There is a need for further research in this area. Another challenge is the transferability of existing approaches to completely different planning systems in the respective countries. Thus, each situation must be considered individually. Nevertheless, water scarcity will drive this, as without new concepts and implementations, sufficient water supply will be jeopardized. Many discussions exist regarding costs for wastewater treatment for water reuse. This can be increasingly ignored, because costs, which are incurred for example from an energetic point of view for the treatment or for the wastewater treatment technology, will no longer play a role at a high water-stress level. In case of a high water scarcity, the focus is rather on being able to provide the water at all.

All in all, regions and cities affected by water scarcity need to analyze all aspects related to the local water supply situation as well as their potential and opportunities for water reuse. In general, the baseline with respect to regulations and technologies is given, but implementation is lacking. Here, planning and water management stakeholders must work closely together to advance water reuse. Only then will it be possible to achieve Goal 6 of the SDGs in the near future.

Author Contributions: Conceptualization, methodology, writing—review and editing, S.B. and M.W. All authors have read and agreed to the published version of the manuscript.

Funding: This research received no external funding.

Institutional Review Board Statement: Not applicable.

Informed Consent Statement: Not applicable.

Data Availability Statement: Not applicable.

Conflicts of Interest: The authors declare no conflict of interest.

References

1. Jodar-Abellan, A.; López-Ortiz, M.I.; Melgarejo-Moreno, J. Wastewater Treatment and Water Reuse in Spain. Current Situation and Perspectives. *Water* **2019**, *11*, 1551. [CrossRef]
2. Burek, P.; Satoh, Y.; Fischer, G.; Kahil, M.T.; Scherzer, A.; Tramberend, S.; Nava, L.F.; Wada, Y.; Eisner, S.; Florke, M.; et al. *Water Futures and Solutions*; Working Paper WP-16-006. Hg. v; International Institute for Applied Systems Analysis: Viena, Austria, 2016.
3. Tortajada, C.; Kastner, M.J.; Buurman, J.; Biswas, A.K. The California drought: Coping responses and resilience building. *Environ. Sci. Policy* **2017**, *78*, 97–113. [CrossRef]
4. WRI–World Resource Institute. Drop by Drop, Better Management Makes Dents in China's Water Stress. 2018. Available online: https://www.wri.org/blog/2018/04/drop-drop-better-management-makes-dents-chinas-water-stress (accessed on 17 May 2019).
5. Drewes, J.E.; Schramm, E.; Ebert, B.; Mohr, M.; Krömer, K.; Jungfer, C. Potenziale und Strategien zur Überwindung von Hemmnissen für die Implementierung von Wasser-wiederverwendungsansätzen in Deutschland. Potentials and strategies to overcome barriers to implementing water reuse approaches in Germany. *Korresp. Abwasser Abfall* **2019**, *66*, 995–1003.
6. Bundesanstalt für Gewässerkunde: Das Niedrigwasser 2018. The Low Tide 2018. Available online: https://www.bafg.de/DE/05_Wissen/04_Pub/04_Buecher/niedrigwasser_2018_dokument.pdf?__blob=publicationFile (accessed on 20 December 2021).
7. Bauer, S.; Linke, H.J.; Wagner, M. Combining industrial and urban water-reuse concepts for increasing the water resources in water-scarce regions. *WER* **2020**, *92*, 1027–1041. [CrossRef] [PubMed]

8. Shoushtarian, F.; Negahban-Azar, M. Worldwide Regulations and Guidelines for Agricultural Water Reuse: A Critical Review. *Water* **2020**, *12*, 971. [CrossRef]
9. Luthy, R.G.; Wolfand, J.M.; Bradshaw, J.L. Urban Water Revolution: Sustainable Water Futures for California Cities. *J. Environ. Eng.* **2020**, *146*, 4020065. [CrossRef]
10. Lyu, S.; Chen, W.; Zhang, W.; Fan, Y.; Jiao, W. Wastewater reclamation and reuse in China: Opportunities and challenges. *J. Environ. Sci.* **2016**, *39*, 86–96. [CrossRef]
11. Dingemans, M.; Smeets, P.; Medema, G.; Frijns, J.; Raat, K.; van Wezel, A.; Bartholomeus, R. Responsible Water Reuse Needs an Interdisciplinary Approach to Balance Risks and Benefits. *Water* **2020**, *12*, 1264. [CrossRef]
12. Melgarejo, J.; Prats, D.; Molina, A.; Trapote, A. A case study of urban wastewater reclamation in Spain: Comparison of water quality produced by using alternative processes and related costs. *J. Water Reuse Desalin.* **2016**, *6*, 72–81. [CrossRef]
13. Klemeš, J.J. Industrial water recycle/reuse. *Curr. Opin. Chem. Eng.* **2012**, *1*, 238–245. [CrossRef]
14. Pintilie, L.; Torres, C.M.; Teodosiu, C.; Castells, F. Urban wastewater reclamation for industrial reuse: An LCA case study. *J. Clean. Prod.* **2016**, *139*, 1–14. [CrossRef]
15. Bauer, S.; Dell, A.; Behnisch, J.; Linke, H.J.; Wagner, M. Sustainability requirements of implementing water-reuse concepts for new industrial park developments in water-stressed regions. *J. Water Reuse Desalin.* **2020**, *11*, 28. [CrossRef]
16. Jaramillo, M.F.; Restrepo, I. Wastewater Reuse in Agriculture: A Review about Its Limitations and Benefits. *Sustainability* **2017**, *9*, 1734. [CrossRef]
17. United States Environmental Protection Agency. Basic Information about Water Reuse. Available online: https://www.epa.gov/waterreuse/basic-information-about-water-reuse (accessed on 20 December 2021).
18. Irvine Company. *Master Planning Saves Water*; Irvine Company: Newport Beach, CA, USA, 2015; Available online: https://www.goodplanning.org/files/MasterPlanningSavesWater.pdf (accessed on 7 April 2022).
19. EU Water Directors. Common Implementation Strategy for The Water Framework Directive and The Floods Directive. In *Guidelines on Integrating Water Reuse into Water Planning and Management in the Context of the WFD*; WFD Reporting Guidance 2016; EU: Brussels, Belgium, 2016.
20. Radcliffe, J.C.; Page, D. Water reuse and recycling in Australia–history, current situation and future perspectives. *Water Cycle* **2020**, *1*, 19–40. [CrossRef]
21. United States Environmental Protection Agency. Summary of the Clean Water Act. Available online: https://www.epa.gov/laws-regulations/summary-clean-water-act (accessed on 12 August 2021).
22. United States Environmental Protection Agency. State-Specific Water Quality Standards Effective under the Clean Water Act (CWA). Available online: https://www.epa.gov/wqs-tech/state-specific-water-quality-standards-effective-under-clean-water-act-cwa (accessed on 12 August 2021).
23. Li, W.; Sheng, G.; Zeng, R.J.; Liu, X.; Yu, H. China's wastewater discharge standards in urbanization: Evolution, challenges and implications. *Environ. Sci. Pollut. Res. Int.* **2012**, *19*, 1422–1431. [CrossRef]
24. US EPA; Smith, C.D.M. *2017 Potable Reuse Compendium*; Living Rivers Europe: Brussels, Belgium, 2017.
25. California Water Boards. Recycled Water Policy. Hg. v. State of California. 2021. Available online: https://www.waterboards.ca.gov/water_issues/programs/water_recycling_policy/ (accessed on 8 March 2021).
26. European Commission. Water Reuse. 2021. Available online: https://ec.europa.eu/environment/water/reuse.htm (accessed on 9 March 2021).
27. Bonvin, D. The Dark Side of Chlorine. 2011. Available online: https://actu.epfl.ch/news/the-dark-side-of-chlorine/ (accessed on 30 March 2022).
28. Sedlak, D.L.; von Gunten, U. The Chlorine Dilemma. *Science* **2011**, *331*, 42–43. [CrossRef]
29. Ghernaout, D.; Elboughdiri, N. Is Not It Time to Stop Using Chlorine for Treating Water? *OALib* **2020**, *7*, 1–11. [CrossRef]
30. Long Beach Water. Reclaimed/Recycled Water. Available online: https://lbwater.org/water-sources/reclaimed-recycled-water/ (accessed on 4 March 2022).
31. Long Beach Water. Recycled Water Filling Stations. 2020. Available online: https://lbwater.org/wp-content/uploads/2019/07/LBWDRecycledWaterFillingStations.pdf (accessed on 4 March 2022).
32. Long Beach Water. Recycled Water System. 2019. Available online: https://lbwater.org/wp-content/uploads/2019/07/REC44_web.pdf (accessed on 4 March 2022).
33. Forsyth, A. Who Built Irvine? Private Planning and the Federal Government (13). *Urban Stud.* **2002**, *39*, 2507–2530. [CrossRef]
34. Irvine Company. 450+ Miles of Purple Pipe. 2022. Available online: https://www.goodplanning.org/sustainability/water-conservation/recycled-water-map/ (accessed on 4 March 2022).
35. Grigg, N.; Rogers, P.D.; Edmiston, S. Dual Water Systems: Characterization and Performance for Distribution of Reclaimed Water. *Water Res. Found.* **2013**, *45*, 75–106.
36. Tortajada, C.; Bindal, I. Water Reuse in Singapore: The New Frontier in a Framework of a Circular Economy? In *Water Reuse within a Circular Economy Contex. 2 Global Water Security Issues Series*; UNESCO: Paris, France, 2020.
37. Department of Statistics Singapore. Population and Population Structure. 2021. Available online: https://www.singstat.gov.sg/find-data/search-by-theme/population/population-and-population-structure/latest-data (accessed on 19 November 2021).
38. PUB, Singapore's National Water Agency. NEWater. 2020. Available online: https://www.pub.gov.sg/watersupply/fournationaltaps/newater (accessed on 2 December 2021).

39. Lee, H.; Tan, T. P: Singapore's experience with reclaimed water: NEWater. *Int. J. Water Resour. Dev.* **2016**, *32*, 611–621. [CrossRef]
40. Cecilia, T.; Pierre, V. Drink more recycled wastewater. *Springer Nat.* **2016**, *45*, 26–28. Available online: https://media.nature.com/original/magazine-assets/d41586--019--03913--6/d41586--019--03913--6.pdf (accessed on 2 December 2021).

Article

Social–Ecological Impact Assessment and Success Factors of a Water Reuse System for Irrigation Purposes in Central Northern Namibia

Martin Zimmermann [1,*] and Felix Neu [2]

1. ISOE—Institute for Social-Ecological Research, 60486 Frankfurt am Main, Germany
2. Independent Researcher, 60486 Frankfurt am Main, Germany; f_neu@web.de
* Correspondence: zimmermann@isoe.de

Abstract: With regard to water supply constraints, water reuse has already become an indispensable water resource. In many regions of southern Africa, so-called waste stabilisation ponds (WSP) represent a widespread method of sewage disposal. Since capacity bottlenecks lead to overflowing ponds and contamination, a concept was designed and piloted in order to upgrade a plant and reuse water in agriculture. Using a social–ecological impact assessment (SEIA), the aim of this study was to identify and evaluate intended and unintended impacts of the upgrading of an existing WSP to reuse water for livestock fodder production. For this purpose, semistructured expert interviews were conducted. In addition, a scenario analysis was carried out regarding a sustainable operation of the water reuse system. The evaluation of the impacts has shown that intended positive impacts clearly outweigh the unintended ones. The scenario analysis revealed the consequences of an inadequate management of the system and low fodder demand. Furthermore, the analysis showed that good management of such a system is of fundamental importance in order to operate the facility, protect nature and assist people. This allows subsequent studies to minimize negative impacts and replicate the concept in regions with similar conditions.

Keywords: casual loop diagrams; livestock fodder production; scenario analysis; semistructured interviews; waste stabilisation ponds

1. Introduction

During the last decades, climate change has proven that water is a scarce resource in many regions of the world, especially in semiarid and arid countries, both now and in the future. Moreover, the Intergovernmental Panel on Climate Change (IPCC) assumes that climate change will increase the hazard of extreme events such as droughts and heavy rainfall even further [1]. However, according to the IPCC, climate change in Africa will have a modest overall impact on future water scarcity compared to other factors such as population growth, urbanisation, agricultural growth and land-use change [1]. Because of these drivers, particularly in developing countries, the pressure on water resources has enormously increased, resulting in limitations of water supply [2]. In these regions, pastoralism has traditionally been of great importance in order to secure the people's livelihood and generate income [3]. Along with the increased water demand, water and fodder for livestock has become rare and of high cost, leading to emergency slaughters during long-term droughts [4].

In order to augment water supply in water-scarce regions, water reuse has created a suitable alternative. By connecting disposal of wastewater, improving hygiene and generating treated water, the multibenefit of water reuse is obvious. Depending on the treatment technology used and the water quality produced, the range of water applications is quite wide [5]. In many developing countries characterised by semiarid climate, so-called waste stabilisation ponds (WSP) or oxidation ponds provide wastewater disposal and treatment by collecting municipal wastewater in a series of ponds [6–8]. WSPs provide several advantages [9,10] and are also used in agriculture [11,12] and industry [13]. Biodegradation processes and sedimentation lead to a reduction in organic matter. Owing to abundant solar radiation, the wastewater discharged finally evaporates into the atmosphere. Associated with low costs, high efficiency and little effort for operation and maintenance, WSPs are therefore a common method of sewage disposal in sub-Saharan Africa [6,14]. In many of these developing countries, the treated water from WSPs has not only been disposed of but also reused for aquaculture and irrigation purposes [6,15,16]. However, in order to reuse the water, certain water quality standards must be met, which can be achieved with different treatment processes [6]. Unfortunately, these plants are often not in good condition due to poor operation and maintenance along with a lack of knowledge and management, representing a high risk for contamination by overflowing, especially during the rainy season. Furthermore, population growth increases the volume of wastewater. At the same time, the lack of financial investments and expertise limit the ponds' capacities.

In central northern Namibia, a transdisciplinary research project established a water reuse system [17–19]. Core idea was to upgrade WSPs for irrigation purposes. Therefore, the pilot project examined different treatment options to reuse municipal wastewater for livestock fodder production in the town of Outapi [20]. By doing so, the project contributed to the Sustainable Development Goals (SDGs) [21], in particular to achieve goal six, namely ensuring clean water and sanitation to all. However, the impacts of such a water reuse system on society and nature have not yet been adequately studied.

Since the 1970s, various types of impact assessments (IA) have been developed as a result of the debate on the impact of human intervention in ecosystems [22]. The most commonly used impact assessment is the environmental impact assessment (EIA) originally developed in the United States in 1970 and based on the US National Environmental Policy Act of 1969 (NEPA). The initial theoretical foundation and application has been widely debated among experts [23–25]. According to the United Nations Environment Programme (UNEP), the EIA considers possible impacts of projects in advance, in order to be able to adapt the planning and design of the projects [26]. The focus on individual projects is in line with the understanding of EIA by the Namibian Ministry of Environment and Tourism, according to which the EIA plays a decisive role in every phase of a project [27]. On the contrary, the strategic environmental assessment (SEA) focuses on the environmental impacts of policies, plans, programmes and other strategic initiatives [26]. This view is closely linked to the definition of IA by the European Commission, which describes IA as a process to suggest to policymakers the advantages and disadvantages of possible policies by assessing their impacts [28]. Technology assessment (TA) carries out scientific and technological forecasts and evaluation processes, thus it is contributing to the public and political debate [29].

The Environmental and Social Impact Assessment (ESIA) was carried out in the early 2000s and is essentially an extension of the EIA. In some cases, the boundaries between these different types of IA are not clear. The European Bank for Reconstruction and Development (EBRD), for example, uses the terms EIA and ESIA as an equivalent [30]. The ESIA emerged from the combination of EIA and Social Impact Assessment (SIA), which was developed within the debate to include a social component [31–33]. The most commonly used definition of ESIA was established in 2012 by the International Finance Corporation (IFC) [34]. According to the IFC policy, an ESIA is a comprehensive document of a project's potential environmental and social risks and impacts. An ESIA is usually prepared for greenfield developments or large expansions with specifically identified physical elements, aspects and facilities that are likely to generate significant environmental or social impacts [35]. Although the ESIA has been well received by multilateral donors, international authorities and private institutions (i.e., WB, AfDB, EU, GIZ), it is rarely used in science, and relatively few scientific articles have been published to date [36,37]. The European Bank for Reconstruction and Development has commissioned several ESIAs, mainly in European countries considering social environmental topics including wastewater treatment plants.

Compared to ESIA, the social–ecological impact assessment (SEIA) addresses a broader regional and overarching context [38]. Furthermore, different starting and future conditions of the region (climate, economy, urbanisation and seasonal migration, sociocultural changes, etc.) can be considered in the analysis. An early example of a SEIA considered rainwater harvesting, sanitation and water reuse, groundwater desalination and subsurface water storage [39]. However, this study is limited to the effects on the water cycle, ecological impacts and land-use changes by using existing empirical data. Brymer et al. [40] conducted a participatory SEIA evaluating the impact of land management in Idaho, USA. Jones and Morrisson-Sanders [41] emphasize that stakeholders must be included in such participatory processes to ensure a long-term success of corresponding interventions. Recent scientific papers focus on the success of societal impacts of transdisciplinary sustainability research [42–44].

The aim of this study is a SEIA of the analysed water reuse system for fodder production in Outapi. This comprises the identification of impacts on society and nature together with an evaluation of these effects by collecting data through expert interviews. A further objective is to identify both positive (intended) and negative (unintended) impacts of the approach by classifying them into ecological, social and economic categories. Based on this, a scenario analysis is carried out to identify success factors and hazards. This contributes to mitigate negative impacts and to assess the transferability of the water reuse concept to other locations in southern Africa. Especially in Namibia, the approach can potentially be transferred to many other towns with WSPs to increase the efficiency of water reuse systems.

2. Materials and Methods

2.1. Namibian Case Study

The location of the studied water reuse system is in the town of Outapi in central northern Namibia (Figure 1). Because of its arid climate, Namibia has no permanently water-bearing rivers except for its border rivers in the south (Oranje) and north (Kunene, Okavango, Linyanti and Zambesi) [45]. Central northern Namibia is characterised by the Etosha pan, a drainless pan measuring 120×50 km. This salty pan is accompanied by the so-called Oshana system in the north, also known as the Cuvelai system. Oshanas describe a complex of shallow, north–south valleys that gradually fill with water during the rainy season depending on the amount of precipitation in northern Namibia and neighbouring Angola [46,47]. After heavy rainfall, these ephemeral systems can turn into a torrential stream in a few minutes, which then only carries water for a few hours or at best a few days. In contrast, due to the extremely low gradient, the water in the Cuvelai system drains off very slowly and in years of high precipitation there are repeated large-scale floods [48].

Namibia possesses groundwater resources that are characterised by a high salt content, especially in the north, so that they cannot be used without treatment. The lack of freshwater resources has an impact on the water supply, especially during dry seasons, which is why there is a risk of excessive utilisation. This process is worsening by population growth and livestock farming, which have traditionally played a major role in Namibia [45,49].

Between 1970 and 1990 in particular, the population in Namibia more than doubled. The country's current growth rate is 2.1% describing a low negative trend over time [50]. The majority of the population is concentrated in the central north due to the high agricultural potential. Approximately 40% of the Namibians live in the regions of Omusati, Oshana, Ohangwena and Oshikoto, reaching population density rates between 4.7 and 22 people per km^2 [51]. In Outapi, the population increased from 2600 to 6500 inhabitants between the last censuses of 2001 and 2011, indicating an annual growth rate of more than 9% [52,53].

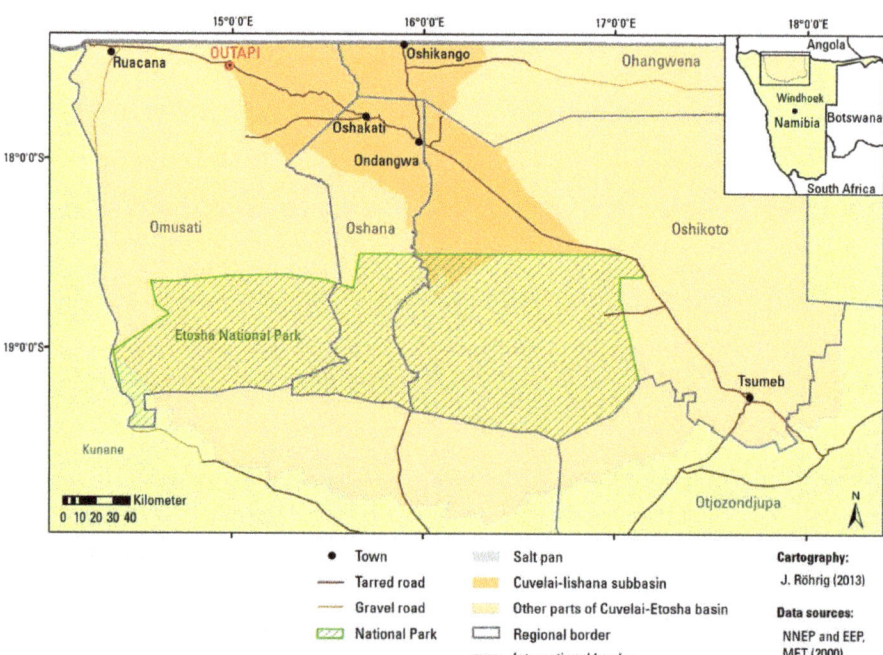

Figure 1. Map of central northern Namibia (map by J. Röhrig, 2013, modified).

Owing to salty groundwater, the water supply is mainly managed with surface water that exists only seasonally depending on the rainy season or in limited accessibility such as the border river Kunene [54]. Hence, in the 1950s, the government of Namibia started an initiative to improve the situation of water supply in the central north by establishing a network of pipelines. Since the 1960s, the so called Calueque–Oshakati Water Supply Scheme has been established by constructing an open canal from Ruacana to Oshakati, several water treatment plants and pipelines to ensure water supply in the north [55].

According to the Outapi Town Council (OTC), 85% of Outapi residents have access to sanitation facilities. Regarding this number, it is not clear whether all informal settlements are included in this specification. Similar to Oshakati and Ongwediva and many other places in northern Namibia, Outapi has WSPs built by Namibian authorities in 2004. The complex consists of two lines with four ponds each that finally lead into an evaporation pond. Each line includes a larger facultative pond, followed by three smaller maturation ponds (total water surface area: 40,000 m^2; total volume: 55,000 m^3), discharging into the evaporation pond (surface area: 41,000 m^2; volume: 20,000 m^3), where the water is supposed to evaporate completely [17,18]. The average total inflow was 753 m^3/d in 2018 [17,18]. The ponds are located about 1.5 km southwest of Outapi's town borders, surrounded by grasslands and Oshanas. The locals use these water areas for fishing purposes and as a source of drinking water for their cattle. Because of urbanisation, people living next to the ponds have been relocated into the town in order to avoid emerging conflicts.

Continuing population growth significantly increased the volume of sewage. This results in the fact that the current capacity of the ponds is not sufficient. Hence, especially during the rainy season, when additional storm water enters the ponds, untreated water breaks through the embankments of the final pond, which leads to contamination and health risks in the Oshanas. Furthermore, inadequate management results in bad discharge qualities due to a lack of knowledge about sustainable operation and maintenance of the WSP. The lack of awareness leads to methane emissions and excessive algae in the ponds. Missing precautions during drainage and drying of the sludge coupled with careless disposal of the sludge cause contaminations [20]. Because of lacking information on the consequences, some locals damage the surrounding fence to allow their cattle to graze and drink in the complex. All these aspects underline the insufficient management and the poor condition of the WSP.

The research project's technical measures to upgrade the WSP can be divided into three parts, namely sludge treatment, pretreatment and post-treatment. The plant consists of two separate lines of ponds (A and B), which can be loaded at the same time (Figure 2). However, only line A is being upgraded, so that line B can ensure the current wastewater disposal. In pretreatment, organic, predominantly suspended constituents (so-called primary sludge) are removed from the raw wastewater to relieve the pollution of the ponds [17,19]. This minimises the chemical oxygen demand and reduces sludge deposits in the ponds. Two separate systems, an upflow anaerobic sludge blanket reactor (UASB) and a microscreen were tested for pretreatment. In the UASB reactor, anaerobic degradation takes place through sedimentation inside the tank, resulting in biogas formation. The deposited sludge of the UASB tank is pumped into the nearby drying bed. To avoid disposal issues of the dried and stabilised sludge, the organic material is used as fertiliser for the nutrient-poor agricultural fields in order to enrich the soil with organic substances. Parallel to this, raw wastewater is pumped to the microscreen, which has a mesh size of 250 µm and can separate approximately 70% of the suspended substances [17,19].

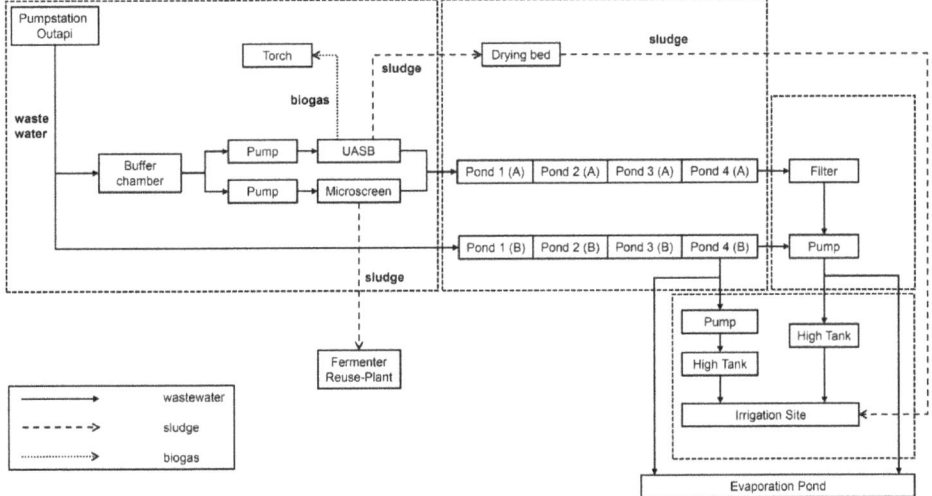

Figure 2. Scheme of upgrading measures (adjusted based on [17]).

The pretreated water flows into pond A1 where installed guide walls control the water flow. This results in an extended retention time in the pond and improved treatment performance. The post-treatment at the end of line A additionally cleans the water by a submerged stone filter. This low-tech approach also reduces filterable substances, in particular, microalgae, which could, for instance, clog the pipes of the drip irrigation system. Pre- and post-treatment lead to a 4.29 log unit reduction in E. coli (3.32 log units for enterococci) and concentrations of E. coli in the effluent of $6.9 \times 10^2 \pm 1.0 \times 10^3$ MPN/100 mL [18]. This does not guarantee drinking-water quality, but agriculture irrigation can use the nutrients in the treated water. In addition, Namibian legislation does not require irrigation water for fodder production to be of drinking-water quality [27,56,57].

The cultivation site for fodder production is located close to the WSP. The test phase included various irrigation systems such as drip, furrow and subsurface irrigation. Considering the local conditions, the focus is on robust low-pressure systems that supply the area via stationary main and overhead lines in order to minimise water losses through evaporation and infiltration while avoiding soil salinisation [20].

Capacity development for sustainable management, operation and maintenance of the system and the establishment of a wastewater treatment plant partnership (WWTPP) support the technical measures. The WWTPP forms a regional network of various operators of wastewater treatment plants and ponds. The purpose of the partnership is to share knowledge and experience regarding sustainable wastewater management and water reuse [58].

2.2. Social–Ecological Impact Assessment (SEIA)

The overall approach to assessing the impacts in this study is the SEIA, which is a relatively new form of impact assessment. Liehr [38] defines the SEIA as a process of evaluating likely impacts of interventions on social ecological systems [59] with regard to consequences for the environment and the people's livelihood. SEIA is therefore able to adequately capture the impacts of a measure on society and nature. By analysing possible impacts, both positive and negative, the SEIA particularly identifies vulnerabilities and risks that must be carefully examined in the future. This results in recommendations for a sustainable adaptation of interventions or further alternatives [38]. By doing so, the SEIA can compensate for the disadvantages of EIA, SEA and TA.

The IFC developed a concept describing the social and environmental impact assessment (S&EA) process [35]. The S&EA and SEIA variants can be regarded as equivalent. The process is described in eight steps. Steps three to five represent the core of the assessment process, which is accompanied by pre- and post-actions. The first two steps comprise the screening and scoping phase that determines the appropriate spatial and temporal system boundaries. After that, baseline studies must be carried out to show the environmental and sociocultural conditions considering the status quo and associated factors (step three). The core of the assessment is the prediction and evaluation of impacts, which requires analysing the impacts identified during scoping and baseline studies in terms of their nature, temporal scale and spatial scale, among other factors (step four). The last step within the actual assessment process is the mitigation (step five), which aims at minimising or even eliminating negative impacts on nature and society. Step seven and eight comprise the social and environmental management plan together with the environmental impact statement.

The SEIA in this study focuses on step four, i.e., identifying impacts and evaluating the operation of the analysed water reuse system. The main criterion regarding the identified impacts refers to the direction of an impact, which can be positive or negative. The SEIA uses the terms intended and unintended impacts to highlight the advantages and disadvantages of the intervention. Other criteria comprise the spatial and temporal extent of an impact. In terms of spatial extent, a distinction can be made between site-specific, small, medium and large. The term site-specific in this context includes the WSP and the cropland used for fodder production. The temporal extent, which describes the duration of the impact, can be divided into temporary, short-term, medium-term, long-term and permanent. Owing to the qualitative data collection method, quantitative criteria such as intensity, probability of occurrence, or significance are not considered.

2.3. Semistructured Interviews

To identify the impacts and be able to transfer the water reuse concept to comparable locations, opinions and views of different stakeholders had to be captured through empirical social research [60–62]. The semistructured interview technique, also called guided interview, represents the most appropriate option in this specific case. This method is based on an interview guide, consisting of a catalogue of open questions allowing the interviewees to freely respond, which increases the chance of receiving new and unexpected information. Questions and ranking are predetermined, but the interviewer can customize both by changing the ranking and even skipping or omitting questions. This form is often referred to as a semistandardised interview guide, which means predefining questions or topics without giving answer options [60]. In semistructured interviews, expert interviews are mostly conducted to obtain expertise of specialists, who have a separate view of certain project components. Expert knowledge refers on the one hand to structural expertise that is easily accessible, but also to practical and action knowledge [60]. In the end, 21 expert interviews were conducted. To facilitate the selection of appropriate interviewees, several stakeholder groups were considered, including science (8 interviews), governmental institutions (5), private sector (4), development cooperation (3) and pastoralists (1). The number of interviews is comparable to similar studies [63–65] and proved to be sufficient, as saturation occurred in terms of statements and identified impacts; i.e., no further impacts could be identified in the course of the field phase.

Concerning the evaluation method, qualitative content analysis was suitable, which aimed to systematically extract the content from qualitative text material by forming categories and, if necessary, to quantify it. The qualitative content analysis thus occupies an intermediate position between qualitative and quantitative research and is often combined in research practice with quantitative content analysis [66]. In contrast to the usual qualitative analysis approach, categorisation was done first and then coding, because an intensive thematic analysis was already carried out during the preparation of the interview guide, from which the themes emerged. These topics or thematic categories comprised (i) the collection and treatment of wastewater, (ii) the reuse of water for fodder production, (iii) alternative water applications, (iv) success factors and hazards and (v) transferability. Within the first two topics, subordinate groups (ecological, social and economic impacts) were developed, which in turn were examined for two characteristics, namely, intended and unintended impacts.

2.4. Scenario Analysis

A scenario analysis was carried out to illustrate possible future developments of the water reuse system. Corresponding to a period of about 10 years, 2030 was chosen as an appropriate time horizon. Storylines were created to describe the circumstances and developments of the water reuse system [66]. In contrast to exploratory scenarios, normative ones allow the formulation of desirable futures (e.g., sustainable operation of the water reuse system) [67–69]. Backcasting can be used to identify measures and pathways to achieve the desirable scenarios. The scenario analysis was divided into four steps [70], including (1) the definition of the problem and the delimitation of the subject, (2) the identification of the most important influencing factors and their development potential, (3) the formation of scenarios and (4) the evaluation of these scenarios.

The factors or uncertainties with the most significant influence on the research object were, on the one hand, the management of the plant and fodder production and, on the other hand, the demand for fodder. This is due to the fact that management is a prerequisite for the operation of the system. Management here refers to the operation and maintenance of the water reuse system, the provision of treated water, the production of fodder and its marketing. In the case of future management, a distinction was made between good management on the one hand and inadequate management on the other, depending on the operator's sense of responsibility. Regarding the demand for fodder affecting the yields of the operating farmers, two variants of low and high demand were assumed. Fodder demand is determined by the need for fodder and the willingness to pay for it. The need for fodder, in turn, depends on the number of animals and the available feeding place.

Finally, four individual scenarios could be derived, namely A1, A2, B1 and B2 (Figure 3). In the best-case scenario (A1), it is assumed that there is a high demand for fodder in the region and, in addition, the operators of the water reuse plant and the cultivation site pursue a very good management. Scenario A2 describes a situation characterised by high demand for fodder and an inadequate management of the treatment and reuse system. The combination of low demand for fodder and good management is assumed in scenario

B1. In the worst-case scenario B2, there is both low demand for fodder and inadequate management of the system.

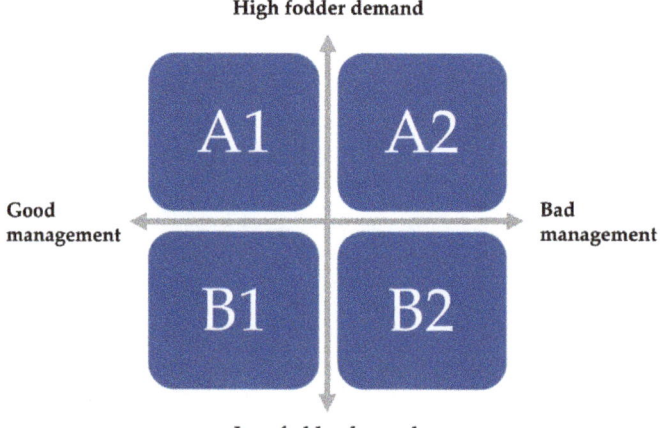

Figure 3. Four scenarios related to the driving forces management and fodder demand: A1 (good management and high fodder demand), A2 (bad management and high fodder demand), B1 (good management and low fodder demand) and B2 (bad management and low fodder demand).

The assessment of the different scenarios took place on a qualitative level by creating a causal loop diagram (CLD). A CLD represents a form of a causal diagram and visually depicts the relations between different variables in a social–ecological system (SES). This type of causal diagram consists of a set of nodes and edges, whereby nodes represent variables and edges the links forming a connection or a relation between two variables [71–73].

3. Results and Discussion

3.1. Ecological Impacts

3.1.1. Intended Impacts

Concerning the intended impacts, improving the effluent quality of treated water is of crucial importance and provides the basis for several of the subsequent aspects (Interviewee UW/H 2018). First, improved water quality represents a lower source of pollution, which reduces the risk of contamination with the surrounding Oshanas in the event of overflowing (cf. [18]). Secondly, the water quality ensures fewer emissions from the WSP, especially methane gas, which is extremely climate-damaging (34 times CO_2 for a period of 100 years; cf. [17]). In this context, the project-related measures contribute to climate change mitigation. At the same time, a method of adaptation to climate change is created by reclaiming municipal wastewater and thus developing a new water resource against the background of water scarcity.

One of the potentially intended impacts is groundwater recharge and groundwater dilution due to irrigation beyond demand (Interviewee TUD I 2018; Interviewee ISOE 2018). Since the dominant soil in the study area is quite sandy, the soils are characterised by a low water storage capacity and a low useable field capacity, i.e., the water would infiltrate quickly contributing to a leaching of salts and also an enrichment of groundwater (cf. [74]). To a large extent, the groundwater in central northern Namibia is very salty (cf. [75]). The additional water from irrigation could therefore help dilute the groundwater and possibly enable groundwater use. Under certain conditions, agricultural irrigation with treated water primarily enhances the value of the soil. Because of the plant growth (e.g., agroforestry) new ecosystems can develop and lead to small isles of biodiversity. Since the intended area of agriculture is withered land, not only the flora benefits, but also the fauna gains access to new habitats through the cultivation (Interviewee ISOE

2018). According to Interviewee HGU I (2018), reuse of dried sludge from the WSP as fertiliser on arable land means that artificial fertilisers might no longer be necessary (cf. [19]). This eliminates the need to produce or import additional fertiliser, reduces emissions and introduces less harmful components into the soil. Furthermore, the reclaimed water contains nutrients, such as nitrogen, phosphorus and potassium. However, by treating the water within the different treatment stages for fodder production, the risk of product contamination cannot be excluded but it can be significantly minimised (Interviewee HGU I 2018).

Concerning the issue of overgrazing, a positive side-effect of fodder production could be the relief and conservation of natural pasture land by providing fodder (Interviewee TUD I 2018). Thus, the illustrated 'transformation of agriculture' in the form of water reuse and fodder production results in the relief of resources of both water resources and grass lands (Interviewee ISOE 2018).

3.1.2. Unintended Impacts

In the case of very intensive, long-lasting rainfall, especially at the transition between the rainy and dry seasons, an overflow of the evaporation pond, which serves as a buffer, cannot be excluded. Owing to the improved water quality and the mixing with rainwater, this would not take on the ecological and social dimensions of an overflow prior to the upgrading, but would have some impacts on nature and society (Interviewee ISOE 2018).

Concerning the soil of the irrigation site, an unintended consequence would be the pollution by organic substances, heavy metals and pharmaceutical residues from the water, and above all, by the use of sludge. Especially heavy metals, which are deposited in the biological phase in the sewage sludge, do not decompose and can therefore potentially enter the cultivation field (Interviewee UW/H 2018; Interviewee HGU I 2018).

Leaching is a potential consequence of excessive irrigation. Besides the leaching of salts, also pollutants from the sludge can be transported into the groundwater. In this context, it must be taken into account that the different soil layers also perform a filtering effect and thus only a part of the pollutants reach groundwater. Shallow groundwater is found at depths of 0 to 20 m, the Ohangwena I aquifer at a depth of 50 to 140 m and the Ohangwena II aquifer at a depth of 220 to 300 m [2].

When managing salts, the risk of salinisation by irrigation must be considered (cf. [76]). As a result of excessive irrigation, salt is dissolved in the soil and rises to the top where it accumulates in the upper soil layer. According to the experts interviewed, the risk of salinisation can be assumed as rather low, as the irrigation is adapted to the plants and additionally during rainy season salts will be washed out (Interviewee TUD I 2018; Interviewee ISOE 2018).

If supplementary fodder is made available to pastoralists, the number of livestock in the surrounding region could increase, leading to overgrazing of natural pastures. As explained above, pastoralists will prefer the natural pasture to the purchase of fodder, so the increase in livestock would accelerate overgrazing. While Interviewee ISOE (2018) and Interviewee UNAM (2018) believe that the volume of fodder production is too small to have an effect, other experts such as Interviewee UW/H (2018), Interviewee KfW II (2018), Interviewee OTC II (2018) and Interviewee MAWF (2018) could imagine that the increase in livestock affects natural pasture.

3.2. Social Impacts

3.2.1. Intended Impacts

By upgrading the WSP system, the hygienic conditions in Outapi have generally improved (cf. [58]). This is partly due to the avoidance of backwater within the sewer system but also to the reduced risk of overflowing ponds and thus contaminating the surrounding Oshanas. Especially fishermen who fish in the Oshana depressions are no longer exposed to the risk of contamination (Interviewee TUD I 2018). Moreover, the fencing of the site, its control and the resulting prevention of human and animal contact with the pond water and their spread of pathogens also contribute to a reduction in health risks. Through the welfare created and improved health conditions, this also gives people back a piece of their dignity, especially in poorly developed regions. This, in turn, strengthens people's work force, which benefits the whole community (Interviewee ISOE 2018).

Almost every expert mentioned job creation and income. This starts with the management in the OTC, continues with the plant's maintenance and safety and ends with the farmer growing fodder and the sellers distributing the products (Interviewee ISOE 2018). Thence, the work of employees could also trigger increased interest in society and thus create awareness for this issue (Interviewee OLBMC 2018). By knowledge sharing in terms of technical and operational solutions within the WWTPP, a large number of municipalities and their inhabitants benefit, too (Interviewee UW/H 2018).

Regarding agricultural production, the availability of fodder can generally be guaranteed at local level throughout the year, making long trips to natural grazing sites avoidable (Interviewee TUD I 2018). As a side-effect, pastoralists can save high costs during droughts for supplementary fodder that has to be provided from distant regions. Moreover, by providing fodder, emergency slaughtering due to the lack of grass can be avoided (Interviewee UW/H 2018). As a further socioeconomic intended impact, the provision of fodder functions as an increase in the yield potential if concentrated fodder is additionally fed before an animal is slaughtered. Since the nutrition in Namibia is very meat-heavy, the sale of meat, and also the additional sources of income in general, can contribute to food security, welfare and rural development (Interviewee TUD I 2018, Interviewee ISOE 2018). If the quality of the treated wastewater would allow the irrigation of vegetables, this could help to achieve a more balanced diet for the population (Interviewee HGU I 2018).

A somewhat broader aspect beyond fodder production is the cultivation of trees with treated water, similar to nurseries. One effect refers to an improved urban atmosphere through trees and parks that provide shade, improve air quality and protect against wind erosion (Interviewee ISOE 2018).

3.2.2. Unintended Impacts

The acceptance of fodder produced with reused water within society is a frequently contested aspect. On the one hand, Interviewee UNAM (2018) believes that there is indeed a lack of acceptance regarding irrigated products due to negative connotations of wastewater. On the other hand, many experts such as Interviewee ISOE (2018), Interviewee NamWater (2018) and Interviewee KfW II (2018) are convinced of the opposite. This is based on the fact that even before the upgrading, sewage of the ponds was drunk by livestock. Furthermore, the quality of the water in the Calueque–Oshakati canal from which the animals drink is of bad quality, because pollutants are introduced when cars are washed at the canal or even some animal carcasses are floating inside the canal (Interviewee MAWF 2018).

If the harmlessness of the products can be ensured by controls, awareness campaigns must be started as already initiated within the project in order to inform about the benefits of the system (Interviewee HGU II 2018; Interviewee GOPA 2018). The dissemination of information would lead to unequal access to information. For example, it matters whether livestock farmers have been informed about the existence of supplementary fodder and how it is marketed so that no stakeholder suffers a disadvantage (Interviewee HGU I 2018).

The displacement of traditional farmers was invalidated by the interviewees. In principle, it would have been possible that farmers specialising in the production of fodder could not have been competitive to the new source of fodder. This effect has been allayed as there is virtually no commercial fodder production in northern Namibia and traditional farmers, in particular, are pursuing subsistence farming (Interviewee TUD I 2018; Interviewee ISOE 2018). However, if there has not been any fodder production so far, a certain dependency on these products can develop and in the event of crop failures, pastoralists dependent on fodder would be severely affected.

Because of the fencing of the plant and the cultivation site, the grazing area will become smaller. In addition, the increase in number of livestock and the associated increase in competition for natural grazing land could have an unintended impact on pastoralists who cannot afford fodder. These pastoralists would have to extend the range where they look for fodder for their livestock and thus take on further distances and additional strains (Interviewee MAWF 2018). The increasing commercialisation of agriculture, pastoralism and marketing could also result in the loss of cultural heritage in the sense of traditional farming and pastoralism methods. Moreover, these effects can contribute to a further widening of the social gap between rich and poor (Interviewee GIZ 2018; Interviewee HGU I 2018). According to the Gini index [50], Namibia is already one of the countries with the greatest disparities between rich and poor, a fact that could worsen due to mismanagement.

Because of the fence around the WSP and its monitoring, it is no longer possible for pastoralists to enter the facility and water their livestock there (Interviewee TUD II 2018; Interviewee UW/H 2018; Interviewee HGU II 2018). In addition, the water volume of the surrounding Oshanas is reduced in order to avoid an overflow of pond water. That change significantly limits fishing during the dry season. This would be an initial negative impact for previous beneficiaries. Yet, the risk of contamination for humans and animals is significantly reduced at the same time. However, the absolute purity of the produced fodder cannot be fully guaranteed, depending on the ingredients of the irrigated water and the sludge applied. This might allow pathogens to spread via contaminated fodder products, which could also have an impact on animal health and thence on meat or milk quality (Interviewee HGU I 2018). In addition to the health consequences, this would also result in financial losses for the animal owners and sellers. Nevertheless, if the water reuse system is well run, there are very low health risks (Interviewee HGU I 2018). Health hazards such as contaminants of emerging concern were not mentioned by the interviewees [8].

Emerging envy, conflicts or even hostilities could arise between farmers. This could take the form of illegally leaving livestock on arable land to feed or maliciously destroying or burning crops (Interviewee KfW II 2018). Out of necessity, pastoralists could also illegally enter the area with their livestock in order to graze or use the ponds as a drinking trough. This in turn could lead to the spread of pathogens by livestock. Envy can also take place at a higher level, for example, between municipalities that feel disadvantaged compared to Outapi (Interviewee NamWater 2018). In extreme cases, this could entail moving residents from a neighbouring town to Outapi, as the conditions in Outapi are more favourable for them. On a larger scale, the town affected by emigration would suffer considerable economic losses and setbacks.

As pastoralism in Namibia is traditionally in the hands of men, women benefit only to a very limited extent or indirectly from fodder production (Interviewee GIZ 2018). When the meat is sold, women are involved again, typically at open markets. Depending on the business model of irrigated fodder production, people already selling fodder could be displaced as well. If the sale of this supplementary fodder is in the hand of the farmers themselves, the traders and vendors in the central north, who procure fodder from distant regions, would be negatively affected and could in the worst case lose their source of income (Interviewee NGO 2018).

Additional fodder could result in an increase in the number of livestock. By purchasing supplementary fodder, selling meat with better quality will be more lucrative, resulting in increased meat consumption (Interviewee HGU I 2018). Depending on whether people can afford to eat more meat and whether they own their livestock, this in turn is associated with an unbalanced nutrition and thus health consequences for humans.

Conflicts could arise regarding the question of how to use the treated water (Interviewee TUD I 2018). For example, farmers could prefer to use the water for the purpose of irrigating vegetables, since the yields to be achieved through this are significantly higher than with another purpose. If water quality does not permit the cultivation of food, this could have drastic consequences on the quality of food and in particular for human health. In this instance the OTC as operator would have to intervene. The responsibility for the plant and its management could also be a burden for the operator, which could lead to a lack of maintenance and thus poorer quality of water (Interviewee ISOE 2018). This is a circumstance that can also be observed in other countries in southern Africa [7].

3.3. Economic Impacts

3.3.1. Intended Impacts

A key economic impact is the relief of the entire WSP system and the reduction in sewage system failures (backwater) due to the technical upgrading (Interviewee TUD I 2018). The increased capacity and thus improved efficiency of the WSP can save costs, which would have been incurred for repair work or an expansion of the plant with additional ponds (Interviewee UW/H 2018). Furthermore, improved management of the plant and the coordination with irrigation contributes to increased efficiency. By providing the treated water and depending on the business model, also the production and sale of fodder, the operator (OTC) generates income to cover the costs incurred for the operation and maintenance of the plant. In addition, the WWTPP contributes to reducing costs significantly by sharing knowledge and experiences among the participating municipalities (Interviewees TUD 2018; Interviewee HGU 2018; cf. [58]). The importance of cost-effective technologies can also be observed in other African countries [77].

Another impact is the creation of value since 'waste' is transformed from a harmful substance into a valuable resource (Interviewee ISOE 2018). On the one hand, this concept illustrates a possibility of sewage disposal that was previously managed via evaporation. On the other hand, a new water resource is created, which can be used for irrigation and fodder production, respectively. In addition, the water contains nutrients such as phosphorus, nitrogen and potassium, which have an added value in agriculture as fertilisers (Interviewee UW/H 2018). At the same time, improved agricultural production and product quality is achieved by using sludge and treated water, which is reflected in rising revenues for farmers and pastoralists (Interviewee UW/H 2018; Interviewee HGU II 2018). A further side effect is that the sewage sludge separated from the wastewater through a microscreen is transported to another plant north of Outapi (Oswin O. Namakalu Sanitation and Reuse Facility), where biogas and electricity can be produced by a fermenter (Interviewees TUD 2018; Interviewee UW/H 2018; cf. [75]).

The evaporation pond serves as an additional buffer to prevent overflowing during the rainy season and to compensate for the lack of precipitation in dry season (Interviewee TUD I 2018). The possibility of year-round cultivation creates economic benefits for farmers and indirectly also for buyers and/or consumers (Interviewee TUD I 2018; Interviewee HGU 2018).

Project-related fodder production can contribute to building a regional market for fodder, which could lead to a strengthening of trade, a development of commercial breeding farms and a general economic upturn in the region (Interviewee TUD I 2018). The Namibian economy also benefited from technical improvements at the WSP that were carried out by Namibian companies, among others. If the water reuse concept is replicated, an economic branch in Namibia focusing on corresponding measures could be established, which may reduce Namibia's dependency on imports (Interviewee HGU 2018; Interviewee NGO 2018).

3.3.2. Unintended Impacts

A question regarding unintended impacts is whether there will be people willing to buy the treated water at all. If the illegal withdrawal of water from the canal is left out, the treated water should not be more expensive than tap water; otherwise, the farmers would not buy it (Interviewee KfW 2018; Interviewee UNAM 2018). Another aspect could be the lack of demand for water during the rainy season due to natural rainfall (Interviewee UNAM 2018). This is invalidated by the buffering function of the evaporation pond since water supply is only provided during dry season when water demand is high. In the case of a tenant farmer, the business model would then have to be designed in such a way that the farmers pay amounts to the operator all year round, such as in the case of a lease.

Furthermore, pipe clogging of the irrigation system by particles could occur, assuming some kind of drip irrigation is applied (Interviewee HGU I 2018; cf. [78]). This would limit plant growth but would also entail repair costs. A significantly greater damage would result for the farmers if livestock entered the field and destroyed the harvest or damaged the irrigation system (Interviewee TUD I 2018).

A major problem associated with the fodder production is the possible lack of fodder demand due to multiple reasons. On the one hand, a reason could be the origin of the treated water (sewage), which could deter buyers (Interviewee OLBMC 2018). On the other hand, the lack of demand could result from high fodder prices due to irrigation, which pastoralists cannot afford. The price of the fodder produced should therefore not be higher than the market price (Interviewee UNAM 2018). Fodder prices that can be charged are likely to be low especially at the beginning of the dry season, which affects the farmers' income (Interviewee ISOE 2018; Interviewee UNAM 2018).

Assuming the tenant model, the conditions described result in a mutual dependency between the operator and the farmer. The farmers rely on treated water, especially in the dry season, and the operator in turn depends on the yields of the farmers in order to cover their running costs and plant maintenance (Interviewee OLBMC 2018). Events such as crop failures could therefore lead to financial losses for farmers and thus move the equilibrium point of the system into another state.

3.4. Success Factors and Hazards for Sustainable Operation

3.4.1. Success Factors

The focus regarding success factors is on the underlying business model of the water reuse system, in particular on management and maintenance (Table 1) (Interviewee KfW 2018; Interviewee UW/H 2018). A prerequisite for good management is that the operators develop an awareness of the system and a willingness to carry out the work involved (Interviewee UW/H 2018). Clear responsibilities, monitoring and control structures must be established within the operator [75,79]. In this context, qualified personnel with a certain sense for responsibility is crucial, e.g., to anticipate possible weaknesses in the system (Interviewee HGU I 2018; Interviewee TUD II 2018). The training of workers is indispensable not only in relation to the WSP system, but also in agriculture related to cultivation and the use of irrigation systems (Interviewee HGU II 2018; Interviewee MAWF 2018). In order to facilitate the maintenance of the plant, it is advantageous to pursue low-tech approaches that simplify the procurement of spare parts and repair (Interviewee UW/H 2018). The WWTPP contributes to this (Interviewee TUD I 2018; Interviewee HGU I 2018).

Table 1. Success factors and hazards for sustainable operation of the water reuse system.

Success Factors	Hazards
Business model, good management	Vandalism (stealing, fence, burning fields, grazing fields)
Awareness, sense of responsibility	Clogging of irrigation pipes
Qualified personnel	Lack of demand for water and fodder
Training of workers	Water usage competition / conflicts in general
Maintenance	Lack of awareness
Low-tech approach	Poor maintenance, no monitoring
Wastewater treatment plant partnership	Missing spare parts
Communication between stakeholders	Decreasing water quality affecting farmers cultivation
Public participation	Lack of specific education (curricula)
Overcoming prejudices	Incorrect irrigation practices
Ensuring product safety	Extreme population growth in the town
Functioning supply chains	Flooding of cultivation site, erosion
Equal distribution in income	Switching off pretreatment
	Employers migration, brain drain
	Deterioration in political relations with Angola

A further success factor involves communication between the various stakeholders, primarily between the plant operator and the farmers, in order to coordinate and simplify work steps, but also among local authorities, associations, urban planners and environmentalists to increase acceptance in society through public participation (Interviewee NGO 2018). By overcoming prejudices regarding treatment of sewage and ensuring product safety, it must be proven and demonstrated that the product is harmless to humans (Interviewee HGU I 2018; Interviewee TUD II 2018).

Functioning supply chains are essential for the business model. This refers to the provision of water, the production of fodder and its purchase. When generating income, care must be taken to ensure a fair distribution to avoid conflicts (Interviewee TUD I 2018; Interviewee UNAM I 2018).

3.4.2. Hazards

Hazards to sustainable operation comprise vandalism (Interviewee HGU II 2018), clogging of irrigation pipes, salinisation (Interviewee TUD II 2018), lack of demand for water and fodder (Interviewee ISOE 2018), and water usage competitions (Interviewee ISOE 2018). Further hazards relate to the success factors, such as awareness problems (Interviewee UNAM II 2018) and therefore lack of maintenance, in particular, missing spare parts and controls (Interviewee UNAM II 2018), resulting in decreased water quality that would in turn affect the farmers and consumers (Interviewee TUD II 2018). Another hazard is the lack of specific education (curricula) (Interviewee HGU I 2018) and thus poorly qualified farmers, which leads to incorrect irrigation practices and impairment of societal health. In particular, health risks for both livestock and humans pose a risk, which in the event of incidents would fuel the negative image of water reuse and cause sales to collapse (Interviewee UW/H 2018).

One hazard not previously considered is the extreme population growth in the town, resulting again in capacity bottlenecks of the WSP (Interviewee UNAM I 2018; cf. [53,80]). Given the current population trends in the study area, planning for the upgrade is for a period of approximately 10 to 15 years (Interviewee ISOE 2018; Interviewee UW/H 2018). Natural hazards can occur in form of flooding of cultivated areas due to heavy rainfall, which in turn can lead to soil erosion. This would only affect a very small area of land and due to the low gradient, soil erosion would not be very significant (Interviewee ISOE 2018; Interviewee HGU II 2018). Going along with the burden of operation and associated investments, in the worst case, the operator could switch off pretreatment and return to the original condition before upgrading (Interviewee TUD I 2018). In addition, as part of the brain drain, the operator's employees may leave their current jobs because working conditions are better elsewhere (Interviewee TUD II 2018; Interviewee HGU I 2018).

A hazard that would affect not only the sanitation and water reuse system in the study area but also northern Namibia entirely would be a deterioration in bilateral relations with Angola, which, in the worst case, could lead to no more water provided for the canal. However, owing to good bilateral relations, this can be considered relatively unlikely at present (Interviewee TUD I 2018).

3.5. Significance of Management and Operation

Based on the success factors and hazards, it becomes clear how important management is for sustainable operation of the water reuse system. To illustrate the changes associated with the mode of operation, the following summary tables show intended and unintended impacts, taking into account good and poor management. To assign the impacts to the different groups and perspectives, a subdivision into stakeholders or spheres of influence was made, which include operators of the system, society, economy, ecosystem and ecosystem services. Table 2 shows the intended impacts occurring in terms of poor operation (only black) and good operation (additionally blue). Table 3 shows the unintended impacts that occur with good operation (black) and poor operation (additionally red).

Table 2. Intended impacts of the upgrading process related to poor (black) and good (blue) operation, allocated to spheres of influences.

	Operators			Society						
	Employees	Farmer	Fisher	Pastoralists	Seller	Consumers	Residents	Municipalities		Economy
OTC	income	income (tenant)	decreased health risks	fodder assurance	income	food security	improved hygiene	wastewater treatment plant partnership		increased efficiency
relief of ponds/minimising overflowing/sewage disposal	welfare/livelihood	higher yields	bigger and healthier fish	avoiding emergency slaughtering, reputation, capital reserve	welfare/livelihood	decreased health risks	triggering interest	strengthening cooperation		promoting regional fodder production
prestige/recommendations		welfare/livelihood	higher yields	avoiding high costs for fodder			awareness	reducing costs		trade
financing of treatment plant		independent to climate variability		avoiding long trips			dignity	decreased health risks downstream		reducing import dependency
revenues		reducing use of artificial fertiliser		welfare/livelihood higher yields, better meat quality			improved town atmosphere			establishing new industries
maintenance										power generation
reduction in repair effort										increased buying power
increased efficiency										
			Ecosystem					Ecosystem Services		Cultural
	Groundwater	Pastures	Biodiversity	Oshanas	Atmosphere	Provisioning	Regulating	Supporting		maintaining traditions
Soil										
less degradation	recharge	conserving pastureland	increasing biodiversity (agroforestry)	minimising overflowing/less contamination	less emissions from ponds	water	climate regulation	soil formation		recreational
increasing moisture	dilution	less desertification	relief of resources	improved water quality	less smell	sludge/fertiliser	purification of water reducing pathogens/diseases	nutrient cycling		science
humus enrichment, additional nutrients						fodder				
reducing use of artificial fertilizer						food				
						power generation				

Table 3. Unintended impacts of the upgrading process related to good (black) and poor (red) operation, allocated to spheres of influences.

	Operators			Society						
	Employees	Tenant farmer	Fisher	Pastoralists	Seller	Consumers	Residents	Municipalities		Economy
OTC	brain drain	contamination	less water in Oshanas, no fishing in ponds contamination in case of overflow	less grass areas due to agriculture no water trough at the ponds	competition for existing seller	health risk increased meat consumption	women do not benefit opening of the social gap displacement due to new farming site	migration of inhabitants		lack of demand for fodder
burden		dependency		lack of acceptance						
information inequity		lack of demand for fodder		cultural losses						
lack of demand for water		pipe clogging		lower quality of meat						
higher costs for maintaining		damaging of pipes lower yields								
less revenues										
			Ecosystem					Ecosystem services		Cultural
	Groundwater	Pastures	Biodiversity	Oshanas	Atmosphere	Provisioning	Regulating	Supporting		commercialising pastoralism
Soil										
pollution (heavy metals)	pollution	more livestock, overgrazing	loss	risk of overflowing in heavy rain	more methane due to livestock		health risks			loss of tradition
salinisation	groundwater rise, dissolving salts	less grass areas	displacement of wild animals	increased risk of overflowing						
damage of soil quality										

3.6. Scenario Analysis

As described above, the two main drivers of the water reuse system comprise management and fodder demand. Figure 4 shows a causal loop diagram of the various system components linked with the identified drivers. The diagram combines the three categories, ecology, society and economy, with the system of water reuse for fodder production. The causal links between the variables are represented by arrows with a positive or negative polarity indicating the type of influence. Furthermore, the diagram shows reinforcing (R) and balancing feedbacks (B), feedbacks consisting of at least two variables forming a loop.

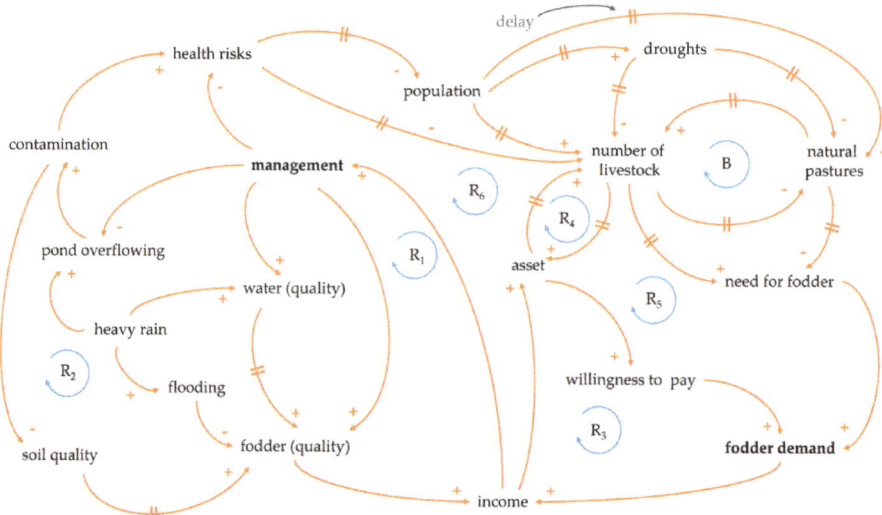

Figure 4. Causal loop diagram of the different system components linked with the main driving forces, management and fodder demand (highlighted). Two strokes on a causal link symbolize a time delay between the current and perceived state of a process. Blue circles containing either an R or a B represent reinforcing or balancing feedbacks.

The most important reinforcing feedback consists, among others, of the variables management and water quality, which are connected by a positive causal link. Water quality, in turn, indicates a positive causal link with the quality of the fodder, which generates income. As a result of the yields achieved from the sale of fodder, the management is again strengthened, which completes the reinforcing effect (R1). Another reinforcing feedback arises from the causal links of the management to pond overflowing and from there to the contamination. Contamination is linked to the soil quality and therefore to the fodder. Similar to R1, fodder generates income, which in turn leads to initial management. Since this loop comprises two negative causal links, a reinforcing feedback is formed (R2). Additional reinforcing feedback concerns income, since income from fodder creates an increase in assets in all their facets. Therefore, the willingness to pay for fodder and also fodder demand rises due to the positive causal links, resulting in initial income (R3). Besides the willingness to pay, the fodder demand depends on the need for fodder, which in turn relies on the number of livestock and natural pastures. These two variables also form a balancing feedback as they are directly connected to two causal links of different polarity (B). The same applies to the variables asset and number of livestock, but due to the two positive causal links there is a reinforcing feedback here (R4). By combining R3 and R4 with the need for fodder, a new reinforcing feedback is created (R5). Furthermore, different existing loops can be merged, e.g., the combination of R1 and R5 with health risks and population leads to the reinforcing feedback R6.

By identifying the feedbacks, the complexity of the water reuse system in relation to the driving forces becomes clear. The diagram facilitates the assessment of the four identified scenarios A1 (good management and high fodder demand), A2 (bad management and high fodder demand), B1 (good management and low fodder demand) and B2 (bad management and low fodder demand). Since a scenario analysis similar to the CLD model developed would be difficult to cope with and numerous overlaps occur in the content of the scenarios, the focus of the evaluation is on highlighting the differences in order to avoid repetition of impacts.

Figure 5 shows the intended and unintended impacts of the upgrading and assigns them to the four scenarios. If effects only apply to one single scenario, they are located in one of the four corners. In addition, there are four fields in which two scenarios demonstrate the same characteristics of a single driving force, for example, a high fodder demand in the upper middle. The ninth and last field in the middle contains impacts that can occur in all four scenarios.

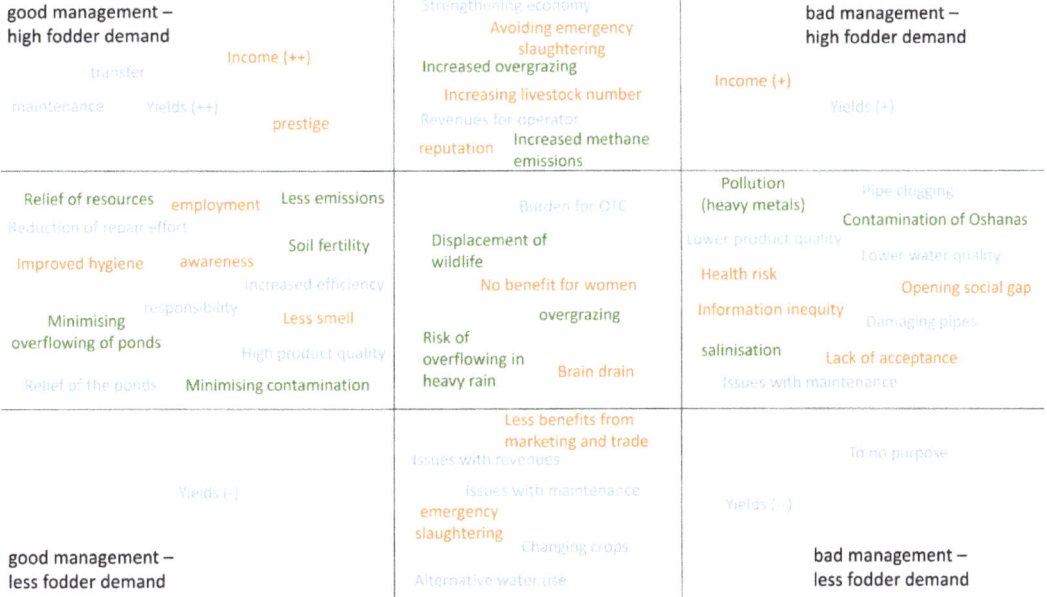

Figure 5. Ecological (green), social (orange) and economic impacts (blue) of the upgrading process associated with the four identified scenarios; corner fields represent impacts that occur only for the specific scenario; middle outer fields symbolise overlaps between two adjacent scenarios with the same driving force, and the field at the centre overlaps of all four scenarios.

Effects that occur in all four scenarios include only unintended impacts, namely, the displacement of wildlife, the risk of ponds overflowing during heavy rainfall, a possible burden on the operator and the fact that women do not benefit from fodder production and pastoralism. In addition, overgrazing is already widespread in northern Namibia. On the one hand, additional fodder could relieve the strain on natural resources. On the other hand, pastoralists would prefer free grass to produced fodder. Another effect that can always occur independently of the driving forces is brain drain, especially if there are wage payment bottlenecks.

Irrespective of the fodder demand, good management leads to relief of the WSP and less repair work for the plant. Furthermore, reusing water relieves the strain on natural resources. With good management, the effluent is characterised by a good water quality, which benefits fodder production with nutrients and a minimised risk of contamination, thereby improving soil fertility. Good management also reduces the risk of ponds overflowing, and in the event of heavy rain, the strain on the surrounding Oshanas is significantly lower than before the upgrade. Society also benefits from the management through improved hygiene and income associated with employment. Additionally, good management reduces emissions and odour nuisance.

In contrast, bad management leads to maintenance issues and thus a lower quality of water. Therefore, an increased risk of land contamination is present, in particular by heavy metals, resulting in lower product quality and health risks for consumers. In addition, the lower water quality and the particles contained in the water lead to pipe blockages and damage to the pipes. With an insufficient management, the risk of ponds overflowing and therefore the risk of contaminating the Oshanas also increases. Concerning society, information inequalities can widen the social divide. In addition, a disregard for the involvement and education of the population can lead to a lack of acceptance.

Assuming that fodder production is guaranteed regardless of the management style, high fodder demand is associated with revenues for the operators. The consideration of a high demand for fodder shows that the number of livestock will increase due to the avoidance of emergency slaughtering. As a result, the increased number of animals will lead to increased methane emissions and overgrazing of natural pastures.

Less fodder demand, however, results in issues with revenues and thus the maintenance of the system. In addition, benefits from marketing and trade are missing out. A low fodder demand based on a lack of willingness to pay or financial resources does not change the state of emergency slaughtering, resulting in losses for pastoralists. Furthermore, a lack of yield by farmers can lead them to reconsider their cultivation system and possibly generate higher margin products that do not meet water quality standards. Otherwise, a complete reorientation towards water reuse alternatives could be initiated.

Concerning the scenario-related impacts, it is noticeable that the number of such effects is relatively small. A dominant factor is the generated yields for fodder and the associated revenues for the operator, which vary from scenario to scenario. In the worst-case scenario (B2), inadequate management coupled with low demand for fodder will generate the lowest yields. In this case, the operation of the underlying system consisting of water treatment and fodder production would not make sense. In terms of yields, this is similar in scenario B1, where there is good management and thus good water quality, but low fodder demand is the basis. If these circumstances do not change in the near future, it would be advisable to rethink this kind of water reuse, as the efforts extend beyond the benefits.

The situation is different in the case of inadequate management combined with high demand for fodder (A2), where income is generated from the yields. However, due to the lower product quality coupled with possible crop losses, the yields to be achieved are lower. Here, more conscious management could significantly increase yields and thus incomes. Optimal conditions are given in scenario A1, where good management meets high fodder demand. This is where the highest yields are achieved, which benefits the operators themselves and the maintenance of the system. As a prestige object, this would enhance the reputation of the operator and the town. At the same time, in terms of transferability, the concept thus represents a prime example for other regions.

4. Conclusions

Against the background of the current problems in managing WSP systems and water supply constraints in central northern Namibia, this study demonstrated that water reuse based on an upgrade of WSPs represents an extremely sensible and sustainable concept. The impact assessment has shown that intended positive impacts clearly outweigh the unintended ones. Wastewater is not only disposed, but is seen as a new water resource. Thus, a polluting liquid is transformed into a value benefitting the population and the ecosystem. In addition, the concept reveals a new opportunity for water reuse and constitutes a major advantage for water supply and relief regarding limited water resources. The relatively new approach of the SEIA has proven useful and insightful in assessing the social–ecological impacts of the water reuse system and the corresponding success factors and hazards of the transformation process. In this way, insights for a long-term enhancement of living and ecosystem conditions could be achieved.

Following conclusions can be drawn:

- Water reuse systems based on WSPs have great transfer potential due to the large number of WSPs in northern Namibia, but also generally in southern Africa. Most WSPs are facing capacity problems with numerous ecological and social consequences, which is an argument in favour of upgrading WSPs.
- The success of a transfer depends on a number of factors rooted in local conditions. These factors comprise, for instance, the local availability of water, the amount of wastewater collected through a sewage system, potential agricultural areas close to the ponds, demand for fodder, the willingness to pay for it and capacities to manage, operate and maintain the system.
- A major problem that arises with replication is the financing of the initial investment, which cannot be refinanced by the income from fodder production alone. Therefore, external investors or donors must be found, although public funds may also be considered, such as subsidies from responsible ministries.
- Limitations of the study lie in the purely qualitative nature of the results, which means that quantitative aspects were hardly taken into account. Future research could consider approaches such as Baysian belief networks (BBN) to provide both qualitative and quantitative analysis of the research topic.
- From a more general point of view, alternatives for upgrading WSPs, e.g., activated sludge plants, might have to be considered in future research as well. It remains to be seen whether such an option would involve significantly greater transformation efforts, given that WSPs are already available and both investment and operating costs are assumed to be higher.
- With the increasing global water and food crises, the reuse of water from WSPs will become much more widespread since water reuse in agriculture gains importance in all parts of the world. In this way, water reuse based on WSPs also contributes to global food security.

Author Contributions: M.Z. and F.N. conceived and designed the study; F.N. conducted and analyzed the interviews and wrote large parts of the paper; M.Z. wrote additional parts of the paper and made the final edits; M.Z. contributed to the acquisition of funds, administered and supervised the project. All authors have read and agreed to the published version of the manuscript.

Funding: The project was funded by the German Federal Ministry of Education and Research (BMBF) under the funding measure WavE (funding code 02WAV1401B). Open access fees were covered by the BMBF Post Grant Fund (funding code 16PGF0371).

Institutional Review Board Statement: The study was conducted in accordance with the Declaration of Helsinki, and approved by the Ethics Commission of the ISOE–Institute for Social-Ecological Research, Frankfurt am Main, Germany (2022).

Informed Consent Statement: Informed consent was obtained from all subjects involved in the study.

Acknowledgments: The authors would like to thank all Namibian and German interview partners together with the German project partners, in particular, the institute IWAR of the Technical University in Darmstadt. The article was written as part of the project 'EPoNa—Upgrading wastewater pond systems to generate irrigation water for animal fodder production using the example of Outapi, Namibia'.

Conflicts of Interest: The authors declare no conflict of interest.

References

1. Niang, I.; Ruppel, O.C.; Abdrabo, M.A.; Essel, A.; Lennard, C.; Padgham, J.; Urquhart, P. *Climate Change 2014: Impacts, Adaptation, and Vulnerability. Part B: Regional Aspects. Contribution of Working Group II to the Fifth Assessment Report of the Intergovernmental Panel of Climate Change*; Barros, V.R., Field, C.B., Dokken, D.J., Mastrandrea, M.D., Mach, K.J., Bilir, T.E., Chatterjee, M., Ebi, K.L., Estrada, Y.O., Genova, R.C., et al., Eds.; Cambridge University Press: Cambridge, UK; New York, NY, USA, 2014; pp. 1199–1265.
2. Himmelsbach, T. Tiefe, semi-fossile Grundwasserleiter im südlichen Afrika: Hydrogeologische Untersuchungen im Norden von Namibia. *GMIT–Geowiss. Mitt.* 2017, 67, 8–18.
3. Dabasso, B.H.; Wasonga, O.V.; Irungu, P.; Kaufmann, B. Emerging pastoralist practices for fulfilling market requirements under stratified cattle production systems in Kenya's drylands. *Anim. Prod. Sci.* 2021, 61, 12–24. [CrossRef]
4. Luetkemeier, R.; Stein, L.; Drees, L.; Liehr, S. Blended Drought Index: Integrated Drought Hazard Assessment in the Cuvelai-Basin. *Climate* 2017, 5, 51. [CrossRef]
5. Asano, T. (Ed.) *Wastewater Reclamation and Reuse: Water Quality Management Library*; CRC Press: Boca Raton, FL, USA; London, UK; New York, NY, USA; Washington, DC, USA, 1998.
6. Sinn, J.; Agrawal, S.; Orschler, L.; Lackner, S. Characterization and evaluation of waste stabilization pond systems in Namibia. *H2Open J.* 2022, 5, 365–378. [CrossRef]
7. Edokpayi, J.N.; Odiyo, J.O.; Popoola, O.E.; Msagati, T.A.M. Evaluation of contaminants removal by waste stabilization ponds: A case study of Siloam WSPs in Vhembe District, South Africa. *Heliyon* 2021, 7, e06207. [CrossRef]
8. K'oreje, K.O.; Okoth, M.; van Langenhove, H.; Demeestere, K. Occurrence and treatment of contaminants of emerging concern in the African aquatic environment: Literature review and a look ahead. *J. Environ. Manag.* 2020, 254, 109752. [CrossRef]
9. Mungray, A.K.; Kumar, P. Occurrence of anionic surfactants in treated sewage: Risk assessment to aquatic environment. *J. Hazard. Mater.* 2008, 160, 362–370. [CrossRef] [PubMed]
10. Mara, D.D. Waste stabilization ponds: Past, present and future. *Desalination Water Treat.* 2009, 4, 85–88. [CrossRef]
11. Craggs, R.; Sukias, J.; Tanner, C.; Davies-Colley, R. Advanced pond system for dairy-farm effluent treatment. *N. Zealand J. Agric. Res.* 2004, 47, 449–460. [CrossRef]
12. Gumisiriza, R.; Mshandete, A.; Rubindamayugi, M.; Kansiime, F.; Kivaisi, A. Enhancement of anaerobic digestion of Nile perch fish processing wastewater. *Afr. J. Biotechnol.* 2008, 8, 362–370.
13. Veeresh, M.; Veeresh, A.V.; Huddar, B.D.; Hosetti, B.B. Dynamics of industrial waste stabilization pond treatment process. *Env. Monit Assess* 2010, 169, 55–65. [CrossRef] [PubMed]
14. Mara, D.; Pearson, H.W. *Design Manual for Waste Stabilization Ponds in Mediterranean Countries*; Lagoon Techn. Internat: Leeds, UK, 1998; ISBN 0951986929.
15. Ho, L.T.; van Echelpoel, W.; Goethals, P.L.M. Design of waste stabilization pond systems: A review. *Water Res.* 2017, 123, 236–248. [CrossRef] [PubMed]
16. Lado, M.; Ben-Hur, M. Effects of irrigation with different effluents on saturated hydraulic conductivity of arid and semiarid soils. *Soil Sci. Soc. Am. J.* 2010, 74, 23–32. [CrossRef]
17. Sinn, J.; Cornel, P.; Lackner, S. Waste stabilization ponds with pre-treatment provide irrigation water—A case study in Namibia. In *IWA Water Reuse 2019: 12th IWA International Conference on Water Reclamation and Reuse, Book of Abstracts*; DECHEMA–Gesellschaft für Chemische Technik und Biotechnologie e.V: Frankfurt am Main, Germany, 2019; pp. 10–17.
18. Mohr, M.; Dockhorn, T.; Drewes, J.E.; Karwat, S.; Lackner, S.; Lotz, B.; Nahrstedt, A.; Nocker, A.; Schramm, E.; Zimmermann, M. Assuring water quality along multi-barrier treatment systems for agricultural water reuse. *J. Water Reuse Desalination* 2020, 10, 332–346. [CrossRef]
19. Lackner, S.; Sinn, J.; Zimmermann, M.; Max, J.; Rudolph, K.-U.; Gerlach, M.; Nunner, C. Upgrading waste water treatment ponds to produce irrigation water in Namibia. *Watersolutions* 2017, 2017, 82–85.
20. Cornel, P.; Engelhart, M. *Ertüchtigung von Abwasser-Ponds zur Erzeugung von Bewässerungswasser am Beispiel des Cuvelai-Etosha-Basins in Namibia (EPoNa)*; Technische Universität Darmstadt: Darmstadt, Germany, 2016.
21. United Nations. *Transforming Our World: The 2030 Agenda for Sustainable Development: Draft Resolution Referred to the United Nations Summit for the Adoption of the Post-2015 Development Agenda by the General Assembly*; United Nations: New York, NY, USA, 2015.
22. Dendena, B.; Corsi, S. The Environmental and Social Impact Assessment: A further step towards an integrated assessment process. *J. Clean. Prod.* 2015, 108, 965–977. [CrossRef]
23. Bond, A.J.; Viegas, C.V.; de Souza Reinisch Coelho, C.C.; Selig, P.M. Informal knowledge processes: The underpinning for sustainability outcomes in EIA? *J. Clean. Prod.* 2010, 18, 6–13. [CrossRef]
24. Jay, S.; Jones, C.; Slinn, P.; Wood, C. Environmental impact assessment: Retrospect and prospect. *Environ. Impact Assess. Rev.* 2007, 27, 287–300. [CrossRef]
25. Ortolano, L.; Shepherd, A. Environmental Impact Assessment: Challenges and Opportunities. *Impact Assess.* 1995, 13, 3–30. [CrossRef]
26. Abaza, H.; Bisset, R.; Sadler, B. *Environmental Impact Assessment and Strategic Environmental Assessment: Towards an Integrated Approach*, 1st ed.; UNEP: Geneva, Switzerland, 2004; ISBN 9789280724295.
27. DEA. *Procedures and Guidelines for Environmental Impact Assessment (EIA) and Environmental Management Plans (EMP)*; DEA: Windhoek, Namibia, 2008.
28. European Commission. *Impact Assessment Guidelines–Technical Report SEC No. 92*; European Commission: Brussels, Belgium, 2009.

29. Grunwald, A. *Technikfolgenabschätzung-Eine Einführung, Zweite, Grundlegend überarbeitete und Wesentliche Erweiterte Auflage (Gesellschaft-Technik-Umwelt, 1, Band 1)*; Sigma: Berlin, Germany, 2010.
30. European Bank for Reconstruction and Development. *Finance of Investment Projects/Environmental and Social Impact Assessments (ESIA) and Public Consultation*; European Bank for Reconstruction and Development: London, UK, n.d.
31. Burdge, R.J. Benefiting from the practice of social impact assessment. *Impact Assess. Proj. Apprais.* **2003**, *21*, 225–229. [CrossRef]
32. Esteves, A.M.; Franks, D.; Vanclay, F. Social impact assessment: The state of the art. *Impact Assess. Proj. Apprais.* **2012**, *30*, 34–42. [CrossRef]
33. Pope, J.; Bond, A.; Morrison-Saunders, A.; Retief, F. Advancing the theory and practice of impact assessment: Setting the research agenda. *Environ. Impact Assess. Rev.* **2013**, *41*, 1–9. [CrossRef]
34. Corsi, S.; Oppio, A.; Dendena, B. ESIA (environmental and social impact assessment): A tool to minimize territorial conflicts. *Chem. Eng. Trans.* **2015**, *43*, 2215–2220. [CrossRef]
35. IFC. *IFC Performance Standards on Environmental and Social Sustainability*; IFC: Washington, DC, USA, 2012.
36. Rosa, J.C.; Sánchez, L. Is the ecosystem service concept improving impact assessment? Evidence from recent international practice. *Environ. Impact Assess. Rev.* **2015**, *50*, 134–142. [CrossRef]
37. African Development Bank Group. *Sustainable Development of Abu Rawash Wastewater Treatment System. Egypt. Summary of the Environmental and Social Impact Assessment*; African Development Bank Group: Abidjan, Côte d'Ivoire, 2017.
38. Liehr, S. *Social-Ecological Impact Assessment (SEIA): Concept and Terms of Reference*; CuveWaters Project Report; ISOE–Institute for Social-Ecological Research: Frankfurt am Main, Germany, 2012.
39. Klintenberg, P.; Wanke, H.; Hipondonka, M. *Social-Ecological Impact Assessment of the Rainwater Harvesting, Groundwater Desalination, Sanitation and Water Reuse, and Sub-Surface Water Storage in the Cuvelai Water Basin: Thematic Study on Ecology, Land Use, Hydrogeological Cycle and Eco-Hydrology*; CuveWaters; ISOE–Institute for Social-Ecological Research: Frankfurt am Main, Germany, 2012.
40. Brymer, A.B.; Holbrook, J.D.; Niemeyer, R.; Suazo, A.; Wulfhorst, J.; Vierling, K.; Newingham, B.; Link, T.; Rachlow, J. A social-ecological impact assessment framework for public lands management: Application of a conceptual and methodological framework. *Ecol. Soc.* **2016**, *21*, 9. [CrossRef]
41. Jones, M.; Morrison-Saunders, A. Understanding the long-term influence of EIA on organisational learning and transformation. *Environ. Impact Assess. Rev.* **2017**, *64*, 131–138. [CrossRef]
42. Lux, A.; Schäfer, M.; Bergmann, M.; Jahn, T.; Marg, O.; Nagy, E.; Ransiek, A.-C.; Theiler, L. Societal effects of transdisciplinary sustainability research—How can they be strengthened during the research process? *Environ. Sci. Policy* **2019**, *101*, 183–191. [CrossRef]
43. Tribaldos, T.; Oberlack, C.; Schneider, F. Impact through participatory research approaches: An archetype analysis. *Ecol. Soc.* **2020**, *25*, 15. [CrossRef]
44. Schäfer, M.; Bergmann, M.; Theiler, L. Systematizing societal effects of transdisciplinary research. *Res. Eval.* **2021**, *30*, 484–499. [CrossRef]
45. Mendelsohn, J.M.; Jarvis, A.; Roberts, C.; Robertson, T. *Atlas of Namibia: A Portrait of the Land and Its People*; David Philip Publishers: Cape Town, South Africa, 2002.
46. Woltersdorf, L.; Jokisch, A.; Kluge, T. Benefits of rainwater harvesting for gardening and implications for future policy in Namibia. *Water Policy* **2014**, *16*, 124–143. [CrossRef]
47. Kluge, T.; Liehr, S.; Bischofberger, J.; Deffner, J.; Felmeden, J.; Kramm, J.; Krug von Nidda, A.; Schulz, O.; Stibitz, V.; Woltersdorf, L.; et al. *IWRM-Verbundprojekt CuveWaters: Integriertes Wasserressourcen-Management im zentralen Norden Namibias (Cuvelai Basin) und in der SADC-Region. Phase III: Transfer eines Multi-Ressourcen-Mix, Teilprojekt 1: Schlussbericht: Projektlaufzeit: 01.10.2013–31.12.2015*; ISOE–Institute for Social-Ecological Research: Frankfurt am Main, Germany, 2016.
48. Mendelsohn, J.; Jarvis, A.; Robertson, T. *A Profile and Atlas of the Cuvelai-Etosha Basin*; RAISON & Gondwana Collection: Windhoek, Namibia, 2013; ISBN 9789991678078.
49. UNESCO. *Managing Water under Uncertainty and Risk: The Challenges*; UNESCO: Paris, France, 2012.
50. World Bank Group. GINI Index (World Bank Estimate). Available online: https://data.worldbank.org/indicator/si.pov.gini (accessed on 24 August 2018).
51. National Planning Commission. *Namibia 2011 Population and Housing Census: Preliminary Results*; Namibia Statistics Agency: Windhoek, Namibia, 2012.
52. Ministry of Agriculture, Water and Forestry. *The Augmentation of a Water Supply to the Central Area of Namibia and the: CuvelaiPart II: Cuvelai Area of Namibia*; Ministry of Agriculture, Water and Forestry: Windhoek, Namibia, 2016.
53. Namibia Statistics Agency. *Namibia 2011 Population and Housing: Census Indicators*; Namibia Statistics Agency: Windhoek, Namibia, 2012.
54. Zimmermann, M. The coexistence of traditional and large-scale water supply systems in central northern Namibia. *J. Namib. Stud.* **2010**, *7*, 55–84.
55. Kluge, T.; Liehr, S.; Lux, A.; Moser, P.; Niemann, S.; Umlauf, N.; Urban, W. IWRM Concept for the Cuvelai Basin in Northern Namibia. *J. Phys. Chem. Earth* **2008**, *33*, 48–55. [CrossRef]
56. Republic of Namibia. *Namibia Vision 2030: Policy Framework for Long-Term National Development*; Republic of Namibia: Windhoek, Namibia, 2004.

57. Röhrig, J.; Werner, W. *Policy Framework for Small-Scale Gardening*; CuveWaters Papers No. 8; ISOE–Institute for Social-Ecological Research: Frankfurt am Main, Germany, 2011.
58. Frick-Trzebitzky, F.; Kluge, T.; Stegemann, S.; Zimmermann, M. Capacity Development for Wastewater Management and Water Reuse in In-formal Partnerships in Northern Namibia. *Front. Water* **2022**, *30*, 115.
59. Hummel, D.; Jahn, T.; Keil, F.; Liehr, S.; Stieß, I. Social Ecology as Critical, Transdisciplinary Science—Conceptualizing, Analyzing and Shaping Societal Relations to Nature. *Sustainability* **2017**, *9*, 1050. [CrossRef]
60. Döring, N.; Bortz, J.; Pöschl-Günther, S. *Forschungsmethoden und Evaluation in den Sozial-und Humanwissenschaften: Mit 194 Abbildungen und 167 Tabellen, 5., Vollständig überarbeitete, Aktualisierte und Erweiterte Auflage*; Springer: Berlin/Heidelberg, Germany, 2016; ISBN 978-3-642-41088-8.
61. Mattissek, A.; Pfaffenbach, C.; Reuber, P. *Methoden der Empirischen Humangeographie*; Westermann: Braunschweig, Germany, 2013.
62. Mayring, P. *Einführung in die Qualitative Sozialforschung: Eine Anleitung zu qualitativem Denken, 6. Auflage*; Beltz: Basel, Switzerland, 2016.
63. Galvin, R. How many interviews are enough? Do qualitative interviews in building energy consumption research produce reliable knowledge? *J. Build. Eng.* **2015**, *1*, 2–12. [CrossRef]
64. Sattlegger, L. Negotiating attachments to plastic. *Soc. Stud. Sci.* **2021**, *51*, 820–845. [CrossRef]
65. Kerber, H.; Kramm, J. From laissez-faire to action? Exploring perceptions of plastic pollution and impetus for action. Insights from Phu Quoc Island. *Mar. Policy* **2022**, *137*, 104924. [CrossRef]
66. Kosow, H.; Gaßner, R. *Methods of Future and Scenario Analysis: Overview, Assessment, and Selection Criteria*; German Development Institute/Deutsches Institut für Entwicklungspolitik (DIE): Bonn, Germany, 2008.
67. Kok, K.; van Vliet, M.; Bärlund, I.; Dubel, A.; Sendzimir, J. Combining participative backcasting and exploratory scenario development: Experiences from the SCENES project. *Technol. Forecast. Soc. Chang.* **2011**, *78*, 835–851. [CrossRef]
68. Quist, J.; Thissen, W.; Vergragt, P.J. The impact and spin-off of participatory backcasting: From vision to niche. *Technol. Forecast. Soc. Chang.* **2011**, *78*, 883–897. [CrossRef]
69. Carlsson-Kanyama, A.; Dreborg, K.H.; Moll, H.C.; Padovan, D. Participative backcasting: A tool for involving stakeholders in local sustainability planning. *Futures* **2008**, *40*, 34–46. [CrossRef]
70. Döll, P.; Hauschild, M.; Fuhr, D. Scenario Development as a Tool for Integrated Analysis and Regional Planning. In *Neotropical Ecosystems. Proceedings of the German Brazilian Workshop Hamburg 2000*; Lieberei, R., Bianchi, H.-K., Boehm, V., Reisdorff, C., Eds.; GKSS-Forschungszentrum Geesthacht: Geesthacht, Germany, 2000; pp. 817–822.
71. Delgado-Maciel, J.; Cortés-Robles, G.; Alor-Hernández, G.; García Alcaraz, J.L.; Négny, S. A comparison between the Functional Analysis and the Causal-Loop Diagram to model inventive problems. *Procedia CIRP* **2018**, *70*, 259–264. [CrossRef]
72. Kirkwood, C.W. *System Dynamics Methods: A Quick Introduction*; Arizona State University: Phoenix, AZ, USA, 1998.
73. Senaras, A.E. Causal Loop Diagrams and Feedbacks: A Case Study in Flexible Manufacturing System. *Yönetim Ekon. Ve Pazarlama Araştırmaları Derg.* **2017**, *1*, 1–12.
74. Zimmermann, M.; Woltersdorf, L.; Felmeden, J.; Müller, K. Water reuse for agricultural irrigation. In *Integrated Water Resources Management in Water-Scarce Regions: Water Harvesting, Groundwater Desalination and Water Reuse in Namibia*; Liehr, S., Kramm, J., Jokisch, A., Müller, K., Eds.; IWA Publishing: London, UK, 2018; pp. 42–51.
75. Liehr, S.; Kramm, J.; Jokisch, A.; Müller, K. (Eds.) Integrated Water Resources Management in Water-Scarce Regions: Water Harvesting, Groundwater Desalination and Water Reuse in Namibia. IWA Publishing: London, UK, 2018.
76. Ghassemi, F.; Jakeman, A.J.; Nix, H.A. *Salinisation of Land and Water Resources: Human Causes, Extent, Management and Case Studies*; University of New South Wales Press: Sydney, Australia, 1995; ISBN 0-86840-198-6.
77. Janeiro, C.N.; Arsénio, A.M.; Brito, R.; van Lier, J.B. Use of (partially) treated municipal wastewater in irrigated agriculture; potentials and constraints for sub-Saharan Africa. *J. Phys. Chem. Earth* **2020**, *118–119*, 102906. [CrossRef]
78. Zimmermann, M.; Boysen, B.; Ebrahimi, E.; Fischer, M.; Henzen, E.; Hilsdorf, J.; Kleber, J.; Lackner, S.; Parsa, A.; Rudolph, K.-U.; et al. *Replication Guideline for Water Reuse in Agricultural Irrigation: Upgrading Wastewater Pond Systems to Generate Irrigation Water for Animal Fodder Production Using the Example of Outapi, Namibia*; ISOE-Materialien Soziale Ökologie No. 63; ISOE–Institute for Social-Ecological Research: Frankfurt am Main, Germany, 2021.
79. Zimmermann, M.; Deffner, J.; Müller, K.; Kramm, J.; Papangelou, A.; Cornel, P. *Sanitation and Water Reuse—Implementation Concept*; ISOE–Institute for Social-Ecological Research: Frankfurt am Main, Germany, 2015.
80. Zacharia, A.; Ahmada, W.; Outwater, A.H.; Ngasala, B.; van Deun, R. Evaluation of Occurrence, Concentration, and Removal of Pathogenic Parasites and Fecal Coliforms in Three Waste Stabilization Pond Systems in Tanzania. *Sci. World J.* **2019**, *2019*, 3415617. [CrossRef] [PubMed]

Article

Sustainable Wastewater Management to Reduce Freshwater Contamination and Water Depletion in Mexico

José de Anda [1,*] and Harvey Shear [2]

1. Center for Research and Assistance in Technology and Design of the State of Jalisco, Department of Environmental Technology, C.A. Avenida Normalistas 800, Guadalajara 44270, Jalisco, Mexico
2. Institute for Management and Innovation, University of Toronto Mississauga, 3359 Mississauga Road, Mississauga, ON L5L 1C6, Canada; harvey.shear@utoronto.ca
* Correspondence: janda@ciatej.mx; Tel.: +52-33-3345-5200

Citation: de Anda, J.; Shear, H. Sustainable Wastewater Management to Reduce Freshwater Contamination and Water Depletion in Mexico. *Water* **2021**, *13*, 2307. https://doi.org/10.3390/w13162307

Academic Editors: Martin Wagner, Sonja Bauer and William Frederick Ritter

Received: 4 July 2021
Accepted: 11 August 2021
Published: 23 August 2021

Publisher's Note: MDPI stays neutral with regard to jurisdictional claims in published maps and institutional affiliations.

Copyright: © 2021 by the authors. Licensee MDPI, Basel, Switzerland. This article is an open access article distributed under the terms and conditions of the Creative Commons Attribution (CC BY) license (https://creativecommons.org/licenses/by/4.0/).

Abstract: At present, most rivers, lakes, and reservoirs in Mexico have significant anthropogenic contamination. The lack of sanitation infrastructure, the increase in the number of nonoperational or abandoned sanitation facilities, limited enforcement of environmental regulations, and limited public policies for the reuse of treated wastewater all contribute to the contamination and water availability problem. The reasons for this are identified as (1) the high maintenance and operational costs in sanitation facilities (including electricity consumption); (2) poor planning and practices of wastewater management and reuse by municipalities; (3) national policies that do not favor the reuse of treated wastewater for agriculture, industry, and municipal services instead of using groundwater as at present; (4) failure to adopt a governance model at the three levels of government; and (5) transparency in the management of financial resources. Some measures to improve this situation include (a) transparent decision-making; (b) participation and accountability in budgeting and planning at the national, state, and municipal levels; and (c) planning for the reuse of treated wastewater to reduce groundwater extractions and to reduce discharges to surface waters from the beginning of every WWTP project.

Keywords: municipal wastewater treatment plants; wastewater treatment systems; wastewater reuse; balanced scorecard; developing countries; Mexico

1. Introduction

Apart from their value as water sources for water supply for urban areas and food production, freshwater lakes have always been important to human life, because they serve as a freshwater fishery, recreation sites, and avenues of transport. They also provide other benefits, such as wildlife preservation, the replenishment of groundwater, flood regulation, regulation of the local climate, and enhancement of the beauty of local landscapes [1].

The continuing increase in global population is raising the demand of freshwater supply. One important factor affecting freshwater availability is associated with socioeconomic development and climate change. Another factor is the general lack of sanitation and waste treatment facilities in high-population areas of developing countries [2]. The quality of surface water or groundwater at any point in a watershed reflects the combined effect of many physical, chemical, and biological processes that affect water as it moves along hydrologic pathways over, under, and through the land [2]. Impacts on the water quality of freshwater ecosystems substantially reduce the water availability of regions [3]. Therefore, an efficient sanitation of wastewaters is becoming an issue that must be considered, particularly in developing countries where there is a major lag in providing this strategic service for the population.

At the end of 2016, in Mexico, the national coverage in the treatment of municipal wastewater was about 54% [4]. However, of the total installed municipal WWTPs, in the same year, about 22.47% were out of operation. Some of the reasons for this were identified as the dominance in the country of a wastewater treatment technology demanding high electricity consumption, maintenance and operation costs, poor planning practices of municipal wastewater management and reuse, the failure to adopt a governance model, and transparency in the management of financial resources at the municipal level [5]. As a result, several water bodies in the Mexican hydrographic basins have been reported to be in an advanced process of eutrophication, increase of toxic contaminants, depletion of water levels, loss of habitat and aesthetic values, and impairment with recreational activities, among others [6].

On the other hand, most Mexico suffers under high water stress, and many regions of the country are highly vulnerable to droughts because of climate change; the contamination of its freshwater bodies will accelerate in a few years the process of the loss of water availability in many regions of the country [7–9]. Baseline water stress measures the ratio of total water withdrawals to the available renewable surface and groundwater supplies. Water withdrawals include domestic, industrial, irrigation, and livestock consumptive and non-consumptive uses. The available renewable water supplies include the impact of upstream consumptive water users and large dams on the downstream water availability. Higher values indicate more competition among users [10]. Figure 1 shows the main regions in the country suffering under water stress.

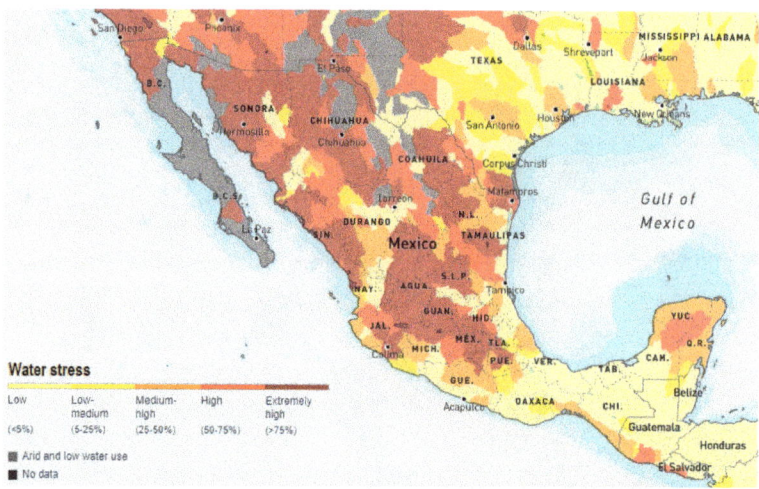

Figure 1. High-water stress in the central and northern areas of Mexico [10].

Baseline water depletion measures the ratio of total water consumption to the available renewable water supplies. The total water consumption includes domestic, industrial, irrigation, and livestock consumptive uses. The available renewable water supplies include the impact of upstream consumptive water users and large dams on the downstream water availability. Higher values indicate a larger impact on the local water supply and a decreased water availability for downstream users. Baseline water depletion is similar to baseline water stress; however, instead of looking at the total water withdrawal (consumptive plus non-consumptive), baseline water depletion is calculated using the consumptive withdrawal only [10]. Figure 2 shows the water depletion in Mexico affecting large central and northern regions of the territory.

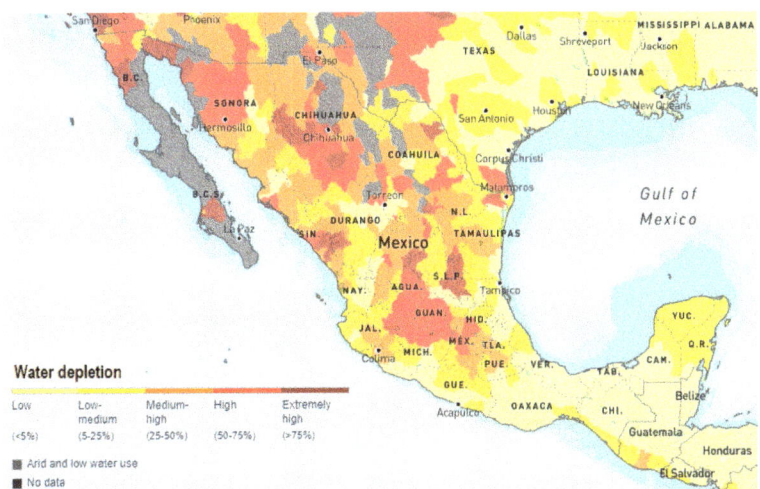

Figure 2. Water depletion in the central and northern areas of Mexico [10].

The purpose of present work is to discuss some of the reasons why, after years of investments in wastewater treatment projects, no substantial progress in sanitation coverage has been made in recent years. The current deficiencies in the national wastewater management model are significantly affecting all surface freshwater ecosystems in the country [4,7].

Some alternatives to improving wastewater management include: (1) strengthening the present water governance model at the three levels of government, (2) the adoption of integrated water resource management (IWRM) practices, and (3) the reuse of municipal treated wastewater for agriculture and the industry. These alternatives are proposed as potential strategies to advance the national commitment in relation to the sixth of the UN Sustainable Development Goals, SDG6: "Ensure access to water and sanitation for all" [11].

2. Materials and Methods

Historical data on the population in Mexico for the period 1950–2018 were obtained from the National Population Council (CONAPO, its acronym in Spanish) [12]. The projected growth of the population until 2050 was also included.

The National Water Commission (CONAGUA, its acronym in Spanish) is a decentralized administrative body of the Ministry of Environment and Natural Resources (SEMARNAT, its acronym in Spanish) created in 1989, whose responsibility is to administer, regulate, control, and protect the national waters in Mexico. Official data from CONAGUA and SEMARNAT were used to analyze the historical evolution of the wastewater and treatment infrastructure and coverage in Mexico for the period 1992–2017 [4,13]. Unfortunately, 2017 was the last date in which official wastewater and sanitation data were published by the federal agencies in Mexico.

The critical problems faced by the treatment sector in Mexico, including the state of municipal wastewater reuse, was previously discussed by de Anda and Shear [14,15], de la Peña et al. [16], and by de Anda-Sánchez [5]. The critical problems in the municipal wastewater treatment programs and the new findings reported in this work are classified according to four perspectives based on an adapted Kaplan and Norton Balanced Score Card (BSC) model applied to measure the performances of the municipal services [17–20]. Later, the most relevant issues in each BSC area were identified. A figure was constructed to visualize the general strategy oriented to improve the provision of the treatment services in the country.

Three generally recognized dimensions of sustainable development have been well-enunciated: ecological, social, and economic [21,22]. In addition to these dimensions, Pawłowski [23] suggested including four more dimensions—namely, moral, legal, technical, and political. For the water and treatment projects, McConville and Mihelcic [24] proposed five dimensions that affect sustainable development: sociocultural respect, community participation, political cohesion, economic sustainability, and environmental sustainability. The authors explained that, in this context, the application of the five sustainability factors throughout the project life cycle will result in the design and implementation of appropriate technology. According to this, the sustainability dimensions proposed by Pawłowski [23] and McConville and Mihelcic [24] are strongly correlated. On the other hand, Saad et al. [25] acknowledged that the sociological factors, including community participation, public involvement, social perception, attitudes, gender roles, and public acceptance, would lead to improvements in wastewater management practices. To discuss the reported results, an analysis of six dimensions of sustainability was included. Finally, the progress in the country to adopt the Integrated Water Resource Management (IWRM) principles was considered [26].

3. Results

Table 1 shows the trend in population growth from 1950 to 2010 and the expected trend to 2050. The official data indicate that the country experienced a gradual decrease in the population growth rate from 2.7% in 1960 to 2.0% in 2010 [12]. According to the official population projections, it is estimated that the population growth will decrease by 2050 until it reaches 0.23%, and the total population will be around 150 million people. The equation used to estimate the average annual growth rate (GR) for every ten-year period (n), where P_n and P_{n-1} are the populations at the start and end of each decade, respectively, was:

$$GR = \frac{1}{10}\left(\frac{P_n - P_{n-1}}{P_{n-1}}\right)100 \qquad (1)$$

Table 1. Historical growth of the Mexican population from 1950 to 2010 and the expected growth in 2050 [12].

Year (n)	Total Mid-Year Population (P)	Average Annual Growth Rate (%)
1950	27,026,573	
1960	36,786,543	3.61%
1970	50,778,729	3.80%
1980	67,561,216	3.31%
1990	84,169,571	2.46%
2000	98,785,275	1.74%
2010	113,748,671	1.51%
2020	127,792,286	1.23%
2030	138,070,271	0.80%
2040	144,940,511	0.50%
2050	148,209,594	0.23%

In Mexico, a population center is considered urban when it has more than 2500 inhabitants; where there are fewer than this, it is considered rural [4]. Figure 3 shows the historical evolution of the urban population in Mexico. At the beginning of the 1960s, the relationship between the urban and rural population in Mexico was close to 50% each. However, like global trends, Mexico has been urbanizing. In 2018, 80.2% of its citizens lived in urban areas. This figure is well above the world average of 55.3% in 2018 [27].

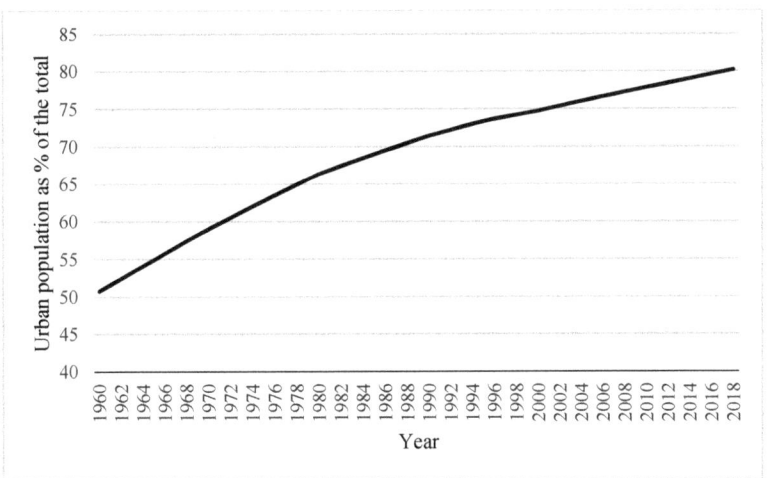

Figure 3. Historical growth of the urban population in Mexico from 1960 to 2018 [12].

3.1. Consumptive Water Use

'Granted' water in Mexico is defined as the concessional volume of water (underground and surface) for public supply. It corresponds to the volume authorized for use or exploitation of the water resource for public supply and domestic use. It includes the population that has a connection to a domestic, public distribution system, a protected well, or a rainwater collection network (Figure 4) [4]. In 2017, the volume granted by CONAGUA for consumptive use in Mexico was 87.84 hm^3. Approximately 60.9% of the consumptive water use in Mexico came from surface sources (rivers, streams, and lakes), while the remaining 39.1% came from aquifers [4]. The largest volume granted for consumptive use was for agriculture, mainly for irrigation, as shown in Table 2. It is important to note that, in Mexico, because of inefficient irrigation technology, about 60% of the water used in agriculture is lost through evaporation or subsoil infiltration. On the other hand, approximately 80% of the drinking water that is used by the population for domestic or commercial use turns into wastewater normally ending in the urban sewage network system [4].

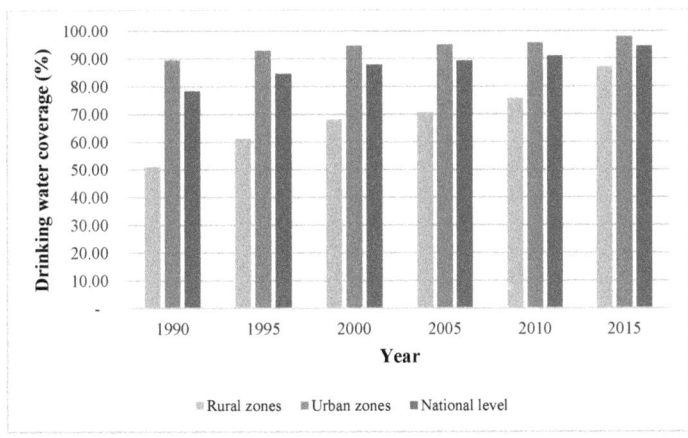

Figure 4. Evolution of the drinking water coverage in rural and urban areas in Mexico in the period from 1990 to 2015 [4,13].

Table 2. Consumptive uses grouped by source type for 2017 [4,13].

Water Use	Origin		Total Volume	Extraction Percentage
	Surface Water	Groundwater		
	Thousands of hm^3	Thousands of hm^3	Thousands of hm^3	%
Agriculture	42.47	24.32	66.80	76.04
Public supply	5.25	7.38	12.63	14.38
Self-supplying industry	2.04	2.23	4.27	4.86
Electric power (without hydroelectric power)	3.70	0.45	4.15	4.72
Total	53.46	34.38	87.85	100.00

3.2. Sewage and Treatment Coverage

In Mexico, the sewage coverage includes people who have a connection to the sewage network or to a septic tank; it does not necessarily include the conduct of wastewater to a treatment plant. According to CONAGUA [4], the evolution of sewage coverage in the country has been improved from 80% in 2000 to close to 92% in 2017. As shown in Figure 5, the trend has been almost asymptotic since 2012. The registered data on the number of municipal WWTPs in operation and out of operation in Mexico are described in Figure 6 for 1991–2017 [4,13]. An out-of-operation WWTP in Mexico includes any facility that the municipality has partially or totally closed because of a lack of financial resources to keep it operating. With the increase in population, the number of municipal WWTPs has increased by an average of about 7% per year. However, as shown in Figure 4, at the beginning of the 1990s, almost 32% of the total number of municipal WWTPs were out of operation. This trend reversed until 2009, when, once again, there was a steady increase in the plants out of operation, reaching 22.5% in 2016, the last year when the government published such records.

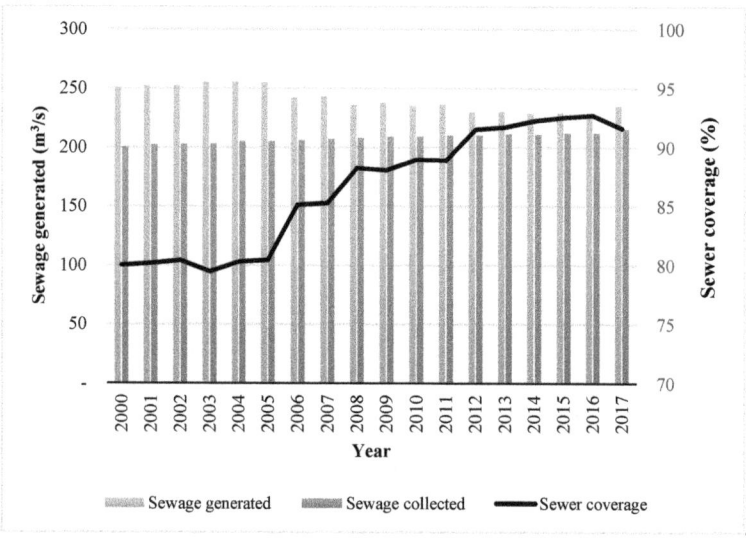

Figure 5. Historical trends of the sewer coverage in Mexico from 2000 to 2017 [4,13].

In Mexico, the installed capacity of a wastewater treatment plant means the design capacity, while the operating capacity is the record of the wastewaters received in the treatment system. In 2017, the total amount of treated municipal wastewater was 137,699 L/s (4342.42 hm^3/year). According to Figure 7, the percentage of the unused capacity of the municipal WWTPs has remained relatively low. In contrast, Figure 8 shows that there is still a significant gap between the wastewater generated in the country and the volume of treated wastewater. As can be seen in the same figure, the percentage of the treatment coverage has been exponential, going from 17% in 1998 to almost 60% in 2017, the last year reported in the official figures.

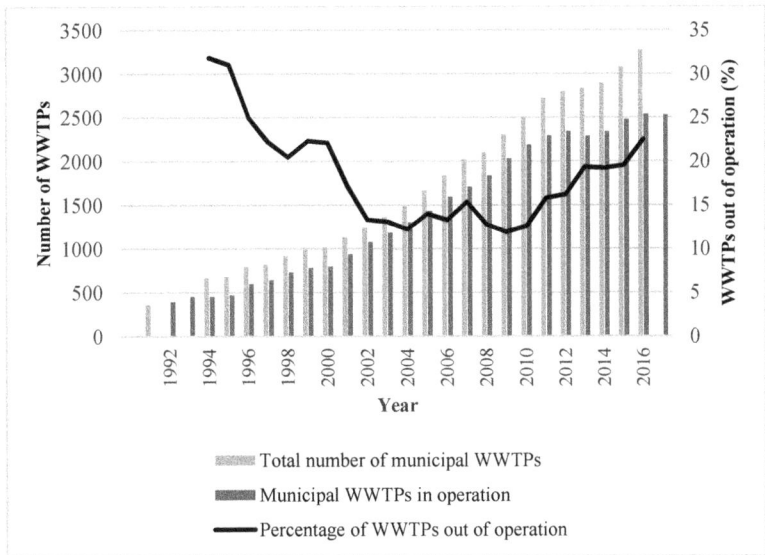

Figure 6. Historical trends of the municipal WWTPs in Mexico from 1991 to 2017 [4,13].

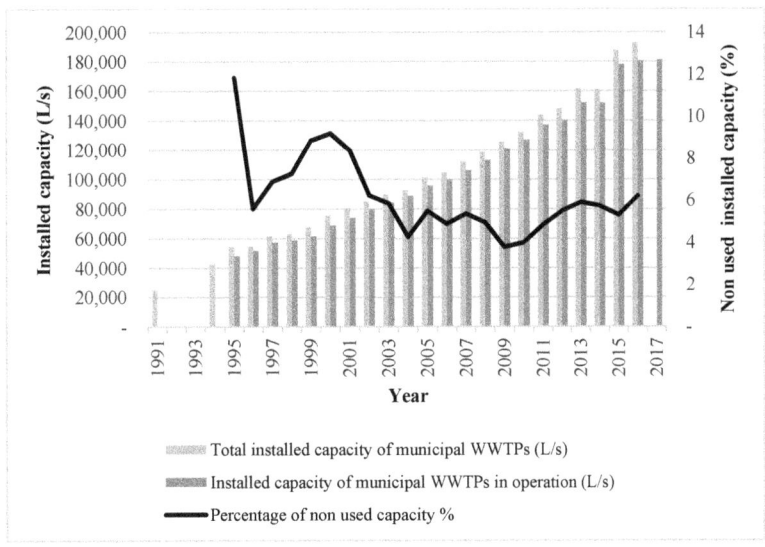

Figure 7. Installed capacity of the WWTPs in Mexico from 1991 to 2017 [4,13].

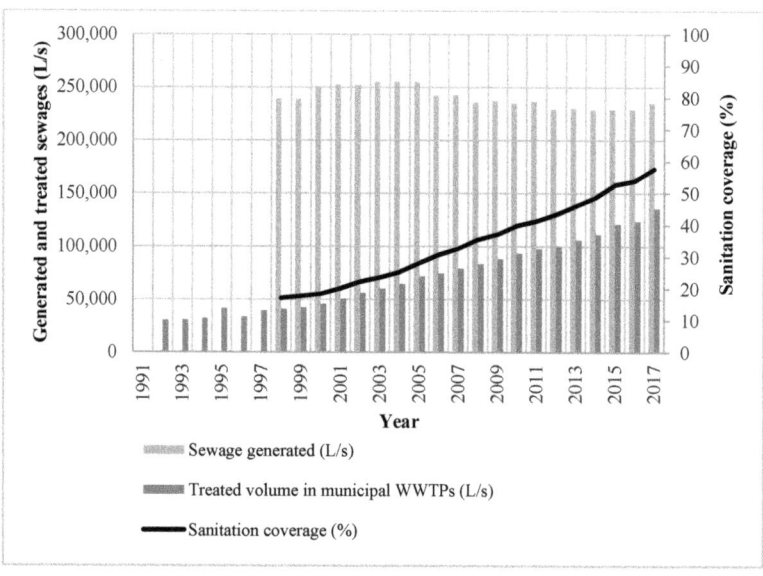

Figure 8. Gap between the wastewater generated and treated in municipal WWTPs, and the percentage of the treatment coverage from 1998 to 2017 [4,13].

3.3. Reuse of Treated Wastewaters

According to CONAGUA [4], in 2017, 39.8 m^3/s (1255.13 hm^3/year) of treated wastewater was reused directly from WWTPs and 78.8 m^3/s (2485.04 hm^3/year) was reused indirectly after first being discharged into a water body. Figure 9 explains the main paths of the water, wastewater, and reused treated wastewaters according to the Mexican water management model. As explained in Table 2, agriculture is by far the most demanding sector for water resources in the country, followed by public supply. The self-supplying industry and thermal electric power generation represent less than 10% of the entire water budget. In Figure 9, the dotted box for the water purification plants (WPP) indicates that, normally, in urban areas, there is the infrastructure to treat the water, but in rural areas, groundwater is used directly to supply the water needs of the population. The dotted box for the municipal and industrial WWTP means that, in some cases, the plant does not exist, exists but does not operate, or operates but does not meet the regulatory requirements. The dotted box in rainwater harvesting means that this infrastructure exists in some areas, since the collection of rainwater is not a priority in most urban municipalities in the country. The grey dotted lines represent diluted raw wastewaters or wastewaters after treatment, which are recycled for agriculture or the industry. The red lines represent the main sources for water pollution of the waterbodies in the country. Figure 9 shows how the discharge of raw wastewaters, municipal WWTPs, industry, energy production, and agriculture increase the pollution of water bodies in the country and how this polluted water is then used as a source of drinking water, thus increasing the public health risks and environmental pollution.

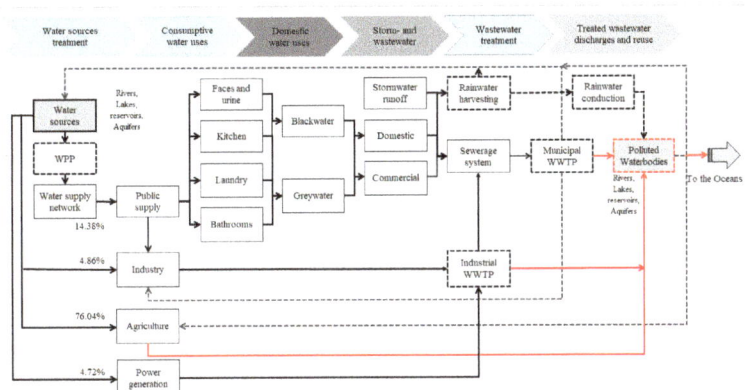

Figure 9. Main paths of the supplied water and wastewater in Mexico (Source: J. de Anda).

4. Discussion

The population projections for the next thirty years in Mexico estimate that the overall growth rate will decrease dramatically. The country must, however, cover the treatment needs of the wastewater generated by a growing urban population (Figure 1). According to a predictive forecast carried out by the National Population Council, the urban population in Mexico is going to increase by 2050 to 89.4% with reference to the population reported by 2018 (Figure 3) [12].

Even though sewerage system coverage has been developed significantly in the last twenty years, the treatment coverage of sewage and sanitation continues to be a challenge (Figures 5 and 8). In the last 20 years, the country has made a significant effort to increase the wastewater treatment coverage, going from just over 20% in the late 1990s to almost 60% in 2017, the last year officially reported. Wastewater treatment plants continue to grow in number; however, there are a significant number of facilities that do not operate (Figure 6). Figure 6 also shows that, at the end of 2016, in Mexico, the registry of municipal Wastewater Treatment Plants (WWTPs) showed 2536 facilities in operation, with an installed capacity of 180,569.72 L/s and a treated flow of 123,586.75 L/s, resulting in a national coverage in the treatment of municipal wastewater of about 54%. This figure is above the national average in other countries in Latin America, where most countries reported less than 40% coverage. However, a significant effort was made to reduce the number of facilities that were not in operation from 1990 to 2010. Unfortunately, this number has gradually increased since then, reaching 22.47% in 2016. This figure means that, in this year, there were 2536 treatment plants out of operation in the country.

In addition to these problems, one must note that, since 1996, the regulations regarding the discharge of sewage into national water bodies have not been modified. The official Mexican standard NOM-001-SEMARNAT-1996 does not regulate many of the compounds that today seriously compromise the health of the population and of surface and groundwater ecosystems such as organic micropollutants, pesticides and pesticide metabolites, urban and industrial organic micropollutants, veterinary compounds, and pharmaceuticals, among others [28]. Modifications to the current official regulations would put municipal treatment services in serious trouble because the technologies of the plants in operation are not capable of meeting even the more demanding requirements of these types of pollutants. That is why, to date, the official draft regulations have seen no progress for their approval in the Chamber of Deputies [29]. Consequently, at present, many waterbodies in Mexico exceed the limits of the most basic water quality parameters [4].

According to the previously mentioned methodology, the problems faced in the municipal wastewaters treatment in the country were organized based on the Balance Score Card (BSC) model. Figure 10 summarizes the results.

4.1. Internal Process Perspective

a. The urban concentration in Mexico today exerts a high to very high level of water stress and water scarcity in most cities located in Central and Northern Mexico [30].
b. There is still an important gap between the sewage generated by the population and the percentage of sewer coverage to collect this sewage [5].
c. Due to the lack of infrastructure for the management of urban rainwater, wastewater and rainwater are mixed in the same municipal sewage systems, meaning that opportunities for the aquifers to be recharged and for rainwater storage and reuse are lost [5].
d. Most municipal WWTPs are bypassed during the rainy season because they were not designed to handle the volumes associated with heavy rainfalls. As a result, during the rainy season, many of the receiving water bodies are contaminated by excess organic loads and other pollutants (personal communication with State Water Commission in Jalisco).
e. The leaks in the sewerage network are as high as 40% in urban areas [31].
f. More than 70% of the municipal wastewater treatment technologies installed in the country require intensive energy use and high Maintenance and Operational (M&O) costs, such as activated sludge, aerated lagoons, and dual anaerobic–aerobic systems [5].
g. There are limitations to the current regulations to control the nutrients and emerging contaminants [5,28].

4.2. Users' Perspectives

a. The interruption of drinking water services in cities during the dry season is more and more frequent.
b. There is increasing social discontent due to the high level of contamination of surface waters in Mexico and the resultant consequences for public health.
c. Farmers refuse to reduce their rights to water extraction volumes from their wells, because they could lose their government-approved water concessions if they agree to use treated water from the municipality.
d. Municipal politicians fear the loss of voter support if water metering is installed. At present, a single fee per user is applied, which is generally insufficient to pay for potable water treatment and for sewage treatment services. The consequence of accepting such a measure is the overuse and overexploitation of the local water resources and the contamination of surface- and groundwaters due to lack of, or inefficient, treatment infrastructures.
e. Water managers of municipalities with conventional WWTPs consider that the government incentives to cover part of their Maintenance and Operating (M&O) costs are insufficient to keep the facilities in good operating condition and that, in any case, the process of getting such incentives is highly bureaucratic.

4.3. Financial Perspective

a. The ground- and surface water volumes given in concession by CONAGUA for agriculture in Mexico cost nothing for farmers if they do not exceed the volume granted by CONAGUA, so, in most cases, farmers prefer to use well water than to use treated water from the municipality [32].
b. In some municipalities, the aquifers are overexploited, but the extraction of the water given in concession by CONAGUA to the farmers is still preferred, because, as noted above, there is no cost for the water, and the energy costs of pumping for agriculture are subsidized by the Federal Electricity Commission.
c. The centralization of the municipal wastewater treatment service involves the commitment to invest, build, and operate a complex facility requiring regular high expenditures for M&O.

d. Medium-sized or large treatment infrastructure projects usually require development banking funds. With this, the municipality acquires long-term debts that, usually, it cannot pay.
e. In most cases, the wastewater treatment plants do not have a wastewater reuse plan.
f. Most municipalities do not have a decentralized water operating agency with financial independence from the municipality. As a result, the expected annual municipal budget for treatment is generally insufficient to maintain the WWTP M&O costs.
g. Municipalities with a low population normally do not have water consumption meters and pay an annual fee for water services for housing independent of water consumption. This practice has led to the overuse of water, overexploitation of aquifers, and insufficient income to provide adequate potable water and treatment services to users.
h. A lack of long-term political continuity in the municipalities to get financial resources with the state and federal agencies that manage the water resources affects the plans to expand the treatment coverage and the renewal of existing treatment infrastructures.
i. Frequently, the managers of potable water and treatment services do not have enough information about the subsidies from CONAGUA to finance part of the M&O expenditures of the WWTP [33].
j. Many municipalities do not have a treatment service for their wastewater and prefer to pay fines for unregulated discharges, because they do not want to incur financial debts or permanent M&O expenses.
k. Knowing the financial limitations of the municipalities, CONAGUA usually defers the fines to municipalities that do not comply with the Mexican Federal Law of Rights declared in its article 276 [34].

4.4. Learning and Growth Perspectives

a. Frequently, water managers do not have a basic educational background in subjects such as hydrology, laws, and the regulations related to the management of water resources or the basic principles of hydraulic urban infrastructure.
b. Sporadic communication between municipal, state, and federal agencies results in the failure of water and wastewater management plans.
c. In most cases, there is an absence of a vision of IWRM based on the hydrologic resources of the basins and subbasins and a lack of expertise and research programs to accelerate technology development and technology transfer activities that can offer innovative treatment technologies with low energy consumption and a low carbon footprint.
d. Poor community participation in decision-making in the management of water resources results in limited confidence in the decisions of municipal representatives.
e. A lack of transparency of the municipal authorities in the management of financial resources generates distrust in the community, and participation in community initiatives declines.

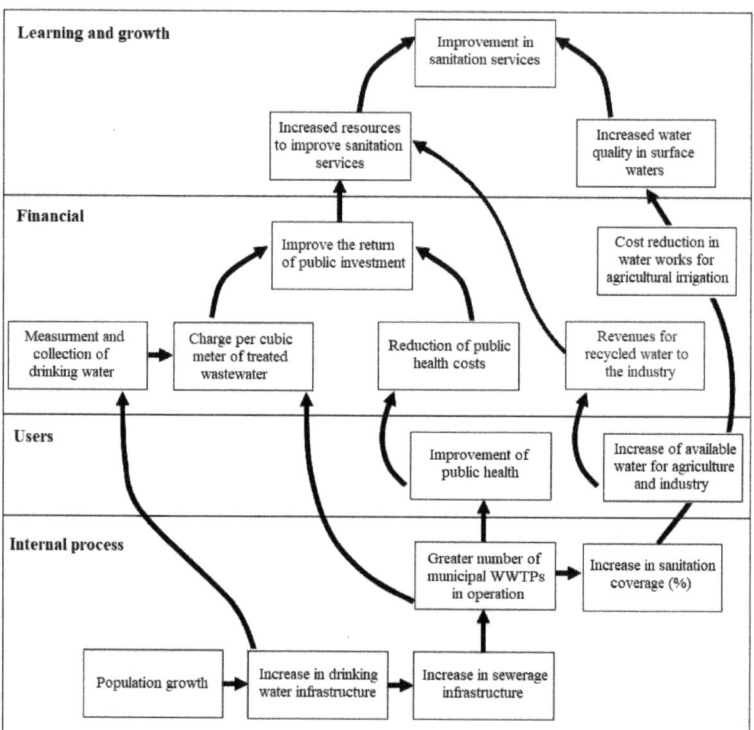

Figure 10. Balanced score card adapted to the sanitation sector (adapted from reference [35]).

4.5. The Sustainable Approach

The sustainable treatment and reuse of municipal wastewater in Mexico was analyzed in six dimensions: social, environmental, technological, economic, legal, and political [23–25,36].

4.5.1. Social Dimension

The work published by Saad et al. [25] explained how to face the social issues related to municipal wastewater management. The publicity of new municipal wastewater-related projects includes advertisement in the media; education; and the inclusion of all stakeholders (politicians, experts, and the public at large). In the decision-making process, these are key elements for the successful design and implementation of wastewater schemes. Public involvement is best achieved through the participation and involvement of users in all parts of the project cycle, from planning and design to implementation and decision-making, thus producing more efficient and sustainable projects/outcomes.

On the other hand, the degree of acceptance of wastewater reuse varies widely, depending on the reuse purposes, and is influenced by many factors, such as the degree of contact, expressions of disgust, education, risk awareness, the degree of water scarcity or availability of alternative water sources, calculated costs and benefits, trust and knowledge, issues of choice, attitudes toward the environment, economic considerations, involvement in decision-making, the source of water to be recycled, and experience with treated wastewater. Other factors that depend on the region and case include the cultural, religious, educational, and/or socioeconomic factors [25]. On the other hand, the engagement of stakeholders in learning, planning, and acting requires a common sense of community, effective formal and informal collaborative processes, and the sharing of power and authority [37] (see Table 3).

Table 3. Pros, cons, and recommendations related to the social dimension in wastewater sanitation management.

	Social Dimension		
Issue	Pros	Cons	Recommended Measures
1. Publicity	Essential to inform the public. Special interest groups may be aware of the plan, but not the public whose ongoing involvement is necessary for success.	Depending on the vehicle (radio, TV, print, etc.), this could be costly in terms of recruiting the public (see 2 below).	Develop a low cost but effective campaign to inform the public about the issues in any wastewater management plan.
2. Public Involvement	Experience throughout the World (e.g., Great Lakes) has shown that public involvement is critical to the success of any sustainability program.	Sustaining public interest and enthusiasm over the long term is difficult. Who pays for travel and other expenses for the public participants?	Establish a Wastewater Management Committee (WMC), with funding, to facilitate public involvement.
3. Degree of Acceptance	Essential for any long-term program. One needs the population to see that the program as necessary for them and for their children, grandchildren, etc. They need to accept the program as benefitting their economy, their health and wellbeing, their spiritual needs, etc.	Provision of costs/benefits may not be easy to translate into concepts that everyone in the population can understand.	The WMC will have the job of analyzing the costs benefits of using wastewater and translating this into language that anyone can understand.
4. Actions of People	Achieving environmental improvement and sustainable communities will depend less on the mandates of government and more on the actions of people, communities, industries, nongovernmental organizations, landowners, and others, working together, often voluntarily. This will lead to a successful design and implementation of a Wastewater Management Plan.	To engage these stakeholders in planning, learning, and acting requires a common sense of community, an effective formal and informal collaborative process, and the sharing of power and authority. Ceding power and authority, when it has long been entrenched, will be difficult.	Experience has shown that when all sectors of society are involved in the development and implementation of a plan, power can be shared. Provide adequate funding for groups (farmers, small scale businesses, etc.) to participate as equals in a WMC.
5. Role of Academia	Essential for the provision of scientific information to make sound decisions.	Academics may not see interacting with the public as part of their research career and may be reluctant to participate.	Once the WMC is established, have its members communicate with senior administrators in the academic community (government and university) to secure participation in the development of the wastewater plan from engineers, scientists, etc.

4.5.2. Environmental and Technological Dimension

Wastewater treatment systems in developing countries like Mexico face several problems to keep them in operation. Most of these systems have been developed in countries with a high level of income per capita; high levels of technical expertise; and without considering the appropriateness of the technology for the culture, land, and climate of Mexico. Often, local engineers trained in the Mexican academic programs continue supporting the choice of conventional systems that later turn out to be inappropriate due to high M&O costs [38]. The successful employment of appropriate technologies requires a deep understanding of the social dynamics of the community in which they are applied [25].

In this sense, it is necessary to consider that the wastewater treatment processes are different in each case due to their different origins (e.g., effluents from agricultural and livestock activities and municipal or industrial discharges). In each case, the most convenient technologies must be clearly identified according to the results of the physical, chemical, and biological characterizations of the water to be treated [5,14]. Additionally, in the case of municipal wastewater, the selection of the appropriate technology to be implemented must be proposed according to the capacity of the community to pay for the M&O expenditures. Some measures to improve the environmental and technological dimensions of the problem are explained in Table 4.

Table 4. Pros, cons, and recommendations related to the environmental and technological dimensions in wastewater treatment management.

Environmental and Technological Dimension			
Issue	Pros	Cons	Recommended Measures
6. Combined domestic wastewater with rainwater	Reduction of urban sewage and rainwater infrastructure construction costs.	When domestic wastewater and rainwater are combined, they increase the demands for treatment capacity and, therefore, the costs of investment, maintenance, and operation [36].	Introduce in municipal construction regulations the separation of domestic wastewater from rainwater. Gradually generate urban infrastructure to separate wastewater from rainwater.
7. In many cases WWTP discharges do not meet environmental regulations	In the absence of complaints and an efficient surveillance system, the government assumes that it is fulfilling its obligations correctly.	Permanent pollution of surface and ground water and harmful effects on the health of people and the ecosystem.	Control the quality of the treated water so that it consistently meets environmental regulations [36].
8. Current wastewater treatment technologies produce important volumes of biological solids that require special treatment and disposal.	Most of the biological solids from WWTPs are managed and disposed in landfills or reused in agriculture.	When biological solids are not properly disposed, they produce offensive odors to closely settled communities, contaminate surface and groundwater and soils, and generate a large amount of greenhouse effect gases.	Ensure that the biological solids generated are treated appropriately for their use as fertilizer (compost), instead of their disposal in landfills that do not comply with environmental regulations [36].
9. Current WWT technologies are intensive in the use of energy and have high operation and maintenance costs.	Current technologies are widely used throughout the country. Wastewater treatment plant staff are familiar with these technologies.	Several municipal WWTPs are out of operation throughout the country because of high M&O costs.	It is necessary to introduce sustainable technologies based on natural processes and low M&O costs [39,40].

4.5.3. Economic and Legal Dimension

In the development of a sustainable scheme of integrated management of municipal water resources, including the management of sanitary wastewater, it is essential that municipalities adopt the scheme of a decentralized public organization that manages the economic resources destined to generate the water services. At present, the fee that users must pay for the supply of potable water and sewerage services, which includes the costs of treatment of the residual water, are integrated into a single fee that varies according to the municipality. Some measures taken in developing countries to improve the economic dimensions of the problem are explained in Table 5.

Table 5. Pros, cons, and recommendations related to the economic and legal dimensions in wastewater treatment management.

Economical and Legal Dimension			
Issue	Pros	Cons	Recommended Measures
10. Most of the technological solutions for municipal WWTP require high energy input and expensive chemical additives which can have a significant impact on the operation costs.	These costs can be offset by the benefits that the community obtains from having clean and safe wastewater discharged into receiving waters.	Municipalities have short administrations (3 years), and they often fail to see the environmental and social benefits of having safe surface and groundwater.	Complying with the condition that the balance of a mass and energy of raw materials, products, and byproducts results in an economic, social, and environmental benefit without compromising the current and future resources of the community [39].
11. In some cases, municipalities acquire debts to build WWTPs and subsequently cannot afford financial and operating expenses.	It is possible to compensate the financial expenses through the sale of the treated wastewater.	In general, sanitation projects are not planned to recover and reuse the treated wastewater and the value of the investment is thereby lost.	Evaluate the real investment capacity of the community and ensure the necessary resources for its maintenance, operation, and reuse [24].
12. In Mexico, the costs of energy have increased in recent years due to reduction of domestic oil reserves and lack of investment in renewable energies.	This situation should promote the use of sanitation technologies based on natural processes.	Not enough national technical capacities to promote sanitation technologies based on natural processes with low energy consumption and low M&O costs.	Consider that the energy, maintenance, and operation costs will increase over time, it will be necessary to build technical capacities in sustainable sanitation technologies [24].
13. Most treated wastewaters are discharged into rivers, lakes, and reservoirs, losing the opportunity to reuse it.	Efficiently treated wastewater protects the quality of surface and groundwater sources of freshwater.	Most of the municipal treated water is not reused in agriculture or industry to lack of incentives.	Rethink the law regarding water use and tax incentives to favor the public, industry, and farmers who reuse reclaimed water in their activities. Implement an education program to make farmers and the public aware of the benefits of reusing water.
14. Sanitation coverage in the country is lower than 60%. It increases few every year.	The programs for the sanitation of wastewater in the country have been maintained in recent years, although not in accordance with the needs of the population.	There is a growing interest from private initiatives to collaborate with government in the cleanup processes of the country's hydrographic basins.	Introduce government programs such as tax incentives for the private sector to invest in the treatment infrastructure for public potable water and treatment to reduce pollution and promote a culture of care and reuse of treated water at all levels of society.

4.5.4. Political Dimension

According to Mestre [41], the Mexican federal administration should reformulate the objectives and treatment strategies, rehabilitate and modernize the large number of installed municipal WWTPs, and accelerate the efforts to improve the water quality and treat 100% of the effluent. To remedy this problem, and to improve the quality of water, Mexico should undertake a major federal–state program to rehabilitate treatment plants and make major investments to achieve universal coverage to treat wastewater. To implement this, some recommended measures are proposed in Table 6 [41].

Table 6. Pros, cons, and recommendations related to the political dimension in wastewater treatment management.

Political Dimension			
Issue	Pros	Cons	Recommended Measures
15. Most national waterbodies in the country face different contamination levels.	There is increasing awareness of the problem in the government, society, and academia.	The existing governance strategies have not been effective in increasing the sanitation coverage throughout the country.	Adjust the roles of federal, state, and municipal governments to improve the water quality. The states must assume the responsibility of reinforcing the governance schemes to accelerate the participation of society and academia in the decision-making process.
16. Most municipalities pay very low or no taxes for sanitation.	People are happy not to pay higher taxes.	The population does not perceive in the short term the environmental damage and its impacts that are being generated by the lack of sanitation.	To convince the municipalities to charge the users appropriate fees for wastewater treatment services.
17. Investment in sanitation services remains limited.	Public resources could be being applied to other higher-priority programs, such as insecurity and fighting poverty.	Gradual loss of water security in different regions of the country.	Make a significant effort to attract funding to invest in new WWTPs, including tertiary treatment.
18. Lack of transparency in the use of public resources to keep sanitation infrastructure in operation.	This practice has allowed some politicians to use public resources for other programs without being held accountable.	The sanitation infrastructure is gradually being abandoned.	Make the origin and destination of subsidies transparent and accountable.
19. Accountability at the municipal level often remains highly controversial and ineffective.	This has allowed politicians to make use of public resources to improve their political position.	The population's confidence in the government's ability to provide basic water supply and sanitation services is gradually being lost.	Create regulatory bodies at the state level and create a national coordinating entity, which has oversight and can regulate the subsector.
20. With the change of municipal government every three years, the employees of the drinking water and sanitation services change.	A new work team enters in which the new municipal government has confidence in its performance.	Capacities created by the previous government are often lost.	Depoliticize the water management organizations so that the permanence of the managerial personnel is not connected to the renewal of municipal administrations.

The strategy proposed by Mestre [41] to improve the wastewater treatment in Mexico includes measures focused on strengthening local government decisions and increasing the federal–state budget to subsidize the construction and operation of municipal WWTPs.

In the National Hydraulic Program 2014–2018, CONAGUA outlined the main lines of action to implement IWRM in the states; however, today the level of implementation of the program has reached a medium-low level, according to a UNEP [26] assessment. One year after starting the new Federal administration, the new 2019–2024 National Hydraulic Program is still under discussion [42].

5. Conclusions

The main issues to solve in terms of wastewater treatment and reuse in Mexico should be focused on (a) increasing the coverage of wastewater treatment facilities and the sewerage system in urban areas and sewage treatment in rural areas; (b) improving the operational state of wastewater infrastructures; (c) planning the wastewater treatment system to suit the conditions of each municipality; (d) favor the reuse of treated wastewater in the Federal Water Law for agriculture, industry, and municipal services instead of groundwater use; (e) building capacities in wastewater management; (f) increasing the efforts to secure funds specifically for wastewater treatment; (g) increasing the capacities of overloaded treatment facilities; (h) ensuring the compliance of wastewater discharges with the required regulation for agricultural reuse; and (i) monitoring for compliance with the recommended guidelines.

The challenges that the sector must face in Mexico are not only in public funding and the environmental, technical, and organizational perspectives but also in the existing legal framework, lack of transparency, accountability, and public participation. Therefore, some additional measures to those mentioned above to improve the management of municipal wastewater and treatment are making transparency, participation, and accountability in budgeting and planning a reality at the national, state, and municipal levels. Reusing treated wastewater should be planned from the beginning of every project, and the energy and M&O requirements to facilitate decision-making should be considered in each project.

In terms of transparency, it is necessary to (a) publish the federal and state funds available for the wastewater treatment program, (b) to improve the transparency of procurement processes, (c) to improve the staff recruitment processes in the water and treatment sector, (d) to improve the management of municipal resources for the water and treatment sector, and (e) to report the annual environmental indicators related to water and treatment to the community.

To increase the sustainable use of water resources, it is also necessary to train municipal politicians and employees in the creation and maintenance of their own operating agencies and to strengthen governance models for the self-management of resources at the municipal level.

Author Contributions: Conceptualization, J.d.A. and H.S.; methodology, J.d.A. and H.S.; software, J.d.A.; validation, J.d.A. and H.S.; formal analysis, J.d.A. and H.S.; investigation, J.d.A.; resources, J.d.A.; data curation, J.d.A.; writing—original draft preparation, J.d.A.; writing—review and editing, H.S.; visualization, J.d.A.; supervision, H.S.; project administration, J.d.A.; funding acquisition, J.d.A. All authors have read and agreed to the published version of the manuscript.

Funding: Centro de Investigación y Asistencia en Tecnología y Diseño del Estado de Jalisco, A. C.

Acknowledgments: The authors acknowledge the Centro de Investigación y Asistencia en Tecnología y Diseño del Estado de Jalisco, A.C. and the University of Toronto–Mississauga for providing the time and information resources to write this work.

Conflicts of Interest: The authors declare that there are no conflict of interest in the present publication.

References

1. UNEP. *The Pollution of Lakes and Reservoirs*; UNEP Environmental Library No. 12; United Nations Environment Programme: Nairobi, Kenya, 1994; p. 35. Available online: https://wedocs.unep.org/handle/20.500.11822/32410 (accessed on 4 July 2021).
2. Peters, N.E.; Meybeck, M. Water Quality Degradation Effects on Freshwater Availability: Impacts of Human Activities. *Water Int.* **2000**, *25*, 185–193. [CrossRef]
3. Edokpayi, J.N.; Odiyo, J.O.; Durowoju, O.S. Impact of Wastewater on Surface Water Quality in Developing Countries: A Case Study of South Africa. *IntechOpen* **2016**, *18*, 401–416. [CrossRef]

4. CONAGUA. *Estadísticas del Agua en México*; Comisión Nacional del Agua. Secretaría de Medio Ambiente y Recursos Naturales: Mexico City, México, 2018; p. 306. Available online: http://sina.conagua.gob.mx/publicaciones/EAM_2018.pdf (accessed on 4 July 2021).
5. De Anda-Sánchez, J. Saneamiento descentralizado y reutilización sustentable de las aguas residuales municipales en México. *Soc. Ambiente* **2017**, *5*, 119–143. [CrossRef]
6. Cotler-Ávalos, H. *Las Cuencas Hidrográficas de México. Diagnóstico y Priorización*; Secretaría de Medio Ambiente y Recursos Naturales, Instituto Nacional de Ecología: Mexico City, Mexico, 2010; p. 231. ISBN 978-607-7655-07-7.
7. Spring, Ú.O. Water security and national water law in Mexico. *Earth Perspect.* **2014**, *1*, 7. [CrossRef]
8. Arreguin-Cortes, F.I.; Saavedra-Horita, J.R.; Rodriguez-Varela, J.M.; Tzatchkov, V.G.; Cortez-Mejia, P.E.; Llaguno-Guilberto, O.J.; Sainos-Candelario, A. State level water security indices in Mexico. *Sustain. Earth* **2020**, *3*, 1–14. [CrossRef]
9. Anda-Sánchez, J. Precipitation in Mexico. In *Water Resources of Mexico*; Raynal-Villaseñor, J.A., Ed.; World Water Resources; Springer: Cham, Switzerland, 2020; Volume 6, pp. 1–14.
10. Aqueduct Tools. Aqueduct Water Risk Atlas. Available online: https://www.wri.org/aqueduct (accessed on 3 June 2021).
11. United Nations. Department of Economic and Social Affairs. Sustainable Development. Sustainable Development Goal 6: Ensure Availability and Sustainable Management of Water and Treatment for All. Available online: https://sustainabledevelopment.un.org/sdg6 (accessed on 3 June 2021).
12. Indicadores Demográficos de la República Mexicana. Consejo Nacional de Población. México D. F., México. Available online: http://www.conapo.gob.mx/work/models/CONAPO/Mapa_Ind_Dem18/index_2.html (accessed on 3 June 2021).
13. El Medio Ambiente en México 2013–2014. Agua, Calidad. Secretaría de Medio Ambiente y Recursos Naturales. Available online: https://apps1.semarnat.gob.mx:8443/dgeia/informe_resumen14/06_agua/6_2_3.html (accessed on 3 June 2021).
14. De Anda, J.; Shear, H. Challenges facing municipal wastewater treatment in Mexico. *Public Work. Manag. Policy* **2008**, *12*, 590–598. [CrossRef]
15. De Anda, J.; Shear, H. Searching for a sustainable model to manage and treat wastewater in Jalisco, Mexico. *Int. J. Dev. Sustain.* **2016**, *5*, 278–294. Available online: https://isdsnet.com/ijds-v5n6-3.pdf (accessed on 14 August 2021).
16. De la Peña, M.E.; Ducci, J.; Viridiana, Z. *Tratamiento de Aguas Residuales en México, Nota Técnica # IDB-TN-521*; Banco Interamericano de Desarrollo: Washington, DC, USA, 2013; p. 42. Available online: https://webimages.iadb.org/publications/spanish/document/Tratamiento-de-aguas-residuales-en-M%C3%A9xico.pdf (accessed on 3 June 2021).
17. Kaplan, R.S.; Norton, D.P. Using the balanced scorecard as a strategic management system. *Harv. Bus. Rev.* **1996**, *74*, 75–85.
18. Epstein, M.J.; Wisner, P.S. Using a Balanced Scorecard to Implement Sustainability. *Environ. Qual. Manag.* **2001**, *11*, 1–10. [CrossRef]
19. Guimarães, B.; Simões, P.; Marques, R.C. Does performance evaluation help public managers? A Balanced Scorecard approach in urban waste services. *J. Environ. Manag.* **2010**, *91*, 2632–2638. [CrossRef] [PubMed]
20. Abdelghany, M.; Abdel-Monem, M. Balanced scorecard model for water utilities in Egypt. *Water Pract. Technol.* **2019**, *14*, 203–216. [CrossRef]
21. Harris, J.M. *Basic Principles of Sustainable Development. Global Development and Environment Institute, Working Paper 00-04*; Tufts University: Medford, MA, USA, 2000; p. 26. Available online: https://sites.tufts.edu/gdae/files/2019/10/00-04Harris-BasicPrinciplesSD.pdf (accessed on 3 June 2021).
22. Holden, E.; Linnerud, K.; Banister, D. Sustainable development: Our Common Future revisited. *Glob. Environ. Chang.* **2014**, *26*, 130–139. [CrossRef]
23. Pawłowski, A. How Many Dimensions Does Sustainable Development Have? *Sustain. Dev.* **2008**, *16*, 81–90. [CrossRef]
24. McConville, J.R.; Mihelcic, J.R. Adapting Life-Cycle Thinking Tools to Evaluate Project Sustainability in International Water and Treatment Development Work. *Environ. Eng. Sci.* **2007**, *24*, 937–948. [CrossRef]
25. Saad, D.; Byrne, D.; Drechsel, P. Social perspectives on the effective management of wastewater. *IntechOpen* **2017**, *12*, 253–267. [CrossRef]
26. Progress on Integrated Water Resources Management. Global Baseline for SDG 6 Indicator 6.5.1: Degree of IWRM Implementation. Available online: https://www.unwater.org/publications/progress-on-integrated-water-resources-management-651/ (accessed on 3 June 2021).
27. Urban Population (% of Total Population)—Mexico. World Bank. Available online: https://data.worldbank.org/indicator/SP.URB.TOTL.IN.ZS?locations=MX (accessed on 3 June 2021).
28. Normas Oficiales Mexicanas. Secretaría de Medio Ambiente y Recursos Naturales. Comisión Nacional de Agua. Available online: http://www.conagua.gob.mx/CONAGUA07/Publicaciones/Publicaciones/SGAA-15-13.pdf (accessed on 3 June 2021).
29. Proyecto de Modificación de la Norma Oficial Mexicana NOM-001-SEMARNAT-1996. Diario Oficial de la Federación. 2018. Available online: https://www.dof.gob.mx/nota_detalle.php?codigo=5510140&fecha=05/01/2018 (accessed on 3 June 2021).
30. Oswald-Spring, Ú. (Ed.) *Water Research in México. Scarcity, Degradation, Stress, Conflicts, Management, and Policy*; Springer: Berlin/Heidelberg, Germany, 2011; p. 527. [CrossRef]
31. Desigualdad, Fugas, Costos y Concesiones Han Puesto en Jaque el Acceso a Este Vital Líquido. Problemáticas Económicas del Agua en México. Ciencia UNAM. Available online: http://ciencia.unam.mx/leer/775/problematicas-economicas-del-agua-en-mexico (accessed on 3 June 2021).

32. Recaudación de la CONAGUA (Nacional). Comisión Nacional del Agua. Available online: http://sina.conagua.gob.mx/sina/tema.php?tema=recaudacion (accessed on 3 June 2021).
33. Programa de Agua Potable, Drenaje y Saneamiento. Comisión Nacional del Agua. Available online: https://www.gob.mx/conagua/acciones-y-programas/proagua (accessed on 3 June 2021).
34. Ley Federal de Derechos. Disposiciones aplicables en materia de aguas nacionales 2016. Secretaría de Medio Ambiente y Recursos Naturales. Comisión Nacional de Agua. 2016. Available online: https://www.gob.mx/cms/uploads/attachment/file/105138/Ley_Federal_de_Derechos.pdf (accessed on 3 June 2021).
35. Bianchi, C.; Montemaggiore, G.B. Enhancing strategy design and planning in public utilities through "dynamic" balanced scorecards: Insights from a project in a city water company. *Syst. Dyn. Rev.* **2008**, *24*, 175–213. [CrossRef]
36. Christ, O. Decentralized waste water treatment system. In *Water in China. Water & Environmental Management S. (Wems)*; Wilderer, P.A., Zhu, J., Schwarzenbeck, N., Eds.; IWA Publishing: London, UK, 2003; pp. 187–196.
37. Randolph, J. Collaborative environmental planning and learning for sustainability. In *Environmental Land Use Planning and Management*, 2nd ed.; Island Press: Washington, DC, USA, 2011; pp. 80–104. ISBN 9781597267304.
38. Abdel-Halim, W.; Weichgrebe, D.; Rosenwinkel, K.H.; Verink, J. Sustainable sewage treatment and re-use in developing countries. In Proceedings of the Twelfth International Water Technology Conference IWTC12 2008, Alexandria, Egypt, 1 January 2008; pp. 1397–1409. Available online: http://www.iwtc.info/2008_pdf/15-2.PDF (accessed on 3 June 2021).
39. Heijungs, R.; Gjalt, H.; Guinée, J.B. Life Cycle Assessment and Sustainability. Analysis of Products, Materials and Technologies. Toward a Scientific Framework for Sustainability Life Cycle Analysis. *Polym. Degrad. Stab.* **2010**, *95*, 422–428. [CrossRef]
40. Crites, R.W.; Middlebrooks, E.J.; Bastian, R.K. *Natural Wastewater Treatment Systems*, 2nd ed.; IWA Publishing: London, UK, 2014; p. 480, ISBN 13 9781780405896.
41. Mestre, E. *Reingeniería. El Agua en Mexico: Lineamientos Propuestos para el Periodo 2018–2024*; Colegio de Ingenieros Civiles de Estado de Jalisco: Guadalajara, Mexico, 2018.
42. Programa Nacional Hidráulico 2020–2024 Comisión Nacional del Agua. Available online: https://www.gob.mx/conagua/articulos/consulta-para-el-del-programa-nacional-hidrico-2019-2024-190499 (accessed on 3 June 2021).

Review

Characterization and Treatment Technologies Applied for Produced Water in Qatar

Hana D. Dawoud, Haleema Saleem, Nasser Abdullah Alnuaimi and Syed Javaid Zaidi *

Center for Advanced Materials, Qatar University, P.O. Box 2713, Doha 2713, Qatar; hanadawoud@qu.edu.qa (H.D.D.); haleema.saleem@qu.edu.qa (H.S.); anasser@qu.edu.qa (N.A.A.)
* Correspondence: szaidi@qu.edu.qa

Abstract: Qatar is one of the major natural gas (NG) producing countries, which has the world's third-largest NG reserves besides the largest supplier of liquefied natural gas (LNG). Since the produced water (PW) generated in the oil and gas industry is considered as the largest waste stream, cost-effective PW management becomes fundamentally essential. The oil/gas industries in Qatar produce large amounts of PW daily, hence the key challenges facing these industries reducing the volume of PW injected in disposal wells by a level of 50% for ensuring the long-term sustainability of the reservoir. Moreover, it is important to study the characteristics of PW to determine the appropriate method to treat it and then use it for various applications such as irrigation, or dispose of it without harming the environment. This review paper targets to highlight the generation of PW in Qatar, as well as discuss the characteristics of chemical, physical, and biological treatment techniques in detail. These processes and methods discussed are not only applied by Qatari companies, but also by other companies associated or in collaboration with those in Qatar. Finally, case studies from different companies in Qatar and the challenges of treating the PW are discussed. From the different studies analyzed, various techniques as well as sequencing of different techniques were noted to be employed for the effective treatment of PW.

Keywords: produced water treatment; produced water characterization; Qatar

Citation: Dawoud, H.D.; Saleem, H.; Alnuaimi, N.A.; Zaidi, S.J. Characterization and Treatment Technologies Applied for Produced Water in Qatar. *Water* **2021**, *13*, 3573. https://doi.org/10.3390/w13243573

Academic Editors: Martin Wagner and Sonja Bauer

Received: 7 November 2021
Accepted: 10 December 2021
Published: 13 December 2021

Publisher's Note: MDPI stays neutral with regard to jurisdictional claims in published maps and institutional affiliations.

Copyright: © 2021 by the authors. Licensee MDPI, Basel, Switzerland. This article is an open access article distributed under the terms and conditions of the Creative Commons Attribution (CC BY) license (https://creativecommons.org/licenses/by/4.0/).

1. Introduction

Produced water (PW), or oilfield wastewater, are terms used to describe the water produced from oil and gas industries in the extraction process [1,2]. It contains brine, as well as a combination of various organic and inorganic compounds [3]. Remarkably, the volume of wastewater all over the world is increasing, which leads to significant attention of the harmful effects of discharging PW on the environment [4,5]. Moreover, this discharge causes a polluted surface in underground water as well as in soil [6]. It is important to mention that the oil extraction process is a physical process consisting of different individual steps [7]. Initially, the oil is extracted by drilling the oil reservoir using subsea pipelines. In this process, the oil, gases, and water from seawater are extracted [8]. To achieve maximum oil recovery and sustain the pressure of the reservoir, two techniques are employed [9,10]. The first technique is water flooding or water injection, in which water is injected into the reservoirs for the purpose of adding extra force into the reservoir [11,12]. The injected water ultimately reaches the reservoirs, and in the later stages of water-injection, the produced water proportion leads to the production of more oil and recovery of the lost pressure in the reservoir due to oil extracting [13]. The second technique is gas flooding or gas injection, where the gas must improve the lifting of the fluid from the reservoir to the manifolds, which aids in reducing the density of the extracted fluid [14,15]. Nonetheless, due to the generation of a large volume of produced water from oil/gas fields in Qatar, different treatments are applied to treat this PW in an efficient and economical way [16]. Currently, there are different technologies used for the purification of wastewater [17–20], and the treatment methods are divided into chemical treatment [21,22], physical treatment [23–25], and biological treatment. Physical treatment includes gravity separation and adsorption [26,27], usage of hydrocyclones separator, membrane filtration-based techniques (sand filtration, membrane distillation, ceramic membranes, membrane bioreactors, hybrid, and asymmetric membranes), application of hydrate inhibitors, demulsification, coalescing, thermal evaporators, forward osmosis, etc. [28–33]. Moreover, chemical treatments consist of photocatalytic treatment, chemical oxidation, chemical precipitation, electrochemical process, Fenton process, treatment with ozone, demulsifiers, and room temperature ionic liquids [34–36]. Presently, the membrane treatment processes such as nanofiltration (NF), ultrafiltration (UF), microfiltration (MF), and reverse osmosis (RO) are gaining importance in reducing the contamination from PW [37]. The biological treatment of produced water is efficient; however, it is rarely utilized for the oil- and gas-produced water [38]. Some of the most commonly used biological treatment configurations were fixed-film treatment, membrane bioreactors, and constructed wetlands and ponds. The fixed-film reactors consisted of tanks covered with higher-surface area media, rotating biological contactors, and granular activated carbon filters, as well as aerobic filters. Table 1 presents the summary of different produced water treatment technologies [39–41]. Table 2 presents the commercial treatment processes worldwide.

Table 1. Summary of different produced water treatment technologies [39–41].

Treatment Technologies	Advantages	Disadvantages
Membrane separation	1. Removal of dissolved organic substances could be achieved by selecting an appropriate membrane. 2. Advanced purification. 3. Have separation impacts on suspended solids.	1. Pollution chances are high. Hence, backwashing is needed. 2. Polymer membrane degrades at higher temperature (greater than 50 °C). The PW temperature is generally greater than 50 °C.
Combined fiber coalescence	1. It can handle PW with higher oil content. 2. The suspend solids could penetrate the fiber module with no blockage.	1. Solid particles having millimeter-scale dimensions can cause blockage. 2. Fiber parameters can influence the separation efficiency. The device must be designed as per the PW characteristics.
Tubular separation	1. Small and could be fixed under water.	1. The anti-fluctuation performance as a pretreatment should be proved.
Media filtration	1. Advanced purification. 2. Demonstrates separation effects on suspended solids.	1. Pollution chances are high. Hence, backwashing is needed. 2. The influent must not have higher oil content. 3. Media replacement is required.
Hydrocyclone	1. Compact with no moving parts. 2. Can handle PW with higher oil content.	1. Limited range of higher efficiency operation region. 2. Separation effects are influenced by blockage as well as the wear of one or more cyclone tubes.
Gravity and enhanced gravity sedimentation	1. Anti-fluctuation ability of the gas concentration, oil content, and flow rate. 2. Essential pretreatment device having extensive operation flexibility. 3. Easy equipment with minimal operational expenses and maintenance needs.	1. Higher occupation of space resulting from the separation mechanism. 2. Lower separation accuracy.

Table 2. Commercial treatment processes worldwide.

Technology	Commercial System	Treatments	Water Recovery	Reference
Reverse Osmosis	CDM Technology	Combination of 3 major processes such as ion exchange process, RO, and evaporation	50–90%	[40]
Reverse Osmosis	Veolia:OPUSTM	Acidification, Degasification and followed by MultifloTM chemical softening and Reverse Osmosis	Higher than 90%	[41]
Reverse Osmosis	Eco-sphere: OzonixTM	Activated carbon cartridge filtration and RO	75%	[40]
Reverse Osmosis	GeoPure water technologies	Combination of pretreatment, UF and RO	50%	[40]
Ion-Exchange (IX) based processes	EMIT: Higgins Loop	Continuous counter current ion exchange contactor for liquid phase separations of ionic components.	99%	[37]
Ion-Exchange (IX) based processes	Drake: Continuous selective IX process	3-phase, continuous fluidized bed system	97%	[40]
Ion-Exchange (IX) based processes	Eco-Tech: Recoflo® compressed-bed IX process	Extension of standardpacked bed IX processes	-	[40]

During the COVID-19 crisis, the worldwide market for PW treatment was assessed to be USD 8.1 billion in the year 2020, and is expected to achieve a revised size of USD 11.2 billion by the year 2027, increasing at a compound annual growth rate of 4.8% over the assessment period 2020–2027 [42]. The PW treatment market in the United States of America is estimated at USD 2.2 billion in 2020. In Qatar, value oil production in the year 2020 was 1343.14 thousand barrels/day, relative to 1550 thousand barrels/day in 2011 [43]. This clearly confirmed the increase in the production volume of produced water also.

Treated produced water can be reused in various applications, such as in increasing oil production by underground injection, irrigation, livestock, or wildlife watering, and various industrial uses (e.g., fire control, vehicle washing, dust control, and power plant makeup water) [41,44,45]. Adopting various treatment technologies is important to assess the adverse environmental impacts, as well as to reach the standards demanded reinjection, reusing, and discharging the PW [46]. All of these will support a harmless release or reinjection of PW in plant irrigation to share in the sustainability of the Qatar environment [47,48].

In this paper, a comprehensive review of produced water generated and treated in Qatar is attempted. To the best of our knowledge, there are very few studies discussing this topic. Since the focus in this paper is on PW, three areas will be highlighted through the review: characterization of produced water; treatments methods, challenges and future developments related to using efficient treatment technology; and cost-effectiveness of treating produced water. The main aim of the current review paper is to promote the appropriate characterization and treatment of PQ in Qatar for reducing the dependence on the very limited freshwater resources. All of the available data in the open literature related to PW in Qatar are summarized and presented in Table 3.

Table 3. Comparison of treatment methods, characterizations, and applications of produced water from different entities in Qatar.

Sl. No.	Treatments Used	Methods Used	Merits/Demerits	Main Characterization	PW Characteristics/Target Contaminants Removed	Significance of the Study	References
1	Membrane processes (UF, NF, and RO), thermal evaporations and advanced oxidation process	Membrane-based	-	TDS, COD, KHL, salinity, conductivity	Monovalent and divalent ions (Calcium, magnesium, potassium)	PW use in irrigation	[49]
2	Forward osmosis	Membrane-based	FO offers ecofriendly dilution before discharge.	pH, TOC, TC, inorganic carbon, water flux, anion and cations, ions, metals, salinity, organic and inorganic contents	Inorganic carbons, Monovalent and divalent ions	Efficiency of FO process	[50]
3	Hollow Fiber Forward Osmosis Membranes	Membrane-based	FO process leads to environmentally friendly dilution before discharge	Conductivity, pH, anion and cation, TDS, TN, TOC, IC, and osmotic pressure	Chloride, sodium, calcium, magnesium, bromide, sulfate, potassium, phosphate, TOC, TN	Efficiency of FO process	[51]
4	Membrane bioreactors (MBRs)	Membrane-based	Economical and good separation performance	COD, conductivity, pH TSS, VSS, TN, TPH, TOG, DO, and TOC	Organic carbons, potassium, ammonium, phosphate	Efficiency of MBRs	[52]
5	Forward osmosis membranes sing thin-film composite FO hollow fiber membranes	Membrane-based	FO process leads to environmentally friendly dilution before discharge	TDS, TOC, inorganic carbon, conductivity, alkalinity, turbidity, pH	Organic carbons	PW use in irrigation	[53]
6	MBRs	Membrane-based	Economical and good separation performance	COD, TOC, TN, anion and cation, oil and grease, TPH, thiosulfate, conductivity	Anion and cation, oil and grease, TPH, thiosulfate	Efficiency of MBRs	[54]
7	Membrane processes, membrane bioreactors, membrane distillation and ozonation	Membrane-based	Economical and good separation performance	TDS, COD, KHL, salinity, conductivity, ions, and cations	Cations, TDS	Efficiency of MB process, MD, ozonation, and membrane bioreactors	[55]
8	Membrane distillation	Membrane-based		TDS, DOC, TOC, COD, conductivity, phenol, oil and grease	Grease, oil, phenol	MD efficiency for PW treatment	[56]
9	Direct membrane filtration, biological such as MBRs, advanced oxidation processes (AOPs) and thermal evaporators	Membrane-based	Economical and good separation performance	TDS, DOC, TOC, COD, conductivity, phenol, oil and grease	Oil, grease, phenol	Efficiency of MBRs and AOPs in PW treatment	[57]
10	Crossflow multi-channel ceramic membrane (TiO$_2$ and SiC)	Membrane-based	Permits produced water re-Injection even in difficult reservoirs with no loss in injectivity	pH, conductivity, sulphide, TS, TOC, hardness, iron, O and G	Iron, oil, grease	Membrane process effectiveness in PW treatment	[58]
11	Chemical cleaning in place (CIP) between ceramic membrane	Membrane-based		pH, calcium, barium, and iron alkaline, oil and inorganic reagents, turbidity and oil and grease	Oil, grease, calcium, barium, inorganic reagents	Ceramic microfiltration membrane	[59]
12	Ceramic membrane	Membrane-based	Permits produced water re-Injection even in difficult reservoirs wit no loss in injectivity	Particulate solids, Feed and permeate oil concentration, TOC, COD, turbidity	Organic carbons and suspended solids	Crossflow ceramic microfiltration (CFCMF) to the removal of emulsified oil produced water	[60]
13	Hybrid Separator-Adsorbent Inorganic Membrane (Al$_2$O$_3$ and AC)	Membrane-based		Salinity, oil removal efficiency, water flux	Oil removal, Salt removal	Inorganic Membrane for PW management	[61]

Table 3. Cont.

Sl. No.	Treatments Used	Methods Used	Merits/Demerits	Main Characterization	PW Characteristics/Target Contaminants Removed	Significance of the Study	References
14	Crossflow membrane filtration (CMF), media (nutshell) filtration (NSF induced gas flotation (IGF), and Hydrocyclones (HCs)	Membrane-based		Efficiency of removing the suspended matters-	Suspended matters removal	application of an effective chemical clean	[62]
15	Hydrogel Forward Osmosis Membrane	Membrane-based	FO process leads to environmentally friendly dilution before discharge	TOC	Organic carbons, oil and grease	Efficiency of FO process for PW treatment	[63]
16	Biotreatment of Hydrate-Inhibitor with activated sludge process	Membrane-based	Most economical approach for organics removal	Ammonium, phosphate, potassium, COD, conductivity, pH TSS, VSS, TN, TPH, TOC, DO and TOC	Organic carbons, Ammonium, phosphate, potassium	PW use in irrigation	[64]
17	Flocculation flotation unit, biotreatment, membrane filtration (UF, RO units), evaporation and crystallization processes	Membrane-based	Permits produced water re-Injection even in difficult reservoirs with no loss in injectivity	pH, conductivity, TOC, ion chromatography, metal, COD, TDS, Ca, Mg, Ba and heavy metal	Heavy metals, organic carbons, dissolved solids, divalent ions	Cooling and power generation among different uses	[65]
18	Coagulation, dissolved air flotation and Evaporation technology	Coagulation	Energy saving process	COD, KHI, ions and cations, Total hardness, TKN, TOC, O and G, TSS, Cl, TDS	Grease, oil, organic carbons	Removal of KHI co-polymers application	[66]
19	Electrocoagulation	Coagulation	Highly efficient and energy saving process	Ammonium, phosphate, potassium, COD, conductivity, cations and anions, ionic-liquid, pH TSS, VSS, TN, TPH, TOC, DO and TOC	Phosphate, potassium, organic carbons	Efficiency of electrocoagulation for PW treatment	[67]
20	Electrocoagulation	Coagulation	Energy saving process	COD, TOC, TPH, O and G and sludge	Oil, grease	-	[68]
21	Electrocoagulation and steel slag	Coagulation	Energy saving process	Oil and grease removal, turbidity, TSS	Suspended solids, oil and grease	Efficiency of electrocoagulation and steel slag for PW treatment	[69]
22	-	Biological treatment	Produces huge amount of biomass that could be employed as feedstock for many products	Bacterial colony-forming units (CFU)	Chloride, sulfate, bromide, sodium, magnesium, calcium, and potassium, strontium and boron	1—Irrigation of 2 turfgrass species, *Paspalum* sp. and *Cynodon dactylon*. 2—Studying the impact of PW irrigation on established grasses, heavy metal accumulation, microbial succession, and germination tests for weeds and turf grass seeds	[70]
23	Microalgae strains	Biological treatment	Most economical approach for organics removal.	Salinity,pH, TOC, TN, TP	Salts, phosphorus	Use of biomass as feedstock	[71]
24	Microscopic microalgae; screening, 5 species of microalgae strains *Dictyosphaerium*, *Scenedesmus*, *Chlorella*, *Monoraphidium*, *Neochloris*.	Biological treatment	Most economical approach for organics removal.	TP, BTEX, Fe, Al, TOC, TN, COD, TKN, turbidity, salinity, pH, and ammonium	Organic carbons, phosphorus, salts, ammonium	Application of microalgae for PW treatment	[72]

Table 3. Cont.

Sl. No.	Treatments Used	Methods Used	Merits/Demerits	Main Characterization	PW Characteristics/Target Contaminants Removed	Significance of the Study	References
25	Five microalgae strains used for water treatment: *Monoraphidium*, *Chlorella*, *Neochloris*, *Scenedesmus*, *Dictyosphaerium*, *Chlorella* and *Dictyosphaerium* species	Biological treatment	Most economical approach for organics removal.	Organic carbon, nitrogen removal and phosphorus and various metals, removal efficiencies, TOC and BTEX	Nitrogen, metals, organic carbons, phosphorus, salts, ammonium	Application of microalgae for PW treatment	[73]
26	-	Biological treatment	Most economical approach for organics removal.	Salinity, bacterial and fungal CFUs	Salts and microorganisms	Irrigation of turf grass—*Paspalum* sp. and *Cynodon dactylon*	[74]
27	-			TDS, boron, sodium, chloride ions, sodium adsorption, and organic contents	Organic compounds, boron, and salt	Plant irrigation in greenhouse for *Salsola burgosna*, *Phramites australis* *Sorghum bicolor*, *Medicago sativa*, *Helianthus annus* and *Zea mays*	[75]
28	Sand filtration activated carbon filtration (ACF) as well as modified activated carbon filtration.	Activated Carbon filtration	Increased removal of COD	Cations, metals, inorganic anions, BTEX, phenolic, organic acids, oil and grease, sulfides, hardness, alkalinity, conductivity, BOD, TOC, COD, and pH	Phenolic, organic acids, oil and grease, sulfides, cations, metals, inorganic anions, BTEX	-	[76]
29	Activated carbon filtration and microemulsions modified AC	Activated Carbon filtration	Increased removal of COD	Heavy metals, salts, toxic organic components, and TDS, BTEX, pH, COD, TOC, TN, TDS, conductivity, alkalinity, hardness and sever all metals	Toxic organic components, heavy metals, salts, dissolved solids	Irrigation application of PW	[77]
30	Series of inclined multiple arc coalescence plates	Coalescing	-	Salinity, oil removal	Oil	Removal of stable oil emulsions from PW	[78]
31	Electrochemical methods	Chemical method	Highest TPH and COD removal efficiency	Corrosion study and scaling study		Examine the impact of PW from the Ras Laffan (North Oilfield) Qatar on corrosion as well as scaling of carbon steel.	[79]
32	Chemical demulsification	Chemical method	-	Cations and anions, ionic liquid	Anions and cations	Efficacy of chemical demulsification for PW treatment	[80]
33	Site 1: 2 phase separation tanks combined with filtration unit as well as chemical injection, and finally the large gravitational separation tanks. Site 2: begins with 2-stage separation with chemical injection, 2 phases succeeded by 3 phase separation tanks combined with hydrocyclone succeeded by surge drum.	Chemical method	-	Total sulfides, dissolved CO_2, concentration of ions, phosphates, ammonia nitrogen, concentration of total Kjeldahl nitrogen, concentration of metals, total dissolved solids, total suspended solids, biodegradable COD, total COD, Phenol concentration, BTEX concentration, total amount of hydrocarbons, conductivity, pH, and oil droplet size distribution	Phosphates, ammonia nitrogen, sulfides	-	[81]
34	Anionic polyacrylamide (PAMs) with electrolyte of aluminum sulphate and ferrous sulphate	Chemical method	-	Turbidity, viscosity, and COD		-	[82]

2. Methodology for the Literature Review

Figure 1 presents the number of articles (by year), where the expression 'produced water Qatar' was found in the title, abstract, or keywords, in the last 10 years. This figure was created from scopus.com (accessed on 5 December 2021). The database showed 96 total studies, with eight studies in 2021 (reported until December), and only four studies during 2012. The entire studies were published from 1993 to 2021. There is a slow growth in studies based on produced water noted in Qatar in the past ten years. Of the different studies, the majority studies were based on the membrane-based PW treatment processes such as membrane distillation, forward osmosis, reverse osmosis, etc. Studies were considered in this work only if the water studied represented produced water. Different project motives were identified by the study authors, even though environmental challenge was the most predominant. It was highly evident from the research works that scientists are interested in the beneficial reusing as well as recycling of PW. The majority of the PW samples were obtained from oilfields, although a few PW samples were obtained from gas production wells.

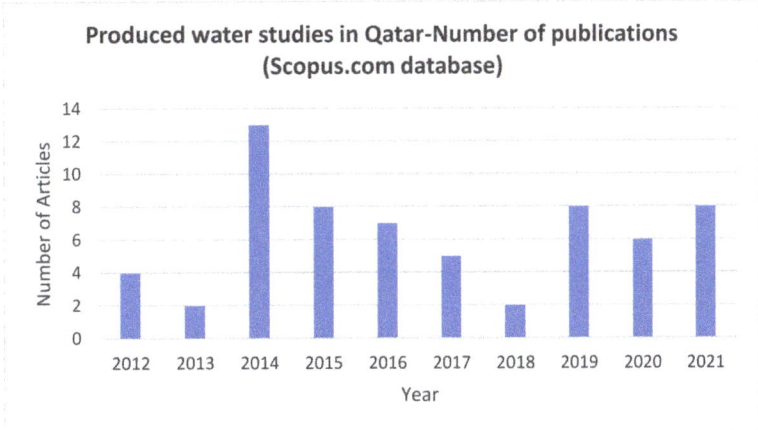

Figure 1. The number of articles (by year), where the expression 'produced water Qatar' was found in the title, abstract, or keywords, in the last 10 years. Data extracted from the scopus.com database (5 December 2021).

3. Produced Water in Qatar

Due to the increasing demand for fresh water, there is a need to develop new water sources in Qatar. With appropriate treatment of PW, it can serve as a new water supply in Qatar. There are three main sources of PW, namely aquifer, formation, and injection water [81]. PW quality varies significantly from field to field, but generally, the total dissolved solids concentration can range from a few thousand to over 400,000 mg/L [83]. It is worth mentioning that the PW produced from the NG production system in the north field offshore is counted as the major volume of wastewater in Qatar, and this could be used for beneficial applications. This water source from the industry can be used for domestic uses if it is treated properly [84,85].

The global estimation of PW production is approximately 250 million barrels per day, relative to a global almost 80 million barrels per day of oil [86]. Based on the global estimation, the water to oil ratio is approximately 3:1. On the contrary, for NG production in the Qatari north field, the ratio of water to gas is approximately 1.20 based on Qatar Petroleum research [87]. Furthermore, Qatar has made a significant investment in liquefied natural gas (LNG), and it produces 77 million tons per year. Moreover, the leading petroleum company in Qatar has added a fourth LNG production line, and is planning to increase its production capacity from the north field to reach 110 million tons a year. Currently, liquefied natural gas companies in Qatar operate 14 LNG trains with a total annual production capacity of 78 million tons [88]. However, several regulations in Qatar are set by the Ministry of the Municipality and Environment to control and regulate the production of produced water. To comply with the regulations of the Ministry of the Environment and ensure sustainable long-term disposal, the liquefied natural gas company in Qatar created 14 trains between 1999 and 2011, producing 78 million tons per year of LNG, and in 2015 they exported 78.4 million tons. Additionally, liquefied natural gas companies in Qatar constructed two advanced wastewater recycling and reduction (WRR) plants for numerous LNG trains at Ras Laffan, Qatar. However, the reverse osmosis (RO) process is the main advanced technology, since it will produce permeate for feeding the boilers, whereas all other supplementary treatments serve as pretreatment stages to eliminate other pollutants, for instance, hydrogen sulfide (H_2S), dispersed & emulsified oil, organics, and suspended solids. The produced water will be combined with the RO brine and injected into disposal wells, causing a reduction of total water disposal volume. On the other hand, in 2018, the major petroleum company in Qatar was awarded a FEED contract for three new LNG trains and added a fourth LNG mega train to reach a capacity of 110 million tons per annum (MTPA), which will be used to service expansion of the North field development.

4. Onshore and Offshore Produced Water Production

Qatar has the world's third-largest reserve of natural gas after Russia and Iran. Moreover, Qatar is the second-largest exporter of natural gas, owning 14 % of the total global gas reserves. Moreover, Qatar's Supreme Council for the Environment and Natural Reserves (SCENR) has assigned all oil production facilities to treat the PW in the state, in order to achieve the recent maximum oil/water concentration limit of 40 ppm, while maintaining an average of 20 ppm [89]. In Qatar, seven main offshore production stations (PS) operate on eight oil production fields. The first PS1 is located in the northeast of Qatar, within a 45 km distance from Al-Rayyan city, and it operates on two fields that were first discovered as oil wells in 1960, Idd- El Shargi north dome and Idd- El Shargi south dome. The second and third PS are operated by Qatar Petroleum, and have two fields in the northeast in Maydan Mahzam (MM) and Bul Hanine (BH) fields of Qatar. These two fields are popular in producing high-quality crudes and associated gas, which began production in 1965 and 1972, respectively. Moreover, Idd-El Shargi and Al-Rayyan fields are operated by Occidental Petroleum of Qatar Ltd., Al-Shaheen field by Maersk Oil Qatar, Al-Khalij field by TOTAL Exploration & Production Qatar [86], and finally, Al-Karkara field by QP Development Company. Another field is El-Bunduq field, which is operated in cooperation with United Arab Emirates and is operated by Bunduq Company Ltd. (Abudhabi, United Arab Emirates) [81]. Nevertheless, Dukhan oil field has onshore field operated by Qatar Petroleum, which is a long narrow anticline over the north-south for around 70 km. The production in the Dukhan field started in 1940 after its discovery in 1939 [90]. Dukhan covers four reservoirs: Fahahil, Khatiyah, Jaleha, and Diyab, one of which contains non-associated gas, and the other three are oil reservoirs. The oil and gas fields are divided into four degassing stations: Khatiyah (north, main, and south), Fahahil Main, and Jaleha. Dukhan field produces up to 335,000 barrels per day [81]. Total is the operator, as well as a shareholder of AlKhalij offshore oilfield, sited on Block 6 in Halul, which was established in1991. Production from this geographically complicated oilfield began in 1997, and currently, approximately 22,000 barrels of oil/day are taken through two subsea pipelines to a treatment plant on Halul Island. In 2015, a breakthrough 200 million barrels of oil was produced from this field. In 2016, Total gained a bid for Qatar's Al-Shaheen offshore oilfield, with a production capacity of 300,000 barrels/day. North Oil Company (NOC), which was founded as a partnership between Qatar Petroleum (70%) and Total (30%), began the operation of the massive oil field starting in 2017. This field generates 300 thousand barrels of oil/day. Table 4 presents the PQ characteristics obtained from a Qatar-based natural gas field [91].

Table 4. PQ characteristics obtained from a Qatar-based natural gas field [91].

Different Parameters	PW Characteristics	
	Filtered Water	Raw PW
Xylene (mg/L)	3.11	3.43
Ethyl benzene (mg/L)	1.05	1.22
Toluene (mg/L)	3.21	3.8
Benzene (mg/L)	16.1	21
Total phosphorus (µg/L)	180	277.78
Total Nitrogen (mg/L)	27.6	35.77
Total organic carbon (mg/L)	317	389.1

5. Factors Affecting Production Volume of Produced Water

The production volume of produced water is affected by several factors. Drilling type is considered as an important factor, for example, the horizontal well produces PW at a higher rate, as compared to the vertical well at the same drawdown. Furthermore, the next important factor is the position of the well, i.e., whether it is placed within heterogeneous or homogeneous reservoirs. An inappropriately drilled well or one that has been inappropriately positioned inside the reservoir structure can lead to an earlier than expected water production. Moreover, the vertical and horizontal well is affected by the type of the reservoirs, for instance, when the homogeneous reservoirs are using the horizontal wells, the production of water is reduced, and the volume of water injected in the oil recovery is enhanced. Further, a perforated completion proposes a higher degree of control in the hydrocarbon-production zone. Special intervals can either aim for improved hydrocarbon production, or be prevented or plugged for minimizing the water production. It is important to mention that inadequate mechanical integrity of drilling could enhance PW production [81,92]. Moreover, the type of water separation as well as treatment facilities influence the production volume of the produced water. Generally, the surface separation, as well as treatment facilities, are employed for the management of produced water. Nevertheless, this type of operation requires lifting costs for bringing the water to the surface, as well as equipment and chemical expenses for water treatment. Substitutes for surface treatment can be downhole separation apparatus that permit the produced water for remaining downhole, thus preventing certain lifting, surface capability, and corrosion expenses, as well as related challenges. Furthermore, the inadequate volume of produced water intended for water flooding impacts the produced water's production volume. If inadequate produced water is obtainable for water flooding, extra source waters should be acquired for augmenting the injection of produced water. In order to maintain a successful water flood operation, the water employed to inject should be of high quality that does not harm the reservoir rock. Previously, freshwater was frequently employed in water floods. Subsurface communication difficulties are another significant factor influencing the production volume of the produced water. Near-well bore communication issues such as barrier breakdowns, channels behind casing, and completions near or into water could lead to more volumes of produced water. Additionally, reservoir communication challenges such as cresting, coning, fracturing out of the hydrocarbon producing zone, and channeling through high permeability zones or fractures could also promote higher volumes of produced water. All of the above-stated factors could increasingly influence the produced water's volume that is finally managed in the course of the life cycle of a well, as well as a project. With augmented volumes of produced water, the financial feasibility of a project becomes an issue, due to the disposal expenses of water, the enlarged size and expense of water treating facilities, as well as related treatment chemicals, the extra cost of lifting water versus hydrocarbons, and loss of recoverable hydrocarbons.

6. Produced Water Characterization

The chemical and physical properties of PW differ in accordance with geographic location, depth, the geologic formation of the production well, geochemistry of the component with the hydrocarbon, and the reservoir lifetime [81,93–95]. Moreover, the chemical properties vary based on the different chemicals added within the production process and the composition of oil and gas in the reservoir. It consists of organic and inorganic materials, metals, and impurities such as radioisotopes [5,96,97], in addition to inorganic anions (chlorides, sulfates, and phosphates), sulfide, metals, cations, total suspended solids (TSS), heavy metals, chemical oxygen demand (COD), biochemical oxygen demand (BOD), total organic carbon (TOC), dissolved and dispersed oil compounds, dissolved gases, and conductivity. Table 5 illustrates the characteristic of oilfield PW content [81].

Table 5. Typical composition of produced water from Qatar. Reproduced from reference [98].

Parameter	Concentration (mg/L)
Total dissolved solid	1000–400,000
Total suspended solid	98–116
Potassium	10–12
Sodium	5462–5836
Chlorine	8475–9219
Total organic carbon	45–71
Magnesium	114–118
Calcium	356–372
Sulfate radical	61–68
Total nitrogen	23–26

In this section, the characteristics of produced water generated in Qatar are initially defined, and then discussed in detail.

6.1. pH

pH is used to find the acidity and alkalinity of the solution. In 1992, Jacobs et al. [99] presented the acidic nature of the PW solution obtained from gas fields. The team stated that the level of pH in PW obtained from gas operations grounds (ranging from 3.5 to 5.5) is more acidic than PW from oil fields (ranging from 6 to 7.7).

6.2. Chemical Oxygen Demand (COD)

COD is a basic method to discover the quantities of contaminants that cannot be oxidized naturally in the produced water. It is measured as the milligrams of O_2 per liter of the sample that is consumed by the chemical demand. Being conscious of the COD amount in the sample leads to the determination of the amount of suspended contaminants, as well as dissolved contaminants present in water [76].

6.3. Total Organic Carbon (TOC)

TOC is used for determining the total amount of organic compounds in PW in mg/L. It is also important to measure the level of pollution in the wastewater. TOC accurately measures the concentration of carbon found in an organic compound, and is usually implemented as a non-precise indicator of water quality. Practically, all TOC analyzers determine the CO_2 formed when organic carbon is oxidized and when inorganic carbon is acidified. The concentration of TOC in PW differs significantly from one well to another [76].

6.4. Biochemical Oxygen Demand (BOD)

BOD is a bioassay process used to determine the concentration of oxygen consumed in the disintegration procedure of organic matter by bacteria. The process includes the measurements of dissolved oxygen mass for a specific volume of solution required for the biochemical oxidation procedure. There are specific conditions to perform this process, such as a temperature 20 °C, full darkness, and monitoring should be carried out over a period of five days. It is measured in mg/L. It has been found that PW from the gas field has higher concentrations of BOD than from the oilfield. The high amount of BOD in PW leads to reduced water quality. Therefore, the PW must be significantly oxidized to avoid the ejection of higher BOD materials into the receiving streams [76].

6.5. Conductivity and Salinity

Conductivity is the measure of the capability of water to pass the electrical flow. Conductivity is precisely related to the concentration of ions present in the water. In one research study, it was noted that the conductivity of PW from natural gas fields was in the range of 4200–180,000 µS/cm [100]. The conductivity is dependent on the value of the temperature, pH, and on the amount of carbon dioxide dissolved in the PW to develop ions. There are two types of conductivity: intrinsic conductivity due to the mentioned factors or extraneous conductivity due to ion's concentration already existing in the sample such as chloride, calcium, sodium, magnesium, and other ions [81]. On the other hand, salinity is the measurement of salts present in the solution determined by the electrical conductivity (EC) of a liquid. The salinity concentration could range from a very small amount of salt to an extreme content that can be higher than that of seawater [76].

6.6. Ions and Inorganic Constituents

Produced water contains various dissolved ions and inorganic elements. Ions are charged particles, such as sodium chloride (NaCl) salt dissolved, and formed Na^+ and Cl^-. To measure the hazardous nature of soil and water, the sodium adsorption ratio (SAR) scale is used [101]. It determines the suitability of water for irrigation, and represents the relationship of the sodium with magnesium and calcium concentration. Overall, higher concentrations of magnesium, calcium, and sodium, and the lesser SAR value are considered as better water for irrigation. In PW, the main inorganic constituents are sodium, calcium, magnesium, bicarbonate, aluminum, arsenic, barium, chloride, sulfate, and potassium. There are minor ions present in inorganic constituents, such as metals consisting in a different range of concentrations alongisde the non-metals, likely boron and fluoride. However, the major ions are found in moderately higher concentrations than minor ions.

6.7. Total Suspended Solids (TSS)

Suspended solids or TSS are found in a smaller size from reservoir rocks such as quartz and clays. Moreover, PW contains various types of solids or dissolved impurities that cannot pass through the filter. The type of TSS differs depending on the size such as in hydraulic fracture, the proppant size ranges from 1.0 mm to larger, whereas iron sulfide particles range from 0.1 µm to smaller than this. Predominately, TSS is denser as compared to the PW or oil, thus it typically sinks to the base of the vessels, tanks, or pipes driving many operating challenges [76,81].

6.8. Heavy Metal

The concentration of heavy metals depends on the formation of geology and the age of the wells. PW tend to contain lead (Pb), iron (Fe), zinc (Zn), barium (Ba), selenium (Se), strontium (Sr), and manganese (Mn) in wide concentrations. In addition, chromium (Cr), vanadium (V), copper (Cu), cadmium (Cd), mercury (Hg), and nickel (Ni) are some popular heavy metal pollutants present in trace amounts in the PW. The concentration of the metals depends on the surroundings, where high concentrations cause toxicity and bioaccumulation [81].

6.9. Total Kjeldahl Nitrogen (TKN)

Total Kjeldahl Nitrogen is the total concentration of organic nitrogen and ammonia. TKN is a method to determine the total nitrogen in an organic substance, involving inorganic compounds such as ammonia and ammonium (NH_3/NH_4^+) in water. This method was established in 1883 by Johan Kjeldahl. The nitrogen constituents and their relationship are shown in Figure 2 [81].

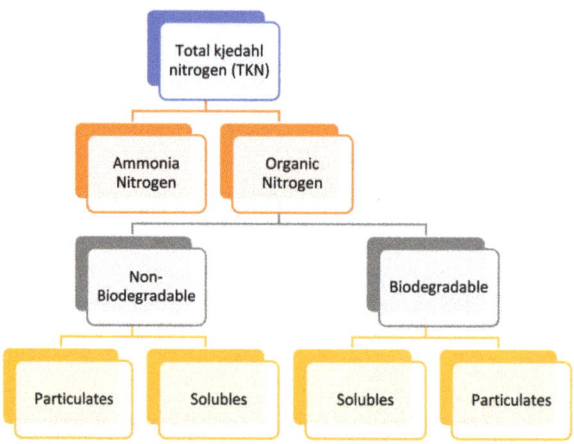

Figure 2. Total Kjedahl nitrogen (TKN) fractions.

6.10. Total Petroleum Hydrocarbon (TPH)

Total petroleum hydrocarbons represents the number of hydrocarbons in the sample, mainly consisting of carbon and hydrogen. It is present either in suspended petroleum hydrocarbon or dissolved form. Petroleum hydrocarbon is a combination of hydrocarbon containing mostly four groups: BTEX (volatile aromatic compounds: benzene, toluene, ethylbenzene, and xylene), phenols, polycyclic aromatic hydrocarbons (PAHs), and NPD (naphthalene, phenanthrene, and dibenzothiophene). Generally, hydrocarbons are categorized into three major categories, namely aromatics, unsaturated, and saturated.

BTEX compounds are naturally created in oil and gas, such as gasoline, diesel fuel, and natural gas. BTEX are very volatile, and therefore they are lost rapidly through PW treatment, such as in the early mixing in the sea or by air stripping [102]. Benzene considers as the most abundant compound among all BTEX components, but when the alkylation is increased, its amount decreases. [76].

6.11. Total Nitrogen (TN)

Total nitrogen is the collective quantity of the entire nitrogen compounds (ammonia, nitrates, and nitrites) in the PW. It contains organically bonded nitrogen, nitrate-nitrogen (NO_3-N), ammonia-nitrogen (NH_3-N), and nitrite-nitrogen (NO_2-N). It can be found by monitoring for free-ammonia, nitrate-nitrite, and organic nitrogen compounds individually, and then summing the values of the components together.

In Qatar, Hussain, A., Minier-Matar, J. et al. [49] performed PW treatment, and the PW was obtained from an offshore Qatari gas processing operation. In the study, the PW for the biotreatability test was characterized. Nitrogen and phosphorus were added to guarantee sufficient nutrients [49]. Additionally, the characterization of the process water for the same test is summarized with PW data in Table 6.

Table 6. Characterization of PW and process water from oil-producing companies in Qatar.

Title	Authors/Date/Reference	Parameters	Unit	Value	Parameters	Value
		Produced Water Source and Composition			**Process Water Source and Composition**	
Advanced Technologies for Produced water treatment	Hussain, A., Minier-Matar, J., et al./ 2014 [49]	COD	mg/L	1572	COD	397 mg/L
		TOC	mg/L	491	TOC	114 mg/L
		TN	mg/L	43	TN	31 mg/L
		Oil & grease	mg/L	47	Oil & grease	10 mg/L
		TPH	mg/L	45	TPH	9 mg/L
		Chloride	mg/L	2265	Chloride	17 mg/L
		Sodium	mg/L	1030	Sodium	359 mg/L
		Calcium	mg/L	329	Calcium	3 mg/L
		Sulfide	mg/L	307	Sulfide	307 mg/L
		Magnesium	mg/L	61	Magnesium	0.2 mg/L
		Bromide	mg/L	51	Bromide	<0.5 mg/L
		Sulfate	mg/L	54	Sulfate	9 mg/L
		Potassium	mg/L	44	Potassium	1.5 mg/L
		Thiosulfate	mg/L	14	Thiosulfate	43 mg/L
		Acetate	mg/L	347	Acetate	3.2 mg/L
		Ammonium	mg/L	11	Ammonium	11 mg/L
		Conductivity	μS/cm	7200	Conductivity	1761 μS/cm
		TDS	mg/L	5189	TDS	1491 mg/L
		Produced and process water (PPW)				
Application of forward osmosis for reducing volume of produced/Process water from oil and gas operations Gas field produced/process water treatment using forward osmosis hollow fiber membrane: Membrane fouling and chemical cleaning	Minier-Matar, J., et al./ 2015 [50] Zhao, S., Minier-Matar, J., and et al./2017 [103]	TOC	mg/L	33		
		Chloride	mg/L	286		
		Sodium	mg/L	329		
		Calcium	mg/L	38		
		Sulfate	mg/L	349		
		Magnesium	mg/L	8.7		
		Bromide	mg/L	5.6		
		Potassium	mg/L	4.7		
		Ammonium	mg/L	8.5		
		Alkalinity	mg/L	223		
		PH		8		
		Conductivity	μS/cm	1810		
		TDS	mg/L	1526		
		Turbidity	NTU	32		
		Produced and process water (PPW)				
Application of Hollow Fiber Forward Osmosis Membranes for Produced and Process Water Volume Reduction: An Osmotic Concentration Process	Minier-Matar, J., Santos, A., et al./2016 [51]	TOC	mg/L	120		
		Chloride	mg/L	284		
		Sodium	mg/L	345		
		Calcium	mg/L	38		
		Sulfate	mg/L	347		
		Magnesium	mg/L	8		
		Bromide	mg/L	5		
		Potassium	mg/L	4.5		
		Phosphate	mg/L	<0.1		

Table 6. Cont.

Title	Authors/Date/Reference	Parameters	Unit	Value	Parameters	Value
		Total nitrogen	mg/L	28		
		Inorganic carbon	mg/L	31		
		PH		8		
		Conductivity	µS/cm	1725		
		TDS	mg/L	1550		
		Osmotic pressure (25°)	bar	1		
		Produced water				
		COD	mg/l	1572		
		TOC	mg/L	491		
		TN	mg/L	34		
		Oil & grease	mg/L	47		
		TPH	mg/L	45		
		Chloride	mg/L	2265		
		Sodium	mg/L	1030		
		Calcium	mg/L	329		
Assessing the Biotreatability of Produced Water from a Qatari Gas Field	Janson, A., et al./2015 [52]	Sulfide	mg/L	828		
		Magnesium	mg/L	61		
		Bromide	mg/L	51		
		Sulfate	mg/L	54		
		Potassium	mg/L	44		
		Thiosulfate	mg/L	14		
		Acetate	mg/L	347		
		Ammonium	mg/L	11		
		Conductivity	µS/cm	7200		
		Total Dissolved solids	mg/L	5189		
		PH		4.3		

Another paper published by Minier-Matar, J. and co-authors (2015) [50] studied the reduction of the volume of produced/processed water (PPW) by 50% through using an application of forward osmosis. The PPW samples were received from oil and gas operations. To characterize the sample, ion chromatography was used to measure the anions and cations such as sulfate, sodium, chloride, and bromide. A conductivity detector was used to find the isolated analyte, while the metals, such as boron and strontium, were measured by inductively coupled plasma. A TOC analyzer, NDIR detector, and pH/conductivity meter were used to find Total organic carbon, Total carbon (TC), CO_2, inorganic carbon, conductivity, and pH [9]. In the same year, Janson et al. [52] assessed the bio-treatability of PW extracted from a gas field operation (North field) in Qatar. Researchers characterized 1200 L of PW to ensure the stability of the composition. Moreover, the characterization data of the PW and PPW is presented in Table 6 [52]. Nonetheless, Zhao, S., Minier-Matar, J., et al. (2017) [103] characterized PPW from the same source as that reported by Minier-Matar, J. et al. [50]. The PPW was a combination of gas field produced water extracted from the offshore gas well and process water from onshore operations, with a blending ratio of 1:5 in Qatar. The studies demonstrated the application of hollow fiber FO membranes for produced and process water volume decline [50,103].

Redoua, A. and AbdulHamid, S. (2016) [66] presented an industrial application to remove kinetic hydrate inhibitor (KHI) polymers from re-injected PW streams. The existence of KHI polymers in the injected PW led to long-term damage in the reservoir. The PW in this project contained KHI treated at the onshore gas processing facilities in Ras Laffan industrial city. A recent paper published by Benamora, A. et al. [79] studied the effect of fluid speed and temperature on the corrosion performance of carbon steel pipelines in Qatari oilfield-produced water. In order to understand the chemical composition of the PW, the sample was supplied from a gas oilfield north of Qatar and the water treatment plant in Qatar.

In 2014, Ahan, J. et al. [81] characterized PW samples supplied from an offshore oil field in Qatar. Chemical and physical characterizations were carried out, in addition to the analysis of factors such as pH, conductivity, concentration of heavy metals, and total Kjeldahl nitrogen [81]. The samples were provided from either oil or gas fields in Qatar, including natural gas from the north field and TOTAL company in Halul Island station. The PW samples were analyzed for different metals and chemical constituents. It was characterized based on total and readily biodegradable COD, BOD, pH, alkalinity, salinity, conductivity, the total quantity of hydrocarbons present in the sample, total solids (TS), BTEX concentration, totals dissolved solids (TDS), phosphate, heavy metals, ammonia nitrogen and phenol, the concentration of ions, total sulfide, dissolved CO_2, and oil droplet size distribution. Abdul-Hakim et al. (2016) [72] studied an application of microalgae using PW from different sources of the petroleum industry in Qatar. In the same year, Al-Kaabi et al. [76] characterized PW samples supplied from an LNG plant at the north field in Qatar, and enhanced the quality of the samples using different treatments discussed in the next section. Shaikh et al. [74] and Atia, A. et al. [75] characterized PW samples provided by a Qatar-based company and studied its effect on the plants for irrigation. Although two studies conducted by Aly et al. [68] and Al-Ghoul et al. [69] used synthetic PW samples for electrocoagulation process. TOC, TPH, O and G, SS, and turbidity were obtained for the PW samples.

7. Treatment Processes

The main target of treating the produced water is to eliminate all the toxic constituents present in it. In 2005, Arthur and co-authors [75] studied the different produced water treatment techniques to remove the soluble organics, gases, impurities, oil and grease, dissolved solids and salts, hardness, salinity, and to remove NORM and disinfectants. Currently, various techniques can be used to manage and treat PW to promote water conservation and sustainability. The technologies designed for PW treatment are established based on whether the installation is onshore or offshore. In general, the installation for offshore is more significant than onshore, since the offshore wells function for longer to stabilize the capital investment. Figure 3 shows the different offshore produced water treatment stages. In onshore facilities, the produced water generated will be re-injected into disposal wells. Therefore, the design of treatment facilities is significant in order to remove most of the contamination. However, in offshore facilities, the treatment required is to only reduce the O and G to acceptable levels so it can be discharged into the sea [76]. In Qatar, the process of the PW begins with fluids extraction from the reservoirs. Once this happens, one or more techniques are used to treat and separate the oil droplet from the water. The separation technique is highly dependent on the production station, for instance, several stations use three-phase separators to separate oil from water in the offshore station, while others degasify the fluid and then deliver it to storage tanks in Halul through pipelines. Moreover, the separation of oil from water in the storage tank depends on the gravity and density difference. Then, the water must be treated as it may have dissolved and dispersed the droplets of oil. Later, the PW is either re-injected in the reservoirs or disposed of into disposal jackets in the water. More importantly, the type of treatment is chosen based on the oil quantity present in the PW, where it varies between 0.5 to 200 μm in diameter [81]. It is worth mentioning that, in oil/gas industries, the pretreatment techniques are used as an initial step. The gravity separation used widely depends on the performance of the oil/water separator, the system surface overflow rate, and it increases the velocity of the individual oil droplets. Then, further treatment technologies are potentially performed such as micro and ultra-filtration, biological-based treatments that are generally used for "downstream" refining processes, forward and reverse osmosis (FO and RO), separation by hydrocyclones, polymer membranes, adsorption, wetlands, aerated lagoons, various flotation methods (column flotation, dissolved air, electro and induced air), membrane bioreactors, activated sludge treatment, coalescence, chemical coagulation, and electrocoagulation [61,62,78]. Due to economic and operational limitations such as high operational, low efficiency, and capital cost, sludge has not been widely implemented for PW treatment [61]. Moreover, to minimize the waste of water, a leading Qatar-based petrochemical company has established formal water to aid the producers to reduce the amount of PW, and reduce the expenses of treatment technologies besides looking for obtainable facilities to handle huge volumes of PW [76].

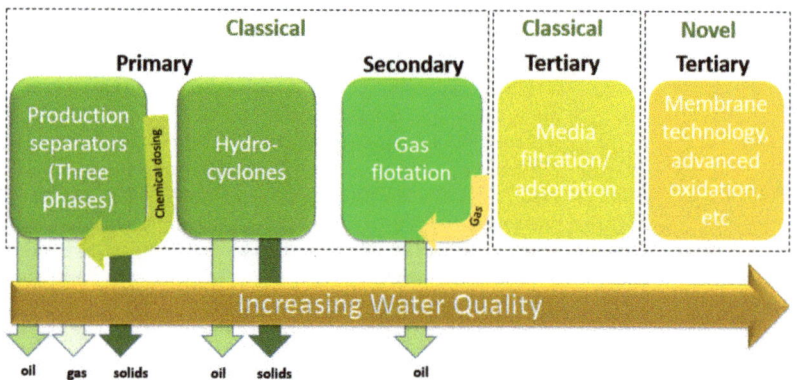

Figure 3. Offshore produced water treatment. Adapted from Reference [62].

Usually, in order to reach the environmental standards of treated water, two or more technologies are necessary to be combined to reach reasonable results. Techniques employing one technology will not be sufficient to treat PW and be acceptable, as per all global environmental standards [104]. Several stages with specified treatment processes are used to treat the pollutants in PW. Initially, physical treatment can be performed where the physical process is performed to remove solid and biomass without using chemicals or bacteria such as filtration. The second stage is a chemical treatment applied to remove the specific chemicals dissolved and suspended particles that are not able to be eliminated from PW through physical treatments. However, chemical treatments require additional costs for the chemicals used in flocculation and coagulation. The last stage is biological treatment using bacteria to remove the biodegradable material.

In this section, the treatment processes of PW conducted in Qatar are discussed in detail, including the final analysis of PW. Figure 4 illustrates the three types of treatments used for PW treatment.

7.1. Gravity Separation and Adsorption

The gravity separation tank is the most technique used to separate oil from water with various concentrations. It depends on the gravity forces to separate the oil from the PW by permitting the oil to float on the water surface, while the suspended solids and particulates are deposited at the bottom. Skimmer tanks, storage tanks, and vessels are examples of equipment used for the gravity separation technique. Centrifuges as well as hydrocyclones are equipment utilized for enhanced gravity separation [81]. The adsorption technique is useful to remove 80% of elements such as Fe, Mg, BTEX, TOC, and heavy metals from PW. Different kinds of adsorbents such as organoclays, zeolites, activated carbon, and activated alumina could also be used [77]. Judd, S., et al. [62] examined the performance of offshore PW oil-separation by employing gravitation plus different technologies for reinjection. The data showed that, for the gravity separator technique, the corrugated plate interceptor

(CPI), de-oiling hydrocyclones (HC), and induce gas flotation (IGF), there is an expected relationship between droplet size and the performance.

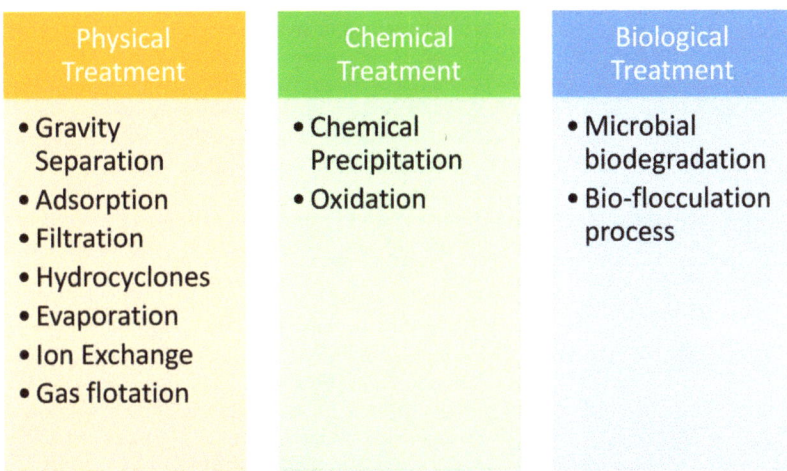

Figure 4. Physical, chemical, and biological treatments for PW.

7.2. Hydrocyclones Separator

Hydrocyclone units are a typical PW treatment device, usually placed downstream of gravity separators. They are considered as a mechanically closed vessel used to sort or separate particles in a liquid suspension based on the ratio of their centripetal force to fluid resistance. It works by directing inflow tangentially and closing the top of the cylindroconical vessel. This will spin the whole contents of the vessel, producing centrifugal force in the liquid [105]. Ahan, J. et al. [81] (2014) presented a comparison of two different stages of PW treatment in two different fields in Qatar. PW samples were obtained from two different fields: field A and field B. In field A, there are two stages of two-phase separation tanks joined with chemical injection and filtration unit, succeeded by gravitational separation tanks, while in field B, similar to field A, this starts with 2-stage separation with a chemical injection, but with three-phase separation tanks including the hydrocyclone, which is then followed by a surge drum. Therefore, the main difference between Fields A and B is the application of the treatment unit, where Field A depends on a gravity separation technology and Field B depends on the hydrocyclones as the main treatment unit for PW. The outcome after the characteristics of PW samples collected before and after the unit operations indicated that hydrocyclones are the more effective device in oil separation with 92.6% separation efficiency for 11 μ oil droplet average size, with 312 h^{-1} as total capacity by volume. The study concluded that field A demanded enhancement in the filtration unit and more treatment units later, even when it minimized the hazardous pollutants better than in Field B [81].

7.3. Filtration and Membrane

Filtration technology is a useful technique to remove and eliminate suspended solids, dissolved salts, TOC, oil, and grease present in PW. The PW passes through layers of porous beads, gravel, anthracite, metal oxides, sand, walnut, ceramic, shells, and others. The efficiency of the filtration can reach more than 90% if it is improved by adding coagulants before the filtration process [75]. Sand, ultra, microfiltration, and nanofiltration (UF, MF, and NF) all are examples of the filtration process used to treat the PW. The leading crude oil producer company in Qatar aimed to limit the oil and TSS from PW using physical separation technologies to reach water quality standards, and then injected it in the disposal wells. Advanced water treatment technologies (AWTTs) such as membrane processes as UF, MF,

NF, reverse and forward osmosis, membrane-bioreactors (MBRs), membrane distillation (MD), advanced oxidation process, and thermal evaporators were utilized to treat the PW. Hussain et al. [49] published their work using advanced technologies for PW treatment, which is implemented in an advanced water laboratory in Qatar. The PW was obtained from an offshore Qatari gas processing operation. The AWTTs were designated to target the main contaminants recognized in PW by using MBRs for organics, membrane distillation for salinity, and membrane process (organics and salinity) for treatment. Results showed a significant flux reduction when MBRs were used, and was then filtered over a reverse osmosis membrane. Moreover, it was found membrane distillation generated high-quality distillate water beside TOC and TDS rejections greater than 87% and 98%, respectively. Overall, the treated PW using AWTTs proved that the process produced effluents that can be used in different applications such as livestock, irrigation, and industrial processes.

7.3.1. Sand Filtration (SF)

Sand filtration treatment is divided into different systems: slow sand filtration, high hydraulic loadings sand filtration (HLR), intermittent sand filtration (ISF), and sand/activated carbon filtration. In the slow sand filter, there is a layer of sand that allows for the pretreatment of water. Moreover, it controls the flow of water, the rate of water to be filtered in $m^3/m^2/h$ ranging between 0.1 and 0.4. The main element of slow sand filtration is the fine grains with a depth summation of almost 1.0 m, and a diameter in the range of 0.15 and 0.35 mm. During the filtration process, the suspended and colloidal materials are stuck at the higher part of the system. The disadvantage of the sand filter is that when the particles are clustered on top of each other, the system will get clogged, and hence the efficiency will be reduced [76]. HLR is produced by placing three identical sand columns parallelly. The column is filled at the bottom and the top with 10 cm of gravel, as well as 80 cm of fine sand. Several studies have proven the effect of the sand filter in reducing and removing the BOD, TSS, COD, TP, TN, and removing the oil & grease. The intermittent sand filter is dependent on the concept of intermitted water levels, as well as its flow. Intermittent sand filtration purification is performed by chemical, physical, as well as biological mechanisms that could be employed to accomplish a remarkable decrease in COD and TSS [76].

The sand/activated carbon filtration system is capable to filter out pyrethroid pesticide residues, organochlorines, and organophosphates to their detection limit. The last type of sand filtration is clay, which is the main element present in the soil. The features of clay are high specific surface area, mechanical and chemical steadiness, high layered structure, and cation exchange capacity. These features offer it a significance in cation exchange capacity (CEC). Clay is capable of trapping metal particles from water when the water flows over soil into the ground, making it useful in eliminating the pollutants from water [76].

7.3.2. Membrane Process (MP)

MP is used to remove the kinetic hydrate inhibitor (KHI) from PW using an efficient membrane such as microfiltration, nanofiltration, ultrafiltration, and reverse osmosis membranes. The advantages of the treatment are either using a combination of numerous membrane processes such as NF or both RO and UF to aid in reducing KHI from the water, or involving other conventional water treatment methods such as clarification or media filtration. Recently, membrane separation has been recognized as a promising approach for handling a variety of oil/water mixtures owing to its low footprint and energy, high separation efficiency, time consumption, and simple operation [106,107]. However, as the inorganic membranes are hydrophilic in the air and convert to being oleophobic after immersion in water, the membranes with such an underwater superoleophobicity can prohibit the oil from touching the membrane surface [108]. Wang, K., et al. [108] reported a stretchable and winnable membrane of ZnO nanorods arrays with a three-dimensional structure conformally grown on woven carbon microfibers for effective oil/water treatment. Outcomes depicted the efficient separation of both oil/saline-water mixtures using the membrane and oil-in-water emulsions, merely driven with gravity, by high separation efficiency over 99%, and extremely high permeation flux of 20,933.4 L m^{-2} h^{-1}. Liu, Z. et al. [109] developed a fast and efficient separation membrane for the treatment of emulsified oil/water mixtures using ultra-long titanite nanofibers/cellulose microfibers. Adding to that, affordable materials and fabrication processes allowed for producing it as a favorable potential for industrial scale-up. Liu, Z. et al. [109,110] found good mechanical flexibility of the new membrane, as well as high separation efficiency up to 99.9% for oil/water emulsions with 3 μm of oil droplet size. Additionally, it has a high water permeation flux at low operation pressure about 6.8×10^4 L m^{-2} h^{-1} bar^{-1}, which is assigned to the interconnected porous structure within the whole membrane, whereas the nanopores selective layer contributes to high oil separation.

7.3.3. Membrane Distillation (MD)

The mass transfer system can be controlled by partial vapor pressure alteration owing to a temperature difference. The difference in the temperature ranges between 10 to 20 °C (warm and cold streams), and is adequate to obtain distilled water at the specific conditions. Membrane distillation is highly important for treating the salinity feed waters and reducing the salinity of PW [49]. Furthermore, MD in comparison with reverse osmosis is better in obtaining good, distilled water quality, which is not affected by high salinity. The RO system requires multiple passes, while MD uses a single pass to reach desired salt rejection, and uses low-grade waste heat, economic system. In the MD system, there are four main module configurations: air gap (AG), vacuum MD (VMD), direct contact MD (DCMD), and sweeping gas MD (SGMD) [56]. Minier-Matar, J. et al. [56] used flat sheet membranes with 0.014 m^2, sandwiched between the feed and distillate plates to treat PW. The results showed that the membrane was not affected by salt concentration, and the flux was stable. In addition, MD produced a high-quality distillate of brines from thermal desalination plants. For better efficiency, the MD process can be combined with some other membrane-based processes (Figure 5).

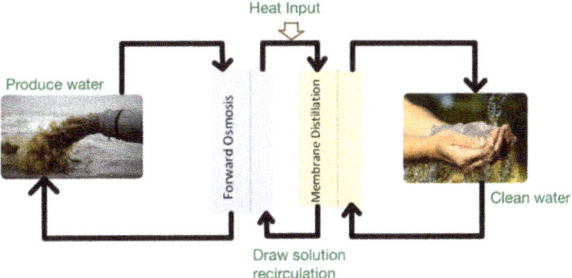

Figure 5. Schematics of the combination of membrane distillation and forward osmosis process.

7.3.4. Membrane Bioreactors (MBRs)

MBR is considered as an excellent method for treating different wastewater streams and industrial wastewaters. The advantage of MBRs is the membrane filter that is used to separate the sludge from the treated water. It is worth mentioning that MBRs have not yet been used to treat PW at upstream gas and oil operations. However, MBRs have offered better treatment for PW than conventional biological techniques, since the process uses an ultrafiltration and microfiltration membrane to isolate the sludge particle and emulsified oil and grease from the treated water [52]. In research done by Janson, A. et al. (2014) [52], the team assessed the biotreatability of PW obtained from the Qatari gas field. They performed a Box–Behnken test to optimize the bio-treatment of PW via a hollow fiber membrane over a range of hydraulic retention time (HRT) of 16 to around 30 h, the temperature of 22 to 38 °C and solids-residence time (SRT) of 60 to 120 days. The outcomes displayed that, after H$_2$S stripping, and over eight months of testing, the chemical oxygen demand of a combination of PPW was approximatley 1300 mg/L, appropriate for supporting the biological activity, with TDS of 5200 mg/L which was sufficiently low. Moreover, it was found that COD removal was between 54 to 63% by desorption, and no significant effect of the removal efficiency appeared in HRT, temperature, and SRT. Moreover, the pH value of MBRs' sludge ranged between 4.9–6, in contrast to the feed water with pH value of 4.3, probably dueto low carbon removal. TOC results consistently followed the COD results with analogous removal values. However, all feedwater acetate and more than 90% of the oil/grease were separated by this technique. The authors concluded that MBRs treatment acted to be carbon limited, for the low volatile suspended solids concentration and accounting both for the absence of nitrification. Furthermore, in the same year by Janson, A. et al. [54], the team evaluated the biotreatability of PW during the summer season under varying conditions of solids retention time (SRT: 60–120 days), hydraulic retention time (HRT: 16–32 h), and temperature (22–38 °C). The authors found that TOC and COD existed in PW from Qatari gas field were removed by approximately 60% through MBR treatment. Moreover, it was discovered that there is no change of TOC% and COD% with any of the input parameters over the different ranges tested (SRT: 60–120 d; HRT: 16–32 h; temperature: 22–38 °C).

7.3.5. Ceramic Membrane

The importance of ceramic microfiltration membranes for treating the PW from an Arabian Gulf oilfield has been conducted by applying a dedicated pilot plant. Moreover, the crossflow multi-channel ceramic membrane process has been studied for a pilot-scale to preserve the required treated water quality while changing the convenient membrane properties for sustaining the flux. In comparison with polymeric materials, the ceramic membrane exhibited the advantage of operation at high temperatures and increased fouling resistance. A significant number of studies for ceramic membrane application to PW have been conducted since the early 1990s [58,59]. Silicon carbide, SiC, titanium dioxide, TiO_2 ceramic membranes with different pore sizes have been used in different studies. In a study by Zsirai, T., et al. (2016) [58], the team used crossflow multi-channel ceramic membrane technology to detect the suitable membrane properties for sustaining the flux through preserving the required treated water quality and minimizing the process footprint. Silicon carbide, SiC, titanium dioxide, and TiO_2 materials were utilized to produce membranes with different pore sizes. PW has been collected from oil platforms operating in the Arabian Gulf. Under the same operating and maintenance conditions of crossflow velocity, transmembrane pressure, and chemical and physical cleaning protocol, the results of SiC membranes were showed to be superior to TiO_2 ones with respect to sustainable permeability. Moreover, after testing the membranes, the results demonstrated that SiC microfiltration membrane showed exceptionally high permeability and high treated water quality, but also the highest fouling propensity with 6.3–7.6 mg/L O&G and 4–8 NTU turbidity for MF (pore size 2 µm) and ultrafiltration (UF, pore size 0.04 µm) membranes. Furthermore, the outcomes indicated that the high fluxes ranging from 1300 to 1800 L m^{-2} h^{-1} are attainable for the technology, but this is conditional upon the application of an efficient chemical clean to sustain permeability and treated water over a long operational time. It was noted that on the chemically cleaned membrane there was a noticeable retrogradation in both permeate water quality and permeability with each consecutive experimental run. Consequently, the need to enhance the efficacy of the chemical clean-in-place (CIP) applied amidst runs to recover both the selectivity and permeability of the membrane. Hence, Zsirai, T., et al. (2018) [59] presented a CIP using a combination of citric acid and caustic soda (NaOH). The results showed that the flux of 700 Lm^{-2} h^{-1} was persistent through the application of 6 wt.% citric acids with 6 wt.% NaOH combined with backflushing at almost double the rate of the filtration cycle flux. Abdalla, M. et al. (2018) [60] published a study about the effect of combined oil/water emulsions and a colloidal particulate suspension (bentonite) on crossflow ceramic microfiltration (CFCMF) fouling and permeability recovery. They investigated the effect on both fouling through the filtration cycle and residual fouling of the ZrO_2-TiO_2 membrane. Outcomes showed that the permeability and selectivity of the membrane were greatly affected by the increase of suspended solids to the o/w emulsion as an increase of 1500 mg· L^{-1} particulate solids to 10% v/v of stabilized emulsion. This led to reducing the permeability by 3.5–5 times over different filtration experiments, compared to the emulsion alone, and a 8–36 times decrease in contrast with the suspension. Oil passage through the 0.45 µm pore size of the microfiltration membrane was concomitantly increased six-fold.

7.3.6. Hybrid and Asymmetric Membranes

In 2018, Fard et al. [61] presented a successful PW treatment technique with a combination of inorganic membrane aluminum oxide (Al_2O_3) in addition to activated carbon (AC) as a low-cost adsorbent material for the adsorption process. Both Al_2O_3 and AC are widely available for use as inexpensive techniques. Despite the fact that inorganic membranes are generally expensive as compared to polymeric membranes, they are more able to be sterilized, autoclaved, have high temperature and wear resistance, withstanding harsh chemical cleaning and frequent backwashing, have a long lifetime, high chemical stability and a stable pore structure. The hybrid membrane exhibited very favorable results for emulsified oil and enhanced permeate flux with 96% removal efficiency. Moreover, the process enhanced the rejection of oil from the water and increased the salinity due to its destabilization of the emulsion [61].

7.3.7. Other Emerging Membrane-Based Processes
Forward Osmosis

FO is an osmotic process that uses a semipermeable membrane to treat PW by separating it from dissolved solutes. The osmotic pressure gradient is the main factor acting as the driving force between a high (draw) and low (feed) concentration solutions (Figure 6). Moreover, the osmotic pressure gradient is important for implementing a net flow of PW within the membrane into the draw, therefore successfully concentrating the feed. The feed solution is either a dilute product stream, seawater, or a waste stream, while the draw solution is a solution particularly tailored for forward osmosis applications or a single or multiple simple salts [111]. Noteworthily, FO is usually preferred as a "pretreatment" process to directly treat feed wastewaters in view of practical water production applications [63]. The membrane support layer controls the antifouling capability of the FO membrane, which also performs the function of allowing water permeation and rejecting pollutants simultaneously. Mainly, two conditions must be included in any super-antifouling support layer, superhydrophilicity and ultra-smooth conditions, in order to reduce the hydrophobic adsorption of foulants on membrane surface. Furthermore, the ultra-smooth conditions will prevent the clogging of foulants [63,112]. Conventional support layers contain a tortuous 1D architecture, known as the internal concentration polarization (ICP) phenomenon. ICP functions as a stable barrier against the diffusion of water molecules and draws solute. Moreover, the construction of 3D architecture with organized pores in the membrane backing the layer will effectively overcome the ICP bottleneck [113].

Figure 6. Forward osmosis process principle.

During 2015, Minier-Matar, et al. [50] studied a novel application of FO for decreasing the volume of PPW from oil/gas facilities. The PPW was disposed of via profound well injection, employing brine from thermal desalination plants as a draw solution. Bench-scale FO tests were performed using flat sheet membranes. The presented results confirmed that using pretreated PPW was successful in the implementation of FO for PW treatment as compared to non-pretreated PPW. Furthermore, FO offered an effective result of 50% volume drop of PPW with an ordinary stable flux of 12 LMH. Different factors affect the flux of feed and draw solutions that lead to the increase in the draw solution osmotic pressure, a reduction on water viscosity, and potential improvement of some membrane properties [50]. Another study by Minier-Matar, J., et al. (2015) [51] analyzed the application of hollow fiber FO membranes for reducing the volume of PPW disposed of in onshore facilities. This work is complementary to previous work where FO was utilized as an osmotic concentration process. Flat sheet and hollow fiber FO membranes were tested, and the latter displayed better flux and rejection. Additionally, the results proved that low-energy osmotic concentration FO has the potential for full-scale utilization to decrease PPW injection volumes. This technique is economically appropriate to decrease PPW injection volumes from Qatari-based gas facilities in an environmentally friendly manner [51]. These two studies proved the concept that osmotic concentration has the potential for full-scale fulfillment with lower energy consumption in contrast to reverse osmosis. After comparing the results, it was found that the hollow fiber membranes had a higher performance regards the constituent rejection and water flux [114]. Furthermore, using hollow fiber membranes showed higher flux with 40% (16.5 $L/m^2 \cdot h$) after treating PPW in contrary to flat sheet membranes (12 $L/m^2 \cdot h$). Temperature showed a high influence in addition to the salinity of the draw solution [50,51].

One of the main difficulties that affect the FO process performance is membrane fouling, which was proved by using hydrophobic organics that have fouled the membranes. However, the fouling might be inhibited with successful pretreatment [50,51,53]. Regarding the quality of water, the lab study confirmed that the forward osmosis membranes are able to reject the field chemicals as well as organics existing in the feed stream, and it will not be transported to the seawater or brine utilized as draw solution. A group of studies of the osmotic concentration concept was performed to illustrate process feasibility in the field [115].

Electrodialysis

Electrodialysis is a separation process whereby charged membranes as well as electrical potential difference are employed for separating the ionic species present in an aqueous solution and other uncharged components [116,117]. The greatest advantage of this electrodialysis technology is that it can separate without any phase change, leading to comparatively lower energy utilization. This technology also has several drawbacks, such as (1) the colloids, organic matter, and silica are not separated by the electrodialysis process; (2) pretreatment of feedwater is needed for preventing the electrodialysis stacks fouling; and (3) elaborate controls are needed in this system, and maintaining them at optimal conditions could be challenging. The study by Sosa-Fernandez et al. [116] evaluated the application of pulsed electric fields during the electrodialysis of polymer-flooding produced water for improving the process performance as well as for reducing the fouling incidences. In another study by Sosa-Fernandez et al. [117], the team evaluated the separation efficiency of divalent ions present in synthetic polymer-flooding produced water by changing the operating conditions. The results confirmed that it is feasible for achieving a preferential separation of divalent cations (magnesium and calcium) by electrodialysis, particularly while using high temperature (40 °C) and lower current densities.

7.4. Hydrate Inhibitors (HI)

Chemical substances can be added to control the formation of hydrates during natural gas production in an oil or gas industry. Qatar is using HI during winter months in natural gas wells to avoid the formation of methane hydrates in the pipes, between the onshore processing facilities and offshore platforms [55,66]. The main two hydrate inhibitors are kinetic hydrate inhibitors (KHI) and thermodynamic hydrate inhibitors such as mono-ethylene glycol (MEG). KHI is usually added at lower concentrations and the chemical composition is proprietary. KHI turns as an antinucleator for delaying hydrate formation. KHI may contain polyvinyl caprolactam, polyvinyl-N-methyl acetamide, poly(vinylpyrrolidone), and polyethylacrylamide. On the other hand, MEG is typically added at higher concentrations, and can reduce the hydrate equilibrium temperature adequately to reduce hydrate formation. Qatar gas has been implemented both KHI & MEG [118]. Janson, A., et al. (2015) [64] published a work related to the biotreatability of kinetic hydrate inhibitor (KHI) and thermodynamic hydrate inhibitor monoethylene glycol (MEG) under set and continuous reactors in aerobic mixed-culture situations without pH control. The results showed that, TOC and COD exhibited more than 80% removal by biological treatment of PW with the addition of 1.5% MEG. However, biotreatment can eliminate about 43% of TOC and COD existing in PW with the addition of 1.5% KHI. The feed water (with either KHI or MEG) to the reactors was at pH 4.5 and this led to stabilizing the reactor, counted very acidic for aerobic activity, and was obviously produced by the inorganic acid via the biological culture [64]. The residual KHI in PW is considered as one of the concerns that can influence the injectivity of disposal wells [66]. In response, various processes were shown in the investigations such as chemical, physical, or biological processes to separate KHI from PW before injection to disposal wells [55]. Between the processes examined, the RO, NF, and UF membranes were estimated to be appropriate via bench-scale testing utilizing synthetic PW containing KHI [119]. KHI removal efficiency was found to be more than 99% for RO, 99% for NF, and 83% for UF membranes. A natural gas producing company in Qatar conducted a project to remove KHI from PW using the evaporation process [18]. Since the PW includes KHI which is treated at the onshore gas operation located in Ras Laffan affects the wells, forming damage within the injection wells besides long-term effects on groundwater pollution. Redoua, A. et al. [66] (2015) used three main steps, including pretreatment, evaporation for the elimination of KHI and then concentrate storage, and handling. This ensured that using KHI products to prevent hydrate formations in the wells is suitable for the NG industry.

7.5. Demulsification

In the petroleum industry, demulsification is a process used to treat water from oil emulsion. Demulsification techniques are categorized as biological, chemical, and physical demulsification. Chemical demulsification is one of the wide separation processes used to separate oil from water. Demulsifiers are referred to as the chemicals used in this process characterized by a strong affinity to the oil-water interface. Demulsifiers are amphiphilic compounds able to abolish the stabilization through adsorbing at the interface. Mechanical strength, interfacial tension, the thickness of interfacial region, and elasticity are examples of interfacial film properties [80]. The physical demulsification processes are microwave irradiation, gravitational settling membrane separation, ultrasonic and filtration. The separation mechanism of oil and water occurs by flocculation and/or coalescence of water droplets and the demulsifiers act as one of them. One of the most important parameters in the demulsifier is hydrophilic-lipophilic balance (HLB) that shows its relative simultaneous attraction to oil and/or water phase [80].

7.6. Coalescing

A coalescer is a technological device using the coalescence that is applied to separate emulsions into their components through numerous processes. Coalescers are divided into two types: electrostatic coalescers, and mechanical coalescers. Electrostatic coalescers use electrical fields, direct current (DC), alternating current (AC) electric fields or both, while mechanical coalescers use filters to develop droplets coalesce. In oil/gas industries mechanical coalescers are used for the removal of water or hydrocarbon condensate [120]. Mixed flow separation, corrugated plate, interceptor, and crossflow separation are examples of equipment implemented for plate coalescence [81]. Qatar-based petroleum companies have applied a technology to de-oil PW to meet the standard requirements and regulations. The work was performed at both PS3 Bul-Hanine offshore field and Dukhan onshore field using TORR technology. Moreover, adsorption, coalescence, desorption, and gravity processes are used to treat the PW from dispersed oil. It has been demonstrated that coalescing elements was appropriate to de-oil most of the PW, despite the challenge in reverse emulsions that demanded modifications of the element. By changing the element's porosity and compressibility factor as well as permeability, the oil removal efficiency can be significantly improved for difficult reverse emulsions. Furthermore, the technology is effective in removing >2 μ of oil droplets [89]. One example of coalescence is the inclined multiple arc coalescence plate, which is the most favorable gravitational oil-water separators among gravitational technology. The performance of the coalescence arc plates was affected by three main factors: size, shape, and geometry [78]. Almarouf et al. (2015) [78] established an effective oil/water separator for the treatment of stable emulsions in PW, combining effects of oil droplet coalescence and chemical demulsification in order to enhance the formation of two phases for further separation. The novel oil-water separator consisted of a series of inclined multiple arc coalescence plates, and exhibited effective results in breaking stable emulsions, therefore enabling their efficient separation from produced water.

7.7. Thermal Evaporators and Advanced Oxidation Processes (AOPs)

Thermal evaporators are an important treatment method of PW, since they are considered as an economically feasible technique. Furthermore, the volumes of freshwater required for makeup are greatly reduced since almost all waste streams are recycled back to the evaporator [49].

7.8. Surfactant Application

Polyacrylamide is a polymeric surfactant, and this simple surfactant along with polymer flooding is greatly improving oil recovery. To separate water from oil, the polyacrylamide with other additives is used as destabilizing agents for water/oil emulsions, such as aluminum and ferrous sulfate. The influence of polyacrylamide on water/oil emulsion was studied through interface electric, interface strength, and interfacial tension property of oil in water wastewater process [121]. Ma, H. et al. (2016) [82] treated the PW/oil emulsions via anionic PAMs with aluminum and ferrous sulfate as an electrolyte. The results presented that the volume of separated water enhanced more than 25% in comparison to PAMs only. Moreover, the COD viscosity and turbidity reduction of separated water improved significantly. The destabilization of water in oil emulsion was improved by using the electrolytes into polyacrylamide in general, as compared to when only polyacrylamide was used [82].

7.9. Activated and Modified Activated Carbon Filtration (AC and MAC)

The advantage of activated carbon is it has a large surface area with changeable pore dimensions and different active sites. Activated carbon is effective in treating water by removing a large range of organic compounds, but it is not capable of removing large molecules such as humic acid that comprise emulsified grease in addition to oil. The reason is that the bigger compounds and particles plug the macroporous space on the activated carbon external surface, which makes it less effective. Activated carbon has several significant characterizations, making it one of the greatest filtration media, for instance, high adsorption ability, thermo-stability, microporous structure, high grade of surface reactivity, low acid/base reactivity, and capability for comprehensive range pollutants removal [76]. In 2010, Al-Ghouti et al. [122] presented research using activated carbon for eliminating the organosulfur compounds from the diesel-non-aqueous medium. The results showed outstanding adsorption skill of granular bead form activated carbon of organosulfur compounds from the diesel-non-aqueous medium. Moreover, the study displayed that the particle size of the activated carbon affected the organosulfur compounds elimination efficiency, which means that the adsorption mainly happened on the external surface area.

On the other hand, despite the fact that activated carbon is economically and obtainable resources from dates, papaya wood, dust, coconut shells, coke as well as rice husk, the modified activated carbon (MAC) has been successfully utilized as adsorbents for the removal of toxic material. Modified activated carbon has a better enhancement in the active surface for adsorbing material. This had been verified by using the waste of pods and husk; it can generate superior-quality microporous activated carbons via stratifying simple steam pyrolysis process [76]. Al-Kaabi et al. [77], in 2016, aimed to study the effect of using sand filtration, AC, and MAC filtration by microemulsions, to remove the major organic and inorganic pollutants, BTEX, and heavy metals from PW samples. In this study, PW samples were received from the north field offshore gas. In a comparison of the three treatments, sand filtration exhibited higher removal efficiency for the TSS (77.5%), TN (63.7%) and corrosion inhibitor (94.1%). Iron and manganese have the highest metals removal efficiency in addition to BTEX with >95%, with the exception of the toluene, which exhibited 26.7%. It is worth mentioning that COD showed the lowest removal efficiency among the other media with only 10.2%. For AC and MAC filtrations, COD was removed by 23.7% via AC, while it increased by 12.6% via MAC. Moreover, MAC was noted to be highly effective in reducing the TOC to 31.1% among the three media. From a comparison between the three treatments SF, AC, and MAC from removal efficiency point of view, it was clear that the treatments used for PW samples efficiently improved the pollutants removal efficiency to utilize it for different applications such as plant irrigation [76].

In a research study by Al Kaabi et al. [123], the chemical and physical characterization of PW was carried out succeeded by treatment by means of sand filtration combined with activated carbon microemulsion modified activated carbon method. Figure 7 presents the schematic representation of PW treatment by means of sand filtration, as well as activated carbon [123]. The study confirmed that the treated water was free from all main contaminants of PW, and it could be considered appropriate for reuse at domestic or industrial level.

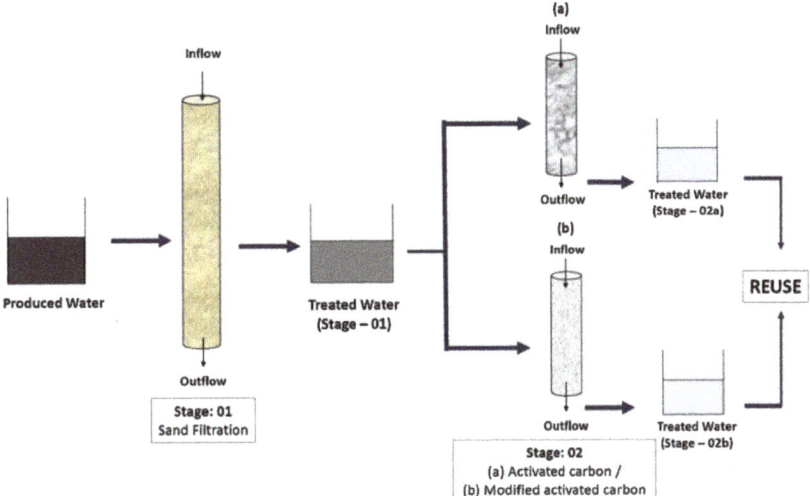

Figure 7. PW treatment by means of sand filtration as well as activated carbon. Reproduced from ref. [123].

7.10. Adsorbents

Application of adsorbent agents for oil-water separation showed good efficiency in the removal of heavy metals from PW, hence it is considered as a new field for further developments and studies. In general, CNT is a new carbon material used in many different fields. It has attracted considerable research attention owing to its thermal, chemical, mechanical stability [124]. Few studies were conducted using CNTs as adsorbent agents for oil-water separation [125]. The CNTs are characterized by exceptionally high adsorption capacity and higher hydrophobicity for oil-water separation and enhanced de-oiling processes. Fard, A. et al. [126] used iron-oxide/CNTs nanocomposites in a study for oil-water separation. The ferric oxide nanoparticles/CNTs showed mass sorption capacities for gasoline oil of up to 7 g/g.

MXene is a new adsorbent material characterized by its high adsorption capacity and efficiency in addition to its superior structural stability, hydrophilic surfaces, availability, flexibility, and high electrical conductivity. There are several studies which have been carried out on the detailed applications of MXenes for water treatment [127,128]. Mxene is material from a family of transition metal carbides used for water purification. It is usually synthesized by etching the first layer from MAX phases. Fard, A. and co-authors utilized Mxene as a two-dimensional (2D) nanosheet adsorbent to remove barium components from produced water. MXene (2D), titanium (III) carbide (II) ($Ti_3C_2T_x$) nanosheets were produced, and the results showed that $Ti_3C_2T_x$ removed 90% of barium within 10 min and the adsorption of barium was pH dependent. MXene exhibited fast kinetics, a large sorption capacity, reversible adsorption properties, and huge trace barium removal that offer a great removal performance of barium with a capacity of 9.3 mg/g [129].

7.11. Biological Treatments

Amongst other traditional treatment methods, biological treatment counts as the cheapest method for the separation of contaminants. In biological treatment, anaerobic or aerobic conditions are maintained, and it is divided into three types, based on the organisms used in eliminating or removing toxic pollutants in PW such as algae, bacteria, and fungi. Virtually, the precise choice of the species, optimization, and maintenance of feeding manners, additives, and environmental conditions are the most important factors that are used to improve the treatment efficiency.

Application of microalgae is one of the eco-technology methods, where the biological treatment process accomplishes a higher rate of removing pollutants from the PW. In general, to bio-remediate produced water effluents, these microalgae are used, where they are capable of employing some of these pollutants as feed source sources of nutrients [72]. Microalgae strains are unicellular organisms employing light as sources of energy to generate biomass and O_2, and due to that, they are grown in an environment with sufficient moisture and sunlight. They also hold chlorophyll-a as a photosynthetic pigment. *Parachlorella Kessler*, *Monoraphidium* sp., *Neochloris* sp., *Chlorella* sp., *Scenedesmus* sp., and *Dictyosphaerium* sp. are some example of microalgae strains used for PW and wastewater treatment. Moreover, some microalgae have a unique property in that they are able to grow in heterotrophic as well as phototrophic conditions [72]. Microalgae require dissolved carbon dioxide, trace metals, nitrogen, and phosphorus for their propagation. Despite the fact that produced water contains these elements at various concentrations, the presence of some heavy metals and toxic organic compounds could be lethal to the microalgae [130]. However, if any microalgal strain could tolerate the toxic compounds of the PW and generate sufficient biomass in it, then the method can be potentially used to supply O_2 to the aerobic bacteria for the degradation of the organics. Qatar has abundant sunlight and adequate unutilized desert land which could be used for microalgal remediation of PW while producing biomass feedstock [130,131].

A study was conducted by Abdul Hakim et al. [72] in a Qatar-based university discussing a solution for pollutants removal from PW using microalgae. The PW samples collected were first filtered using a 0.45 μm Millipore filter to eliminate the maximum of the TSS and other main contaminants. After that, the filtered water was employed to grow in several species of microalgae (*Chlorella*, *Dictyosphaerium*, *Neochloris*, *Monoraphidium*, and *Scenedesmus*) to study their abilities to remove heavy metals [72]. In this study,

Dictyosphaerium sp., *Scenedesmus* sp., and *Chlorella* sp. presented a remarkable quantity of biomass yield among all other concentration of PW, which could be attributed to the low Cr concentration existing in the tested PW [91]. Moreover, the removal efficiency of phosphorus and other metals were high via *Dictyosphaerium* microalgae species, as phosphorus was removed by 88.83%. Despite *Neochloris* sp. having low biomass generation, it removed 41.61% of TOC from the different levels of PW concentrations, and recovered 100% iron and aluminum [73]. Regarding nitrogen and BTEX removal efficiency, the results within the microalgae strains were the same. Nevertheless, the author found that the difference in PW concentration has no meaningful consequence on the pollutant's separation effectiveness of microalgae strains. Therefore, microalgae strains were able to grow and live in PW effluents-deriving from petroleum industries and separate contaminants [72,91]. In 2018, a produced water sample from a local petroleum company was collected by Das, P et al. [71]. The PW samples collected were first characterized and found that PH, TOC, TN, TP, and salinity were 4.17, 720 ppm, 52.5 ppm, 0.21 ppm, and 4.3 ppt respectively. NaOH was added as a pretreatment to raise the pH to 7.1 and permitted to remove 40% of TOC, 38.3% of TN, and 19% of TP [71]. However, the authors found that pretreated produced water was still toxic for some of the local microalgae strains, and even for zebrafish. Thereafter, two conditions were studied for the growing viability of three freshwaters and three marine microalgae strain in the pretreated PW. The first condition was studying the six microalgae strains without additional nutrients, and the second condition was adding N and P. Of these strains, only *Scenedesmus* sp. and *Chlorella* sp. were able to grow in both conditions. *Chlorella* sp. was able to reach the highest biomass yield in the nutrient-supplemented pretreatment PW with 1.2-g L^{-1}. Furthermore, as the pretreated PW was supplemented with nitrogen and phosphorus, *Chlorella* sp. biomass yield increased more and simultaneously removed 73% of TOC, 92% of TN, and other heavy metals. In this research, all the zebrafish managed to survive for at least nine days in the *Chlorella* sp. remediated produced water [71].

7.12. Electrocoagulation (EC)

Electrocoagulation has received extensive attention in the last several years as a green and one of the effective electrochemical techniques for water treatment. It possesses numerous advantages over traditional PW treatment techniques, such as the capability of treating oily water, generating less sludge, and eliminating chemical additives [68]. EC presented its ability in dealing with several pollutants such as organic and inorganic contaminants with higher efficiency without any by-product wastes. This technique combines the advantages of coagulation, electrochemistry, and flotation. In EC, a chemical reaction occurs due to the movement of an electric current through an electrolyte that exists between two electrodes: cathode and anode [67]. EC essentially aims to remove the pollutants from water through electrocoagulation, electro flocculation, electro-oxidation, destabilizing as well as neutralizing the repulsive forces among the suspended particles. When one of the repulsive forces is neutralized, it will lead to forming bigger suspended particles, and thus fall, which makes this technique unique in comparison with other processes [69]. In 2018, Aly, D. [68] developed a new cell to mitigate the cathode passivation problems via using several types of metal (Al and Fe) and perforated hollow cylindrical cathode for the cell electrode. The cathode electrode was used with compressed air allowed to flow from cathode perforations to clean the electrode and provide a sufficient mixing. The results of the new cell design showed a higher removed efficiency of organic contaminants with about 96.8% for TOC, 97.9% for TPH, and 94.6% O&G. As a consequence, this design was found to be more effective in treating PW and minimizing cathode passivation compared with other basic electrocoagulation setups with plate electrodes [68].

7.13. Steel Slag Treatment

Steel slag is considered as a by-product manufactured during the steel separation process. The steel is molten to liquid metals and then solidified, leading to the creation of metal oxides and silicate solids. Furthermore, it contains CaO, Fe_2O_3 and other metals such as silicate, Mn, and Mg materials. Nowadays, steel is made by two technologies, namely electric arc blast and the oxygen steel convertor process. The characteristics and qualities of slag steel are ranging from low to high depending on its disposal location. Qatar-based steel companies manufacture approximately 400,000 tons of slag annually. This high level of production will result in the problem of recyclability or disposal of slag from the company. Hence, one of the solutions is to use steel slag in the electrocoagulation process for treating the produced water [69].

In 2017, Al-ghoul, M and Al Haawari, A. [69] used steel slag as a supplementary coagulant in the electrocoagulation process as a treatment for PW. After the application of 10 mA/cm^2 current density at 10 minutes' reaction time, results showed that the slag sample had a TSS removal efficiency of 90% in comparison with a pure sample of 55.7%. Moreover, the ability of the slag sample to remove the turbidity reached 85.9%, while the pure sample presented 80.1% removal efficiency. It has been demonstrated that increasing the reaction time led to an increase in the removal efficiency both TSS and turbidity to a certain extent. It was found that at a reaction time of 30 min, the optimum removal percentage was obtained. The slag sample showed 94.8% for TSS and 92.5% for turbidity removal percentages, and 90% TSS and 90.3% turbidity for the pure sample. The removal efficiency for oil and grease analysis for the samples with and without steel slag showed almost same result, around 98.9% [69].

8. Case Studies of PW Treatment in Qatar

Several case studies have been carried out in Qatar for the effective treatment of PW. The PW from one of the NG fields based in Qatar was obtained by the group of Al-Ghouti et al. [91]. This PW was employed for examining the separation of heavy metals employing microalgae. As presented in Table 7, complete separation (100%) of Fe and Al from PW was accomplished using microalgae, whereas K demonstrated the least separation efficiency (11.27%).

Table 7. Separation of trace metals from PW employing microalgae. Reproduced from Ref. [91].

Trace Metals	Filtered Water (ppb)	Feed Water (ppb)	% Removal	Microalgae Species
Cd	0.06	0.09	97.37	*Chlorella*
Ni	3.71	7.83	92.29	*Dictyosphaerium*
Cr	17.2	24.09	19.36	*Dictyosphaerium* sp.
Fe	100.19	287.94	100	*Neochloris* sp.; *Chlorella* sp.
Mn	318.56	318.56	87.8	*Neochloris* sp.
Sr	105.73×10^2	111.98×10^2	21.23	*Dictyosphaerium* sp.
K	677.40×10^2	736.18×10^2	11.27	*Scenedesmus* sp.
Ba	43.35	55.69	13.06	*Monoraphidium* sp.
V	1.46	1.87	36.26	*Scenedesmus*
Al	13.68	114.41	100	*Neochloris* sp.
Mg	392.57×10^2	417.15×10^2	13.9	*Dictyosphaerium* sp.
Cu	180.78	224.97	91.65	*Dictyosphaerium* sp.
B	374.7×10^2	425.9×10^2	20.23	*Dictyosphaerium* sp.

Shaikh, S. et al. [74] studied the heavy metal accumulation, microbial succession, and germination tests for turf grass seeds and weeds for evaluating the impacts of PW irrigation. PW was used to irrigate turfgrass—*Cynodon dactylon* and *Paspalum* sp. The samples were collected from Total Qatar, sourced from their station at the Halul Island. According to the *C. dactylon* results, it showed lower tolerance capacity towards PW in comparison with *Paspalum* sp. which exhibited better tolerance by withstanding at least withstand 30% PW and 4.5% salinity. As a consequence, *Paspalum* sp. was more capable of being used in Qatar's areas that are planned to be irrigated with produced water. The Ministry of Qatar has already used *Paspalum* sp. as turf grass around Qatar's parks, roadsides, and golf courses, which can allow for the use of PW for their growth [6]. Moreover, studying the microbial succession depicted that produced water irrigation had resulted in variation in the fungal species, especially in 10% of produced water and 30% of PW-treated soil, and were absent in soil treated with tap water. The effect of concentration of produced water (L-R, 0% (tap water) up to 100% produced water) on turf grass (*Cynodon dactylon*) coverage (%) after being subjected to 14 weeks of treatment was studied. In general, the concentration of PW used for irrigation is the key to determine the effect of weeds and turf grass growth besides their abundance [74].

In a study by Al Kaabi et al. [123], PW samples from gas production process situated in Qatar north field were gathered, characterized, as well as treated. The characterization results demonstrated that the PW had high TOC and chemical oxygen demand (COD) values, i.e., 2405 and 10,496 ppm, respectively, and higher contents of benzene, toluene, ethylbenzene, and xylene (BTEX), and other metals. Table 8 presents the PW characterization from the north offshore gas field, Qatar. The results of organic compounds such as the benzene were available in maximum concentrations (11,170 ppb), succeeded by ethylbenzene (4648.6 ppb), xylene (1156.8 ppb), and toluene (378.1 ppb). After the treatment, sand filtration demonstrated the greatest removal efficiency for corrosion inhibitors (94.1%) and TSS (77.5%), which is due to the straining mechanisms. The highest removal efficiency of metals was for manganese and iron and was also able to separate BTEX.

Table 8. PW characterization from north offshore gas field, Qatar. Reproduced from ref. [123].

Parameters	Mean Values	Parameters	Mean Values
Major parameters		Metals	
TSS (ppm)	21.34 ± 3.51	Strontium (ppb)	13,181 ± 114
pH	4.43 ± 0.01	Vanadium (ppb)	2.55 ± 0.04
Conductivity	7035 ± 56	Sodium (ppb)	1,198,167 ± 16,526
COD (ppm)	10,496 ± 162	Zinc (ppb)	4.97 ± 0.28
Salinity (ppt)	4502 ± 36		
BOD (ppm)	1034 ± 42	Other pollutants	
TOC (ppm)	2405 ± 16	Propionate (ppm)	17.37 ± 1.04
BTEX		% KHI	0.27 ± 0.05
Ethyl benzene (ppb)	4648 ± 688	Phenol (ppm)	1.96 ± 0.07
Xylene (ppb)	1156 ± 88	HEM (ppm)	40.54 ± 4.20
Benzene (ppb)	11,170 ± 4298	Formate (ppm)	0.35 ± 0.04
Toluene (ppb)	278.2 ± 14.3	Corrosion Inhibitor (ppm)	623.3 ± 15.5
Metals		Acetate (ppm)	368.7 ± 4.04
Potassium (ppb)	100,922 ± 122	TN (ppm)	47.41 ± 0.25
Nickel (ppb)	7.08 ± 0.28	% MEG	0.33 ± 0.07
Molybdenum (ppb)	5.52 ± 0.02	Other Ions	
Manganese (ppb)	258.3 ± 2.7	Sulfide (ppm)	326.3 ± 21.1
Iron (ppb)	4144 ± 114	Sulphate (ppm)	46.13 ± 0.19
Aluminum (ppb)	10.28 ± 6.75	Silica (ppm)	2.0 ± 0.1
Arsenic (ppb)	7.24 ± 1.89	Phosphate (ppm)	2.06 ± 0.08
Chromium (ppb)	30.31 ± 0.37	Magnesium (ppb)	45,064 ± 1223
Barium (ppb)	60.51 ± 0.45	Chloride (ppm)	2921 ± 10
Cobalt (ppb)	7.04 ± 0.70	Boron (ppb)	5744 ± 95
Copper (ppb)	0.62 ± 0.05	Calcium (ppb)	285,565 ± 2205
Cadmium (ppb)	0.05 ± 0.01		

In another study by Atia, F. et al. [75], the team collected the soil from the Mesaieed area in Qatar, and seeds of the crop plant species from the local market, seedlings of *Phragmitis*. sp. from El-Khour area (north of Doha) and *Salsola*. sp. from the biology field of a Qatar-based university campus. Their PW samples were provided by an oil and gas company in Qatar, and the team used it at different dilution percentages for plant irrigation (0, 10, 20, 30, 40, 50%) in the greenhouse for *Zea mays, Medicago sativa, Sorghum bicolor, Helianthus Annuus, Salsola baryosma* and *Phragmites australis*. The PW samples were chemically and physically characterized, and the results presented higher content of chloride, boron ions, sodium, sodium adsorption, and TDS ratio as 122, 0.038 g/L, 61, 139.9, and 300 meq/L, respectively. Among all plants, *Medicago sativa* was the only survivor which tolerated 10.0% produced water with a reduction in length, intensity, and biomass to reach to 33%. Furthermore, *Salsola baryosma* tolerated irrigation of PW 20.0% without any remarkable variations in the morphological characteristics. After that level, the heavy metals accumulation and morphological growth were completely disturbed. It has been demonstrated that the soil showed a huge enhancement in salinity, as well as sodium adsorption ratio (SAR) levels which interfered with the soil physical characteristics affecting the permeability and water flow. Figure 8 presents the comparison between the development of root and shoot of alfalfa at 10 % produced water irrigation, having been irrigated using tap water. It was noted that there was no remarkable variation in the root growth under 10 % produced water irrigation, whereas the shoot growth showed remarkable difference. Carbon accumulation percentage is related with the organic accumulation. From control to 10 and 20% PW, it can be noted that the accumulation is smooth and slowly improves, while it has sharpness at 30 % PW (Figure 9). Moreover, in the study, it was noted that the organic contents of produced water were below detection limits of ultra-performance liquid chromatography (UPLC), as well as gas chromatography (GC) instruments after dilutions at altered levels [75].

Onwusogh, U. et al. [65] presented some treatments technologies to treat and reuse the PW from onsite effluent treatment plant (ETP). Among different type of treatments, the biotreatment, crystallization and evaporation processes, flocculation, flotation, membrane filtration (included submerged Ultrafiltration (sUF) and RO units), and catalytic wet air oxidation (cWAO) were used for cooling and power generation. According to the results, all of the treatments enhanced the removal efficiency of most of the pollutants from PW. It is worth mentioning that cWAO is effective in removing KHI and delivers a high quality of suitable water for treatment in onsite ETP. All PW samples were obtained from the gas field. The characterization of organic and inorganic components illustrated that the concentration of chloride (Cl) was the highest reached about 1237 mg/L, followed by sodium (Na), and calcium (Ca) with approximately the same concentration 424 and 386 mg/L, respectively. Moreover, the concentration of sulfur (S), sulfate (SO^{-2}_4), potassium (K) and magnesium (Mg) were almost similar to each other with values almost 45 mg/L. Boron (B), silica (Si), strontium (Sr), and iron (Fe) showed poor values. The GTL plant mentioned in this case studied generated up to 400 m^3/day of PW consisting of 1.5 wt% of KHI solution. The total dissolved solids (TDS) concentration was noted to be between 3000 to 5000 ppm [65].

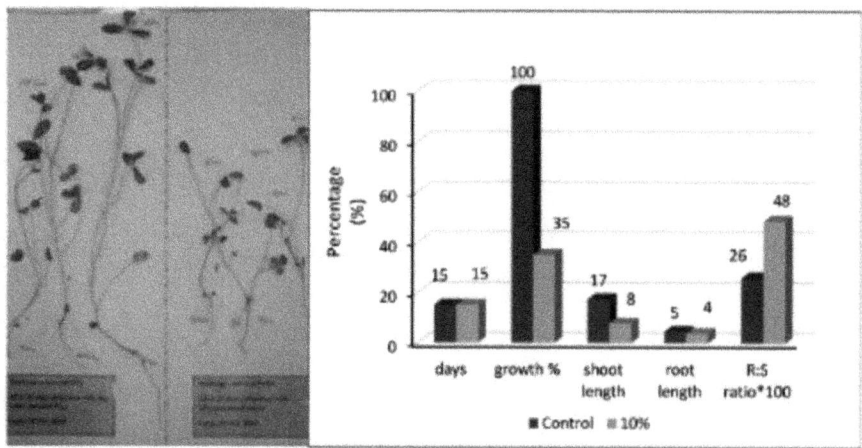

Figure 8. Alfalfa irrigated with tap water and 10% PW after 15 days. Reproduced from Ref. [75].

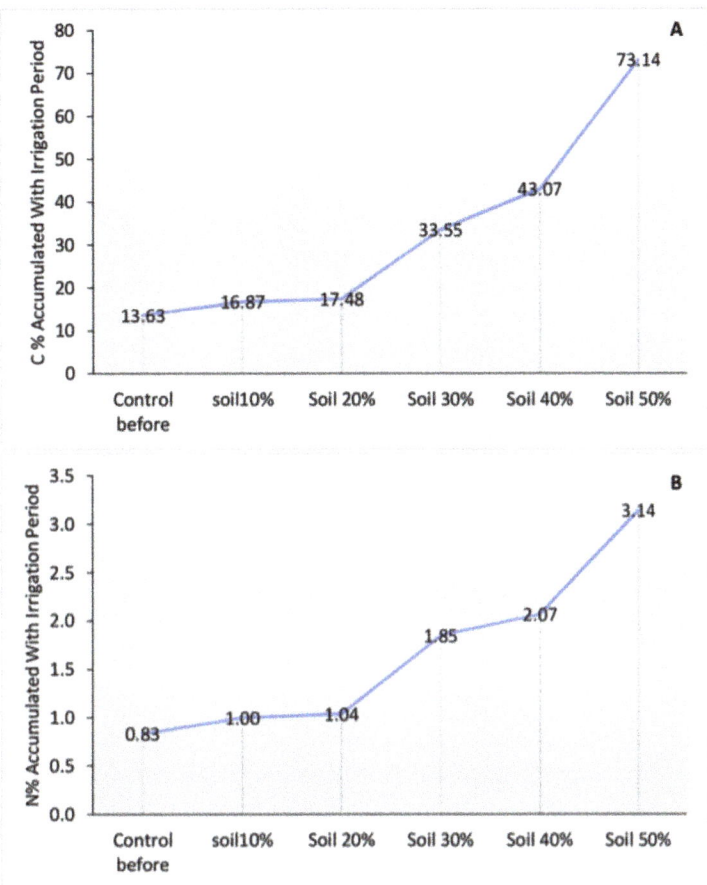

Figure 9. (**A**) Carbon % accumulation in the soil with produced water irrigation and (**B**) nitrogen % accumulation in the soil. Reproduced from ref. [75].

9. Future Outlook

Recently, several advanced technologies or series of technologies developed for treating the PW have been presented. Certain technologies are entirely new, whereas others have been modified from several other municipal or commercial wastewater process operations or treatments. PW management is substantially important for the economic and safe production of gas and oil. The application of online oil-in-water monitors could perform an important role in manufacturing process control, as well as the optimization, discharge reporting, and injection of water quality checks. With the technological developments, an upgraded perception, and a proper understanding of the doubts related to the existing sampling, as well as measurement routine, together with the continuous efforts from industry, an increased use of online oil-in-water monitors can be noted for the PW management.

Moreover, the difference in PW treatment technologies, the effect of numerous constituents of produced water on the agriculture, re-injection in oil/gas industry, or on the growth of microalgae need to be investigated further. The proper characterization step of PW for determining the major components is recommended as the initial step for selecting the finest treatment options. The characterization results will determine whether physical, thermal, or chemical treatments are adopted. Moreover, it is also recommended to adopt the zero liquid discharge concept, thereby avoiding any extra contaminant stream generation and thus waste minimization. Presently, there are specific PW treatment technologies; however, most of the technologies are tailor-made for meeting the specific treatment requirements for each site, and typically concentrate on just a particular group of PW hazardous elements [132].

The National Development Strategy as well as Qatar Vision 2030 both realize the requirement for good water management for sustaining the quick growth of Qatar, in spite of the shortage in natural freshwater. They take active engagement in bringing several diverse water stakeholders collectively for a sustainable future. In 2010, ConocoPhillips started the Global Water Sustainability Center for studying the technology associated with the PW, as well as desalination. This center in Qatar research aimed to develop advanced and economic solutions to treat as well as recycle PW from the oil and gas industries and contributing support to Qatargas.

10. Current Challenges and Environmental Issues of Produced Water

The immense amount of PW water discharged each year offers an ecological problem to the industries. Due to the importance of PW, the adequate treatments of it have been approached to be very critical to reuse or discharge PW to the environment. Therefore, the present major challenge in oil and gas companies have become to be the exploration of an effective and inexpensive treatment method for minimizing the pollutants present in the PW for discharging or/and reusing the constitutes. In the oil and gas industry, approximately 3 to 10 barrels of produced water are brought to the surface for every barrel of oil produced. All of the dispersed oil, heavy metals, large amounts of organic material, inorganic salts, and high levels of sulfur and sulfides present in PW are of particular environmental concern. Therefore, reducing the hazardous waste of PW before discharging it to the environment becomes essential. In the event that the PW was not carefully managed, such as releasing it to neighboring surface water bodies without proper treatments or allowing it to soak into the ground, massive ecological damage will occur. Nevertheless, industries are following reliable programs of activities to prevent the ecological damage. As an illustration, the offshore discharges, even properly treated, will offer certain impacts, although moderate, such as increased pollutant levels and reduced oxygen in the nearby fields. One of the most important objectives of environmental monitoring is to evaluate if the discharge regulations are sufficiently protective. Hence, several physical treatments were used to clean the PW before the discharge in addition to the regulations added to frontier on levels of contaminants, which can be discharged to the sea. Moreover, to moderate the discharge of PW, reinjection has been used for many years. In the case of majority open water situations, the quantity of local dilutions as well as the currents that stimulate dispersion lessens the possible effects very rapidly. In the nearby shore settings, emissions from offshore platforms can offer a significant impact, and generally zero discharge is needed in these regions. However, if some other water managing practice were utilized, it may have a lesser water impact however can have a more energy use impact.

Other major challenges of produced water are the generation of hazardous sludge in chemical treatment, the consequent disposal problems, the sensitivity to the initial concentration of wastewater, higher operation expenses, potential chronic toxicity of the treated PW, high treatment cost, and public acceptance. Moreover, the sensitivity to variation of organic chemicals in biological treatment is another major challenge for treating PQ, besides the salt concentration of influent waste. Furthermore, the quantity and characteristics of the PW varies over time, making a "one size fits all" solution unlikely.

With regards to oil and gas PW, the treatment expense effectively depends on the chemical and physical properties of the PW, which could differ extensively in the production fields and vary over time in a specific field, as well as the monitoring environment. As an illustration, the PW from gas field, particularly coalbed methane production, usually has low total dissolved solids, grease content, and oil, as compared to that from oil production. Hence, the proper treatment of PW can possibly be competitive as compared to other wastewater sources. Although current technologies are proven to meet the present standards of drinking water [132], there are several anxieties concerning the unidentified poisonous effects or undiscovered hazardous compounds. Until the human health impacts of chemical compounds are well understood, toxicity assays are required for addressing any matters on the possible synergy between hazardous compounds and the probability of unobserved hazardous compounds in the treated PW.

11. Reuse of Produced Water

Treated produced water can be reused in various applications, such as in increasing oil production by underground injection, irrigation, livestock or wildlife watering, and various industrial uses. Al-Kaabi et al. [77] mentioned in their study, the demand for a design to adopt an environment-friendly method, for instance, phytoremediation to separate more contaminants from PW and soil to reuse it in the Qatari landscape and biofuel plantation. The technological solutions for drinkable reuse of PW will require it to be adapted as per the properties of the PW and the quantity of water to be treated. Due to the need for desalination as well as the separation of huge quantity of organic compounds, the reverse osmosis technology will most probably be employed for drinking reuse applications. Even though RO is capable of removing many organic compounds with increased efficiency, the combined chronic toxicity of the organic compounds present in the mixture in the reverse osmosis permeate should be thoroughly assessed prior to its direct reuse is executed. A technique for confirming maximal safe pollutant concentrations based on mixed impact of the pollutant mixture is necessary.

The benefit of using membrane process to treat and reuse PW have been studied successfully in many field studies. The study by Al Kaabi et al. [123] confirmed that the treated water was free from all main contaminants of PW, and it could be considered appropriate for reuse at domestic or industrial level. Furthermore, the Gas Processing Center (GPC) in Qatar University contributes to meeting the challenge of huge volume discharge of PW by developing advanced treatment approaches for achieving a zero harmful release in oil and gas production, and to permit the economic reuse of PW generated at the time of field operations.

12. Conclusions

As produced water is wastewater generated during the gas and oil production process, this water is typically trapped during subsurface formations, and is brought to the surface along with oil or gas to maintain reservoir pressure. The presence of pollutants, heavy metals, and toxic organic compounds in PW attract the attention of researchers to find an effective solution to treat this water. Qatar is one of the most water-stressed countries in the world, with relatively very limited freshwater resources, and therefore, the proper treatment of PW has turned out to be critical to reuse or discharged to the environment.

The present study highlighted the generation of produced water in Qatar, the PW characteristics in detail, and the physical, chemical, and biological treatment techniques for the PW. In the end, case studies from different companies in Qatar and the challenges of treating the produced water are discussed. From the different studies analyzed, various techniques, as well as sequencing of different techniques, were noted for the employment for the treatment of PW. In the oil and gas industry, approximately 3 to 10 barrels of produced water are brought to the surface for every barrel of oil produced. Therefore, researchers and companies detailed in their works and literature review that the characteristic of produced water varies from area to area, and thus the treatment technologies also vary for obtaining the fresh water. The proper characterization step of PW for determining the major components is recommended as the initial step for selecting the finest treatment options. It is also recommended that the zero liquid discharge concept is adopted, thereby avoiding any extra contaminant stream generation and thus waste minimization. The sensitivity to variation of organic chemicals in biological treatment is another important challenge to treat PW, alongside the salt concentration of influent waste. Although the present technologies are proven to meet the present standards of drinking water, there are still several anxieties concerning the unidentified poisonous effects or undiscovered toxic compounds. The aims of the largest LNG producer in Qatar are aligned with of the Ministry of Municipality and Environment (MME) in maximizing the use of water and reducing wastewater discharge. There are an increasing number of membrane-based studies in Qatar for effective water treatment. However, different factors such as high initial capital costs of some PW treatment technologies, restrict an overall recommendation for treating produced water. Furthermore, the vast oil/gas and petrochemicals industry production requires a huge amount of water to be used in the industrial processes, driving a further emphasis on water use and resources.

Funding: This research received no external funding.

Institutional Review Board Statement: Not applicable.

Informed Consent Statement: Not applicable.

Data Availability Statement: Some or all data that support the findings of this study are available from the corresponding author upon reasonable request.

Acknowledgments: This publication was supported by Qatar University Grant QUEX-CAM-QP-PW-18/19.

Conflicts of Interest: The authors declare no conflict of interest.

References

1. Khader, E.H.; Mohammed, T.J.; Mirghaffari, N.; Salman, A.D.; Juzsakova, T.; Abdullah, T.A. Removal of organic pollutants from produced water by batch adsorption treatment. *Clean Technol. Environ. Policy* **2021**, *23*, 1–8. [CrossRef]
2. Alzahrani, S.; Mohammad, A.W. Challenges and trends in membrane technology implementation for produced water treatment: A review. *J. Water Process Eng.* **2014**, *4*, 107–133. [CrossRef]
3. Dickhout, J.M.; Moreno, J.; Biesheuvel, P.M.; Boels, L.; Lammertink, R.G.H.; De Vos, W.M. Produced water treatment by membranes: A review from a colloidal perspective. *J. Colloid Interface Sci.* **2017**, *487*, 523–534. [CrossRef] [PubMed]
4. Mao, S.; Deng, Y.; Pelusi, D. Alternatives selection for produced water management: A network-based methodology. *Eng. Appl. Artif. Intell.* **2020**, *91*, 103556. [CrossRef]
5. Alammar, A.; Park, S.H.; Williams, C.J.; Derby, B.; Szekely, G. Oil-in-water separation with graphene-based nanocomposite membranes for produced water treatment. *J. Membr. Sci.* **2020**, *603*, 118007. [CrossRef]
6. Ganiyu, S.O.; Sable, S.; El-Din, M.G. Advanced oxidation processes for the degradation of dissolved organics in produced water: A review of process performance, degradation kinetics and pathway. *Chem. Eng. J.* **2021**, *429*, 132492. [CrossRef]
7. Hendges, L.T.; Costa, T.C.; Temochko, B.; González, S.Y.G.; Mazur, L.P.; Marinho, B.A.; da Silva, A.; Weschenfelder, S.E.; de Souza, A.A.U.; de Souza, S.M.G.U. Adsorption and desorption of water-soluble naphthenic acid in simulated offshore oilfield produced water. *Process Saf. Environ. Prot.* **2021**, *145*, 262–272. [CrossRef]
8. McLaughlin, M.C.; Blotevogel, J.; Watson, R.A.; Schell, B.; Blewett, T.A.; Folkerts, E.J.; Goss, G.G.; Truong, L.; Tanguay, R.L.; Argueso, J.L.; et al. Mutagenicity assessment downstream of oil and gas produced water discharges intended for agricultural beneficial reuse. *Sci. Total Environ.* **2020**, *715*, 136944. [CrossRef]
9. Babu, P.; Bollineni, C.; Daraboina, N. Energy Analysis of Methane-Hydrate-Based Produced Water Desalination. *Energy Fuels* **2021**, *35*, 2514–2519. [CrossRef]
10. Ali, S.; Ijaola, A.O.; Asmatulu, E. Multifunctional water treatment system for oil and gas-produced water. *Sustain. Water Resour. Manag.* **2021**, *7*, 89. [CrossRef]
11. Madadizadeh, A.; Sadeghein, A.; Riahi, S. A Comparison of Different Nanoparticles' Effect on Fine Migration by Low Salinity Water Injection for Oil Recovery: Introducing an Optimum Condition. *J. Energy Resour. Technol.* **2021**, *144*, 013005. [CrossRef]
12. Tale, F.; Kalantariasl, A.; Malayeri, M.R. Estimating transition time from deep filtration of particles to external cake during produced water re-injection and disposal. *Part. Sci. Technol.* **2021**, *39*, 312–321. [CrossRef]
13. Wang, Y.; Yu, M.; Bo, Z.; Bedrikovetsky, P.; Le-Hussain, F. Effect of temperature on mineral reactions and fines migration during low-salinity water injection into Berea sandstone. *J. Pet. Sci. Eng.* **2021**, *202*, 108482. [CrossRef]
14. Zheng, J.; Chen, B.; Thanyamanta, W.; Hawboldt, K.; Zhang, B.; Liu, B. Offshore produced water management: A review of current practice and challenges in harsh/Arctic environments. *Mar. Pollut. Bull.* **2016**, *104*, 7–19. [CrossRef]
15. Torres, L.; Yadav, O.P.; Khan, E. A review on risk assessment techniques for hydraulic fracturing water and produced water management implemented in onshore unconventional oil and gas production. *Sci. Total Environ.* **2016**, *539*, 478–493. [CrossRef]
16. Sosa-Fernandez, P.; Miedema, S.; Bruning, H.; Leermakers, F.; Post, J.; Rijnaarts, H. Effects of feed composition on the fouling on cation-exchange membranes desalinating polymer-flooding produced water. *J. Colloid Interface Sci.* **2021**, *584*, 634–646. [CrossRef]

17. Simona, C.; Raluca, I.; Anita-Laura, R.; Andrei, S.; Raluca, S.; Bogdan, T.; Elvira, A.; Catalin-Ilie, S.; Claudiu, F.R.; Daniela, I.-E.; et al. Synthesis, characterization and efficiency of new organically modified montmorillonite polyethersulfone membranes for removal of zinc ions from wastewasters. *Appl. Clay Sci.* **2017**, *137*, 135–142. [CrossRef]
18. Su, Y.N.; Lin, W.S.; Hou, C.H.; Den, W. Performance of integrated membrane filtration and electrodialysis processes for copper recovery from wafer polishing wastewater. *J. Water Process. Eng.* **2014**, *4*, 149–158. [CrossRef]
19. Eljamal, O.; Jinno, K.; Hosokawa, T. Modeling of solute transport and biological sulfate reduction using low cost electron donor. *J. Environ. Geol.* **2009**, *56*, 1605–1613. [CrossRef]
20. Maamoun, I.; Eljamal, O.; Falyouna, O.; Eljamal, R.; Sugihara, Y.J.W.S. Stimulating effect of magnesium hydroxide on aqueous characteristics of iron nanocomposites. *Water Sci. Technol.* **2019**, *80*, 1996–2002. [CrossRef]
21. Chekioua, A.; Delimi, R. Purification of H_2SO_4 of pickling bath contaminated by Fe (II) ions using electrodialysis process. *Energy Procedia* **2015**, *74*, 1418–1433. [CrossRef]
22. Tristán Teja, C.; Fallanza Torices, M.; Ibáñez Mendizábal, R.; Ortiz Uribe, I. Reverse electrodialysis: Potential reduction in energy and emissions of desalination. *Appl. Sci.* **2020**, *10*, 7317. [CrossRef]
23. Căprărescu, S.; Zgârian, R.G.; Tihan, G.T.; Purcar, V.; Totu, E.E.; Modrogan, C.; Chiriac, A.-L.; Nicolae, C.A. Biopolymeric membrane enriched with chitosan and silver for metallic ions removal. *Polymers* **2020**, *12*, 1792. [CrossRef]
24. Caprarescu, S.; Radu, A.L.; Purcar, V.; Sarbu, A.; Vaireanu, D.I.; Ianchis, R.; Ghiurea, M. Removal of copper ions from simulated wastewaters using different bicomponent polymer membranes. *Water Air Soil Pollut.* **2014**, *225*, 2079. [CrossRef]
25. Obotey Ezugbe, E.; Rathilal, S. Membrane technologies in wastewater treatment: A review. *Membranes* **2020**, *10*, 89. [CrossRef]
26. Eljamal, R.; Eljamal, O.; Maamoun, I.; Yilmaz, G.; Sugihara, Y. Enhancing the characteristics and reactivity of nZVI: Polymers effect and mechanisms. *J. Mol. Liq.* **2020**, *315*, 113714. [CrossRef]
27. Osama, E.; Junya, O.; Kazuaki, H. Removal of phosphorus from water using marble dust as sorbent material. *J. Environ. Prot. Sci.* **2012**, *2012*, 21755.
28. Abd Halim, N.S.; Wirzal, M.D.H.; Hizam, S.M.; Bilad, M.R.; Nordin, N.A.H.M.; Sambudi, N.S.; Putra, Z.A.; Yusoff, A.R.M. Recent development on electrospun nanofiber membrane for produced water treatment: A review. *J. Environ. Chem. Eng.* **2021**, *9*, 104613. [CrossRef]
29. Liang, H.; Zou, C.; Tang, W. Development of novel polyether sulfone mixed matrix membranes to enhance antifouling and sustainability: Treatment of oil sands produced water (OSPW). *J. Taiwan Inst. Chem. Eng.* **2021**, *118*, 215–222. [CrossRef]
30. Weschenfelder, S.E.; Fonseca, M.J.C.; Borges, C.P. Treatment of produced water from polymer flooding in oil production by ceramic membranes. *J. Pet. Sci. Eng.* **2021**, *196*, 108021.
31. Wu, M.; Zhai, M.; Li, X. Adsorptive removal of oil drops from ASP flooding-produced water by polyether polysiloxane-grafted ZIF-8. *Powder Technol.* **2021**, *378*, 76–84. [CrossRef]
32. Qi, P.; Sun, D.; Gao, J.; Liu, S.; Wu, T.; Li, Y. Demulsification and bio-souring control of alkaline-surfactant-polymer flooding produced water by *Gordonia* sp. TD-4. *Sep. Purif. Technol.* **2021**, *263*, 118359. [CrossRef]
33. Husveg, R.; Husveg, T.; van Teeffelen, N.; Ottestad, M.; Hansen, M.R. Variable Step Size P&O Algorithms for Coalescing Pump/Deoiling Hydrocyclone Produced Water Treatment System. *Model. Identif. Control* **2020**, *41*, 13–27.
34. Jin, Y.; Davarpanah, A. Using photo-fenton and floatation techniques for the sustainable management of flow-back produced water reuse in shale reservoirs exploration. *Water Air Soil Pollut.* **2020**, *231*, 441. [CrossRef]
35. Simões, A.J.A.; Macêdo-Júnior, R.O.; Santos, B.L.P.; Silva, D.P.; Ruzene, D.S. A bibliometric study on the application of advanced oxidation processes for produced water treatment. *Water Air Soil Pollut.* **2021**, *232*, 297. [CrossRef]
36. Coha, M.; Farinelli, G.; Tiraferri, A.; Minella, M.; Vione, D. Advanced oxidation processes in the removal of organic substances from produced water: Potential, configurations, and research needs. *Chem. Eng. J.* **2021**, *414*, 128668. [CrossRef]
37. Hammack, R.; Mosser, M.H. Produced water management. In *Coal Bed Methane*; Elsevier: Amsterdam, The Netherlands, 2020; pp. 349–369.
38. Camarillo, M.K.; Stringfellow, W.T. Biological treatment of oil and gas produced water: A review and meta-analysis. *Clean Technol. Environ. Policy* **2018**, *20*, 1127–1146. [CrossRef]
39. Liu, Y.; Lu, H.; Li, Y.; Xu, H.; Pan, Z.; Dai, P.; Wang, H.; Yang, Q. A review of treatment technologies for produced water in offshore oil and gas fields. *Sci. Total Environ.* **2021**, *775*, 145485. [CrossRef]
40. Duraisamy, R.T.; Beni, A.H.; Henni, A. State of the art treatment of produced water. In *Water Treatment*; Books on Demand: Norderstedt, Germany, 2013; pp. 199–222.
41. Scanlon, B.R.; Reedy, R.C.; Xu, P.; Engle, M.; Nicot, J.P.; Yoxtheimer, D.; Yang, Q.; Ikonnikova, S. Can we beneficially reuse produced water from oil and gas extraction in the US? *Sci. Total Environ.* **2020**, *717*, 137085. [CrossRef]
42. Global Produced Water Treatment Industry (2021 Report). Available online: https://www.reportlinker.com/p06032674/Global-Produced-Water-Treatment-Industry.html?utm_source=GNW (accessed on 24 October 2021).
43. The U.S. Energy Information Agency. Qatar: Oil Production. Available online: https://www.theglobaleconomy.com/Qatar/oil_production/ (accessed on 24 October 2021).
44. Kabyl, A.; Yang, M.; Abbassi, R.; Li, S. A risk-based approach to produced water management in offshore oil and gas operations. *Process Saf. Environ. Prot.* **2020**, *139*, 341–361. [CrossRef]

45. Santos, K.A.; Gomes, T.M.; Rossi, F.; Kushida, M.M.; Del Bianchi, V.L.; Ribeiro, R.; Alves, M.S.M.; Tommaso, G. Water reuse: Dairy effluent treated by a hybrid anaerobic biofilm baffled reactor and its application in lettuce irrigation. *Water Supply* **2021**, *21*, 1980–1993. [CrossRef]
46. Rahman, A.; Agrawal, S.; Nawaz, T.; Pan, S.; Selvaratnam, T. A review of algae-based produced water treatment for biomass and biofuel production. *Water* **2020**, *12*, 2351. [CrossRef]
47. Lahlou, F.Z.; Mackey, H.R.; McKay, G.; Onwusogh, U.; Al-Ansari, T. Water planning framework for alfalfa fields using treated wastewater fertigation in Qatar: An energy-water-food nexus approach. *Comput. Chem. Eng.* **2020**, *141*, 106999. [CrossRef]
48. Ahmad, A.Y.; Al-Ghouti, M.A. Approaches to achieve sustainable use and management of groundwater resources in Qatar: A review. *Groundw. Sustain. Dev.* **2020**, *11*, 100367. [CrossRef]
49. Kondash, A.J.; Redmon, J.H.; Lambertini, E.; Feinstein, L.; Weinthal, E.; Cabrales, L.; Vengosh, A. The impact of using low-saline oilfield produced water for irrigation on water and soil quality in California. *Sci. Total Environ.* **2020**, *733*, 139392. [CrossRef]
50. Minier-matar, J.; Hussain, A.; Janson, A.; Wang, R.; Fane, A.G.; Adham, S. Application of Forward Osmosis for Reducing Volume of Produced/Process Water from Oil and Gas Operations. *Desalination* **2015**, *376*, 1–8. [CrossRef]
51. Minier-Matar, J.; Santos, A.; Hussain, A.; Janson, A.; Wang, R.; Fane, A.G.; Adham, S. Application of Hollow Fiber Forward Osmosis Membranes for Produced and Process Water Volume Reduction: An Osmotic Concentration Process. *Environ. Sci. Technol.* **2016**, *50*, 6044–6052. [CrossRef]
52. Janson, A.; Katebah, M.; Santos, A.; Minier-Matar, J.; Hussain, A.; Adham, S.; Judd, S. Assessing the Biotreatability of Produced Water from a Qatari Gas Field. *Int. Pet. Technol.* **2015**, *20*, 1113–1119. [CrossRef]
53. Zhao, P.; Gao, B.; Xu, S.; Kong, J.; Ma, D.; Shon, H.K.; Yue, Q.; Liu, P. Polyelectrolyte-promoted forward osmosis process for dye wastewater treatment—Exploring the feasibility of using polyacrylamide as draw solute. *Chem. Eng. J.* **2015**, *264*, 32–38. [CrossRef]
54. Jansona, A.F.; Santosa, A.; Hussaina, A.; Minier-matara, J.; Juddb, S.; Adhama, S. Application of Membrane Bioreactor Technology for Produced Water Treatment. In *International Gas Processing Symposium*; Elsevier: Doha, Qatar, 2015.
55. Adham, S.; Hussain, A.; Matar, J.M.; Janson, A.; Gharfeh, S.; Hussain, A.; Matar, J.M.; Janson, A.; Gharfeh, S.; Adham, S. Screening of Advanced Produced Water Treatment Technologies: Overview and Testing Results Screening of Advanced Produced Water Treatment Technologies: Overview and Testing Results. *IDA J. Desalin. Water Reuse* **2016**, *5*, 7953.
56. Hussain, A.; Janson, A.; Adham, S. Treatment of Produced Water from Unconventional Resources by Membrane Distillation. In Proceedings of the Internatioal Petroleum Technoloty Conference, Doha, Qatar, 19–22 January 2014.
57. Dores, R.; Hussain, A.; Katebah, M.; Adham, S.S. Using Advanced Water Treatment Technologies to Treat Produced Water from The Petroleum Industry. In Proceedings of the SPE International Production and Operations Conference & Exhibition, Doha, Qatar, 14–16 May 2012.
58. Zsirai, T.; Al-Jaml, A.K.; Qiblawey, H.; Al-Marri, M.; Ahmed, A.; Bach, S.; Watson, S.; Judd, S. Ceramic Membrane Filtration of Produced Water: Impact of Membrane Module. *Purif. Technol.* **2016**, *165*, 214–221. [CrossRef]
59. Zsirai, T.; Qiblawey, H.; Buzatu, P.; Al-Marri, M.; Judd, S.J. Cleaning of Ceramic Membranes for Produced Water Filtration. *J. Pet. Sci. Eng.* **2018**, *166*, 283–289. [CrossRef]
60. Abdallaa, M.; Nasser, M.; Fard, A.K.; Qiblawey, H.; Benamor, A.; Judd, S. Impact of Combined Oil-in-Water Emulsions and Particulate Suspensions on Ceramic Membrane Fouling and Permeability Recovery. *Purif. Technol.* **2018**, *212*, 215–222. [CrossRef]
61. Fard, A.K.; Mckay, G.; Atieh, M.A. Hybrid Separator -Adsorbent Inorganic Membrane for Oil–Water Separation. In Proceedings of the 3rd World Congress on Civil, Structural, and Environmental Engineering (CSEE'18), Budapest, Hungary, 8–10 April 2018; pp. 2–5.
62. Judd, S.; Qiblawey, H.; Al-Marri, M.; Clarkin, C.; Watson, S.; Ahmed, A.; Bach, S. The Size and Performance of Offshore Produced Water Oil-Removal Technologies for Reinjection. *Sep. Purif. Technol.* **2014**, *134*, 241–246. [CrossRef]
63. Qin, D.; Liu, Z.; Liu, Z.; Bai, H.; Sun, D.D. Superior Antifouling Capability of Hydrogel Forward Osmosis Membrane for Treating Wastewaters with High Concentration of Organic Foulants. *Environ. Sci. Technol.* **2018**, *52*, 1421–1428. [CrossRef]
64. Janson, A.; Santos, A.; Hussain, A.; Global, C.; Judd, S. Biotreatment of Hydrate-Inhibitor Containing Produced Waters at Low PH. *SPE J.* **2015**, *20*, 1254–1260. [CrossRef]
65. Onwusogh, U. Feasibility of Produced Water Treatment and Reuse—Case Study of a GTL. In Proceedings of the Internatioal Petroleum Technology Conference, Doha, Qatar, 6–9 December 2015.
66. Al Redoua, A.; Hamid, S.A.; Limited, D.E. Industrial Application for the Removal of Kinetic Hydrate Inhibitor (KHI) Co-Polymers from Produced Water Streams. In Proceedings of the International Petroleum Technology Conference, Doha, Qatar, 6–9 December 2015.
67. Moussa, D.T.; El-naas, M.H.; Nasser, M.; Al-Marri, M.J.A. Comprehensive Review of Electrocoagulation for Water Treatment: Potentials and Challenges. *J. Environ. Manag.* **2016**, *186*, 24–41. [CrossRef]
68. Aly, D.T.A. A Novel Electrocoagulation System for Produced Water Treatment. Master's Thesis, Qatar University, Doha, Qatar, 2018. Available online: http://hdl.handle.net/10576/11221 (accessed on 4 December 2021).
69. Alghoul, M. Treatment of Produced Water Using an Enhanced Electrocoagulation Process. Master's Thesis, Qatar University, Doha, Qatar, 2017.

70. Abu-dieyeh, M.H.; Shaikh, S.S.; Naemi, F.; AlAlghouti, M.; Youssef, T.A. The Influences of Produced Water Irrigation on Soil Microbial Succession and Turfgrass Grass Establishment in Qatar. In Proceedings of the Qatar Foundation Annual Research Conference, Doha, Qatar, 1 March 2018; Hamad bin Khalifa University Press: Doha, Qatar, 2018.
71. Das, P.; Abdulquadir, M.; Thaher, M.; Khan, S.; Chaudhary, A.K.; Alghasal, G.; Al-Jabri, H.M.S.J. Microalgal Bioremediation of Petroleum-Derived Low Salinity and Low PH Produced Water. *J. Appl. Phycol* **2019**, *31*, 435–444. [CrossRef]
72. Abdul Hakim, M.A. Potential Application of Micronalgae in Produced Water Treatment. *Desalin. Water Treat.* **2016**, *135*, 47–58. Available online: https://qspace.qu.edu.qa:8443/bitstream/handle/10576/5087/443354_pdf_F0D5787A-36CD-11E6-9423-32BB4 D662D30.pdf?sequence=1&isAllowed=n (accessed on 4 December 2021). [CrossRef]
73. Abdul, M.A.; Al-ghouti, M.A.; Das, P.; Abu-dieyeh, M.; Ahmed, T.A.; Mohammed, H.; Aljabri, S.J. Potential Application of Microalgae in Produced Water Treatment. *Desalin. Water Treat* **2018**, *135*, 23146. [CrossRef]
74. Shaikh, S. Grass Establishment, Weed Populations and Soil Microbial Succession in Turf Grass System as Influenced by Produced Water Irrigation. Master's Thesis, Qatar University, Doha, Qatar, 2017.
75. Atia, F.A.; Al-Ghouti, M.A.; Al-Naimi, F.; Abu-Dieyeh, M.; Ahmed, T.; Al-Meer, S.H. Removal of toxic pollutants from produced water by phytoremediation: Applications and mechanistic study. *J. Water Process. Eng.* **2019**, *32*, 100990. [CrossRef]
76. Al-Kaabi, M.A.J.I. Enhancing Produced Water Quality Using Modified Activiated Carbon. Master's Thesis, Qatar University, Doha, Qatar, 2016.
77. Al-Kaabi, M.; Ghazi, A.B.; Qunnaby, R.; Dawwas, F.; Al-Hadrami, H.; Khatir, Z.; Yousif, M.; Ahmed, T. Enhancing the Quality of "Produced Water" by Activated Carbon. *Qatar Found. Annu. Res. Conf. Proc.* **2016**, *22*, 16–17.
78. Almarouf, H.M.N.; Al-Marri, M.J.; Khraisheh, M.; Onaizi, S.A. Demulsification of Stable Emulsions From Produced Water Using A Phase Separator With Inclined Parallel Arc Coalescing Plates. *J. Pet. Sci. Eng.* **2015**, *135*, 16–21. [CrossRef]
79. Benamor, A.; Talkhan, A.G.; Nasser, M.; Hussein, I.; Okonkwo, P.C. Effect of Temperature and Fluid Speed on the Corrosion Behavior of Carbon Steel Pipeline in Qatari Oilfield Produced Water. *J. Electroanal. Chem.* **2018**, *808*, 218–227. [CrossRef]
80. Shehzad, F.; Hussein, I.A.; Kamal, M.S.; Ahmad, W.; Sultan, A.S.; Nasser, M.S.; Shehzad, F.; Hussein, I.A.; Kamal, M.S.; Ahmad, W. Polymeric Surfactants and Emerging Alternatives Used in the Demulsification of Produced Water: A Review Polymeric Surfactants and Emerging Alternatives Used. *Polym. Rev.* **2017**, *58*, 63–101.
81. Ahan, J.A. *Characterization of Produced Water from Two Offshore Oil Fields in Qatar*; Environmental Engineering College of Engineering: Doha, Qatar, 2014. Available online: https://qspace.qu.edu.qa/handle/10576/3287/restricted-resource?bitstreamId=0fb376 32-1e3b-4851-a058-80c438f15419 (accessed on 4 December 2021).
82. Ma, H.; Sultan, A.S.; Shawabkeh, R.; Fahd, K.; Mustafa, S. Destabilization and Treatment of Produced Water-Oil Emulsions Using Anionic Polyacrylamide with Electrolyate of Aluminum Sulphate and Ferrous Sulphate. In Proceedings of the Abu Dhabi International Petroleum Exhibition & Conference, Abu Dhabi, United Arab Emirates, 7–10 November 2016.
83. Ray, J.P.; Engelhardt, F.R. *Produced Water: Technological/Environmental Issues and Solutions*; Springer Science & Business Media: Berlin/Heidelberg, Germany, 1993.
84. Karanisa, T.; Amato, A.; Richer, R.; Abdul Majid, S.; Skelhorn, C.; Sayadi, S. Agricultural Production in Qatar's Hot Arid Climate. *Sustainability* **2021**, *13*, 4059. [CrossRef]
85. Echchelh, A.; Hess, T.; Sakrabani, R.; Prigent, S.; Stefanakis, A.I. Towards agro-environmentally sustainable irrigation with treated produced water in hyper-arid environments. *Agric. Water Manag.* **2021**, *243*, 106449. [CrossRef]
86. Hedar, Y. Pollution impact and alternative treatment for produced water. *E3S Web Conf.* **2018**, *31*, 03004.
87. Natural Gas Operation Annual Report. Available online: https://s3.amazonaws.com/rgi-documents/97e32b78209f56efc6d4ec2 ee5458fa89c264783.pdf (accessed on 23 October 2021).
88. James, L.; Ahmed, N. *The Pioneer*; The Public Relations Department, Qatar Gas Operating Company Limited: Doha, Qatar, 2011. Available online: https://www.qatargas.com/english/MediaCenter/ThePioneer/ThePioneer-August2011.pdf (accessed on 3 December 2021).
89. Plebon, M.J.; Saad, M.A.; Al-Kuwari, N. Application of an Enhanced Produced Water De-Oiling Technology to Middle East Oilfields. In *Produced Water—Best Management Practices*; TORR Canada Inc.: Montreal, QC, Canada; Qatar Petroleum: Doha, Qatar, 2007.
90. Khalifa, O.M.; Rais, T.A.M.; El Maadi, A.; Sani, R.A.; Mahrous, E.K.; Al-Marri, B. Coupling Integrated Data Management with Reservoir Surveillance Workflows for Giant Mature Field in Qatar. In Proceedings of the Internatioal Petroleum Technoloty Conference, Doha, Qatar, 6–9 December 2015.
91. Al-ghouti, M.A.; Al-kaabi, M.A.; Ashfaq, M.Y.; Adel, D. Produced Water Characteristics, Treatment and Reuse: A Review. *J. Water Process Eng.* **2019**, *28*, 222–239. [CrossRef]
92. Yaqub, A.; Isa, M.H.; Ajab, H. Electrochemical degradation of polycyclic aromatic hydrocarbons in synthetic solution and produced water using a Ti/SnO$_2$-Sb$_2$O$_5$-RuO$_2$ anode. *J. Environ. Eng.* **2015**, *141*, 04014074. [CrossRef]
93. Macheca, A.D.; Uwiragiye, B. Application of Nanotechnology in Oil and Gas Industry: Synthesis and Characterization of Organo-modified Bentonite from Boane Deposit and its Application in Produced Water Treatment. *Chem. Eng. Trans.* **2020**, *81*, 1081–1086.
94. McLaughlin, M.C.; Borch, T.; McDevitt, B.; Warner, N.R.; Blotevogel, J. Water quality assessment downstream of oil and gas produced water discharges intended for beneficial reuse in arid regions. *Sci. Total Environ.* **2020**, *713*, 136607. [CrossRef]
95. Konur, O. Characterization and properties of biooils: A review of the research. *Biodiesel Fuels* **2021**, *1*, 137–152.

96. Fillo, J.P.; Koraido, S.M.; Evans, J.M. Sources, characteristics, and management of produced waters from natural gas production and storage operations. In *Produced Water*; Springer: Boston, MA, USA, 1992; pp. 151–161.
97. Stephenson, M.T. A survey of produced water studies. In *Produced Water*; Springer: Boston, MA, USA, 1992; pp. 1–11.
98. Ahmad, N.A.; Goh, P.S.; Yogarathinam, L.T.; Zulhairun, A.K.; Ismail, A.F. Current advances in membrane technologies for produced water desalination. *Desalination* **2020**, *493*, 114643. [CrossRef]
99. Jacobs, R.P.W.M.; Grant, R.O.H.; Kwant, J.; Marquenie, J.M.; Mentzer, E. The composition of produced water from Shell operated oil and gas production in the North Sea. In *Produced Water*; Springer: Boston, MA, USA, 1992; pp. 13–21.
100. Fillo, J.P.; Evans, J.M. Characterization and Management of Produced Waters from Underground Natural Gas Storage Reservoirs. *Am. Gas Assoc. Oper. Sect. Proc.* **1990**, *1*, 448–459.
101. Healy, R.W.; Bartos, T.T.; Rice, C.A.; McKinley, M.P.; Smith, B.D. Groundwater chemistry near an impoundment for produced water, Powder River Basin, Wyoming, USA. *J. Hydrol.* **2011**, *403*, 37–48. [CrossRef]
102. Terrens, G.W.; Tait, R.D. Monitoring Ocean Concentrations of Aromatic Hydrocarbons from Produced Formation Water Discharges to Bass Strait. In Proceedings of the International Conference on Health, Society of Petroleum Engineers Australia, New Orleans, LA, USA, 6 June 1996; pp. 739–747.
103. Zhao, S.; Minier-Matar, J.; Chou, S.; Wang, R.; Fane, A.G.; Adham, S. Gas field produced/process water treatment using forward osmosis hollow fiber membrane: Membrane fouling and chemical cleaning. *Desalination* **2017**, *402*, 143–151. [CrossRef]
104. Ray, J.P.; Engelhardt, F.R. Produced Water Technological Environmental Issues and Solution. *Environ. Sci. Res.* **1993**, *46*, 96–230.
105. Dyakowski, T.; Kraipech, W.; Nowakowski, A.P.; Williams, R. A Three Dimensional Simulation of Hydrocyclone Behaviour. In Proceedings of the Second International Conference, Sydney, NSW, Australia, 9–11 November 1999; pp. 205–210.
106. Du, Z.; Ding, P.; Tai, X.; Pan, Z.; Yang, H. Facile preparation of Ag-coated superhydrophobic/superoleophilic mesh for efficient oil/water separation with excellent corrosion resistance. *Langmuir* **2018**, *34*, 6922–6929. [CrossRef]
107. Chen, S.; Lv, C.; Hao, K.; Jin, L.; Xie, Y.; Zhao, W.; Sun, S.; Zhang, X.; Zhao, C. Multifunctional negatively-charged poly (ether sulfone) nanofibrous membrane for water remediation. *J. Colloid Interface Sci.* **2019**, *538*, 648–659. [CrossRef]
108. Wang, K.; Han, D.S.; Yiming, W.; Ahzi, S.; Abdel-Wahab, A.; Liu, Z. A windable and stretchable three-dimensional all-inorganic membrane for efficient oil/water separation. *Sci. Rep.* **2017**, *7*, 16081. [CrossRef]
109. Liu, Z. Efficient and Fast Separation of Emulsified Oil/Water Mixtures with a Novel Micro/Nanofiber Network Membrane. In Proceedings of the Qatar Foundation Annual Research Conference, Doha, Qatar, 19–20 March 2018.
110. Wang, K.; Yiming, W.; Saththasivam, J.; Liu, Z. A flexible, robust and antifouling asymmetric membrane based on ultra-long ceramic/polymeric fibers for high-efficiency separation of oil/water emulsions. *Nanoscale* **2017**, *9*, 9018–9025. [CrossRef]
111. Lee, S.; Boo, C.; Elimelech, M.; Hong, S. Comparison of fouling behavior in forward osmosis (FO) and reverse osmosis (RO). *J. Membr. Sci.* **2010**, *365*, 34–39. [CrossRef]
112. Chen, S.; Li, L.; Zhao, C.; Zheng, J. Surface hydration: Principles and applications toward low-fouling/nonfouling biomaterials. *Polymer* **2010**, *51*, 5283–5293. [CrossRef]
113. Qin, D.; Liu, Z.; Bai, H.; Sun, D.D. Three-dimensional architecture constructed from a graphene oxide nanosheet–polymer composite for high-flux forward osmosis membranes. *J. Mater. Chem. A* **2017**, *5*, 12183–12192. [CrossRef]
114. Minier-Matar, J.; Hussain, A.; Santos, A.; Janson, A.; Wang, R.; Fane, A.G.; Adham, S. Advances in Application of Forward Osmosis Technology for Volume Reduction of Produced/Process Water from Gas-Field Operations. In Proceedings of the International Petroleum Technology Conference, Doha, Qatar, 6–9 December 2015.
115. Awad, A.M.; Jalab, R.; Nasser, M.S.; El-Naas, M.; Hussein, I.A.; Minier-Matar, J.; Adham, S. Evaluation of cellulose triacetate hollow fiber membrane for volume reduction of real industrial effluents through an osmotic concentration process: A pilot-scale study. *Environ. Technol. Innov.* **2021**, *24*, 101873. [CrossRef]
116. Sosa-Fernandez, P.A.; Post, J.W.; Ramdlan, M.S.; Leermakers, F.A.M.; Bruning, H.; Rijnaarts, H.H.M. Improving the performance of polymer-flooding produced water electrodialysis through the application of pulsed electric field. *Desalination* **2020**, *484*, 114424. [CrossRef]
117. Sosa-Fernandez, P.A.; Post, J.W.; Leermakers, F.A.M.; Rijnaarts, H.H.M.; Bruning, H. Removal of divalent ions from viscous polymer-flooding produced water and seawater via electrodialysis. *J. Membr. Sci.* **2019**, *589*, 117251. [CrossRef]
118. Kumar, P.; Mitra, S. Challenges in Uninterrupted Production and Supply of Gas to Mega LNG Trains. In Proceedings of the Internatioal Petroleum Technology Conference, Doha, Qatar, 20–22 January 2014.
119. Adham, S.; Gharfeh, S.; Hussain, A.; Minier-Matar, J.; Janson, A. Kinetic Hydrate Inhibitor Removal by Physical, Chemical and Biological Processes. In Proceedings of the Offshore Technology Conference—Asia, Kuala Lumpur, Malaysia, 18 March 2014.
120. Cobham, A.; Way, H. The treatment of oily water by coalescing. *Filtr. Sep.* **1992**, *29*, 295–300.
121. Al-Shamrani, A.A.; James, A.; Xiao, H. Destabilisation of oil–water emulsions and separation by dissolved air flotation. *Water Res.* **2002**, *36*, 1503–1512. [CrossRef]
122. Al-Ghouti, M.A.; Al-Degs, Y.S.; Khalili, F.I. Minimisation of organosulphur compounds by activated carbon from commercial diesel fuel: Mechanistic study. *Chem. Eng. J.* **2010**, *162*, 669–676. [CrossRef]
123. Al-Kaabi, M.A.; Al-Ghouti, M.A.; Ashfaq, M.Y.; Ahmed, T.; Zouari, N. An integrated approach for produced water treatment using microemulsions modified activated carbon. *J. Water Process. Eng.* **2019**, *31*, 100830. [CrossRef]
124. Upadhyayula, V.K.; Deng, S.; Mitchell, M.C.; Smith, G.B. Application of carbon nanotube technology for removal of contaminants in drinking water: A review. *Sci. Total Environ.* **2009**, *408*, 1–13. [CrossRef] [PubMed]

125. Ali, A.; Ali, K.; Kwon, K.R.; Hyun, M.T.; Choi, K.H. Electrohydrodynamic atomization approach to graphene/zinc oxide film fabrication for application in electronic devices. *J. Mater. Sci. Mater. Electron.* **2014**, *25*, 1097–1104. [CrossRef]
126. Fard, A.K.; Rhadfi, T.; Mckay, G.; Al-marri, M.; Abdala, A.; Hilal, N.; Hussien, M.A. Enhancing oil removal from water using ferric oxide nanoparticles doped carbon nanotubes adsorbents. *Chem. Eng. J.* **2016**, *293*, 90–101. [CrossRef]
127. Rasool, K.; Pandey, R.P.; Rasheed, P.A.; Buczek, S.; Gogotsi, Y.; Mahmoud, K.A. Water treatment and environmental remediation applications of two-dimensional metal carbides (MXenes). *Mater. Today* **2019**, *30*, 80–102. [CrossRef]
128. Ihsanullah, I. MXenes (two-dimensional metal carbides) as emerging nanomaterials for water purification: Progress, challenges and prospects. *Chem. Eng. J.* **2020**, *388*, 124340. [CrossRef]
129. Fard, A.K.; Mckay, G.; Chamoun, R.; Rhadfi, T.; Preud'Homme, H.; Atieh, M.A. Barium removal from synthetic natural and produced water using MXene as two dimensional (2-D) nanosheet adsorbent. *Chem. Eng. J.* **2017**, *317*, 331–342. [CrossRef]
130. Strømgren, T.; Sørstrøm, S.E.; Schou, L.; Kaarstad, I.; Aunaas, T.; Brakstad, O.G.; Johansen, O. Acute toxic effects of produced water in relation to chemical composition and dispersion. *Mar. Environ. Res.* **1995**, *40*, 147–169. [CrossRef]
131. Das, P.; Thaher, M.I.; Hakim, M.A.Q.M.A.; Al-Jabri, H.M.S.; Alghasal, G.S.H. A comparative study of the growth of Tetraselmis sp. in large scale fixed depth and decreasing depth raceway ponds. *Bioresour. Technol.* **2016**, *216*, 114–120. [CrossRef]
132. Fakhru'l-Razi, A.; Pendashteh, A.; Abdullah, L.C.; Biak, D.R.A.; Madaeni, S.S.; Abidin, Z.Z. Review of technologies for oil and gas produced water treatment. *J. Hazard. Mater.* **2009**, *170*, 530–551. [CrossRef]

Article

Assessment of the Effect of Irrigation with Treated Wastewater on Soil Properties and on the Performance of Infiltration Models

Ammar A. Albalasmeh [1,*], Ma'in Z. Alghzawi [1], Mamoun A. Gharaibeh [1] and Osama Mohawesh [1,2]

1 Department of Natural Resources and the Environment, Faculty of Agriculture, Jordan University of Science and Technology, Irbid 22110, Jordan; maen_alghazzawi.just@live.com (M.Z.A.); mamoun@just.edu.jo (M.A.G.); osama@mutah.edu.jo (O.M.)
2 Department of Plant Production, Faculty of Agriculture, Mutah University, Karak 61710, Jordan
* Correspondence: aalbalasmeh@just.edu.jo; Tel.: +962-2720-1000 (ext. 22050)

Abstract: An alternative strategy for saving limited water resources is using treated wastewater (TWW) originating from wastewater treatment plants. However, using TWW can influence soil properties owing to its characteristics compared to conventional water resources. Therefore, assessing the effect of TWW on soil properties and soil water infiltration is crucial to maintain sustainable use of TWW and to increase the water use efficiency of the precious irrigation water. Moreover, several studies were carried out to assess the performance of infiltration models. However, few studies evaluate infiltration models under the use of treated wastewater. Therefore, this study aims to assess the effect of TWW irrigation on soil properties after 2 and 5 years and to evaluate five classical infiltration models with field data collected from soil irrigated by treated wastewater for their capability in predicting soil water infiltration. This study revealed that using TWW for irrigation affects significantly on soil properties after 2 and 5 years. The soil irrigated with TWW had significantly higher electrical conductivity, organic matter, sodium adsorption ratio, cation exchange capacity, and lower soil bulk density compared to control. The basic infiltration rate and cumulative infiltration decreased significantly compared to control (60.84, 14.04, and 8.42 mm hr^{-1} and 140 mm, 72 mm, and 62 mm for control, 2, and 5 years' treatments, respectively). The performance of the infiltration models proposed by Philip, Horton, Kostiakov, Modified Kostiakov, and the Natural Resources Conservation Service was evaluated with consideration of mean error, root mean square error, model efficiency, and Willmott's index. Horton model had the lowest mean error (0.0008) and Philip model had the lowest root mean square error (0.1700) while Natural Resources Conservation Service had the highest values (0.0433 and 0.5898) for both mean error and root mean square error, respectively. Moreover, Philip model had the highest values of model efficiency and Willmott's index, 0.9994 and 0.9998, respectively, whereas Horton model had the lowest values for the same indices, 0.9869 and 0.9967, respectively. Philip model followed by Modified Kostiakov model were the most efficient models in predicting cumulative infiltration, while Natural Resources Conservation Service model was the least predictable model.

Keywords: evaluation; infiltration models; model efficiency; treated wastewater; Willmott's index agreement

1. Introduction

Arid and semi-arid regions are characterized by a short period and low amount of rainfall; therefore, water is a scarce commodity and considered as a limiting factor for agricultural production. This water shortage has compelled the decision-makers to look for other water sources. One of these sources is the treated wastewater (TWW) [1]. Under these conditions, persistent monitoring of the TWW and soil properties are required due to the variation in the TWW properties because of the raw water source as well as advancement of the treatment plant. In comparison to fresh water, TWW has higher content of organic matter as well as nutrients that are required for plant growth especially in arid soils [2]. However, depending on the source and treatment type, TWW may contain some elements that could affect soil and plant adversely [3–5]. One of the main problems of using TWW is the physical clogging by suspended solids and bioclogging facilitated by dissolved organic matter [6–8]. Therefore, understanding and quantifying water movement in the vadose zone is essential for identifying strategies of water conservation, flood or runoff control, erosion control, and assessment of potential aquifer contamination due to migration of water-soluble chemicals present in the vadose zone [9]. In irrigation science, infiltration is a vital and key dynamic process for agricultural activities to be considered for irrigation system design and optimization, irrigation scheduling, and irrigation management [10–12].

Infiltration models have been developed over the years ranging from empirical to physical-based models [13–15]. When these models are used to predict soil infiltration, only a few are satisfactory for field application; therefore, the presentation of variability of soil infiltration properties is a key problem [16]. In this context, new infiltration models have been developed by simplifying the assumptions of Richard's equation by Philip [14] and Green and Ampt [17]; however, the models have some strict restrictions such as soil distribution and ponding depth that imposed limitations on the practical applications [18–20]. Other limitations remain in the practical application of models developed by Horton [13] and Green and Ampt [17] because these models are applicable particularly when rainfall intensity is exceeding the initial water absorption capacity of the soil [21]. Such models will inaccurately predict soil infiltration characteristics in arid and semi-arid regions. On the other hand, multiple factors play a crucial role in the variation of soil infiltration, i.e., soil texture, topography, land use, bulk density, water content, biological activity, etc. [16]. Another factor playing a role in determination of the infiltration and its prediction is the suspended solid in the treated wastewater, which can clog the pores and reduce the infiltration [6,22]. This variation is not well presented by infiltration models. Moreover, the data used as an input for the infiltration models being obtained from field and laboratory experiments are affected by several conditions (soil texture, water quality, spatial heterogeneity, etc.). Hence, the effect of using TWW in irrigation on soil properties and the performance of infiltration models should be evaluated worldwide using field data to identify which infiltration model is best suited. Therefore, the objectives of this study were to (i) assess different infiltration models based on describing and predicting soil infiltration characteristics in areas irrigated with treated wastewater and (ii) evaluate the effect of using TWW in irrigation on soil properties at different periods.

2. Infiltration Models

Soil water infiltration models can be divided into three categories: (1) physically-based, (2) empirical, and (3) semi-empirical models. Physically-based models are approximate solutions of the Richards equation by Philip model. The derivation of these models depends on the considerations of flow dynamics, moisture content, hydraulic conductivity, and initial and boundary conditions [23]. Empirical models are built based on data obtained from field experiments such as Kostiakov model, and semi-empirical models that utilize the simple forms of the continuity equation such as Horton model. Horton and Green and Ampt models provide estimates of the infiltration capacities as a function of time, and they are the most commonly used models [12]. Furthermore, the popularity of these empirical models is due to their simplicity in various water resource applications and reliability in yielding satisfactory results [23]. Infiltration model parameters are typically determined before a rainfall-runoff model is used [24]. A brief description of each of these infiltration models that have been used in this study is given below.

2.1. Philip Model (PH)

Philip model [14] is based on a semi-analytical solution of Richards equation, which results in two term infiltration equation as:

$$I = St^{1/2} + At \tag{1}$$

where I is the cumulative infiltration at time t, S is the soil sorptivity, and A is constant; at long time, A is equal to soil saturated hydraulic conductivity (Ks).

2.2. Horton Model (HO)

Horton [13] presented an exponential infiltration model described as:

$$I = i_f t + \frac{i_o - i_f}{\beta}\left(1 - e^{-\beta t}\right) \tag{2}$$

where i_f is final infiltration rate, i_o is the initial infiltration rate at time (t = 0), and β is constant for soil and initial condition.

2.3. Kostiakov Model (KO)

Among the empirical models, Kostiakov model [15] was the first to propose an infiltration model in which cumulative infiltration can be described as:

$$I = kt^a \tag{3}$$

where k and a are constants and can be evaluated using the measured infiltration data. The limitation of this model is that it predicts zero infiltration rate at long time. However, it is more effective under a relatively short period of water application [9].

2.4. Modified Kostiakov Model (MK)

Due to the limitation of Kostiakov model at long time, it was modified by Smoth [25] through accounting for the final infiltration rate, described as:

$$I = kt^a + i_f t \tag{4}$$

The logic for adding the i_f is that the infiltration rate decreases as more water infiltrates into soil until a constant rate, known as ultimate infiltration capacity, is achieved.

2.5. NRCS Model (SCS)

Natural Resources Conservation Services (NRCS) modified Kostiakov Model to include a term for cracking, commonly known as SCS Model, described as:

$$I = kt^a + d \tag{5}$$

where d is equal to 0.6985 [26].

3. Materials and Methods

3.1. Experimental Site

The experiments were carried out at Jordan University of Science and Technology (JUST) in the province of Irbid, Jordan (32 27′ 57.4″ N latitude, 35 57′ 54.4″ E longitude). The soil is classified as fine, mixed, thermic Typic Calcixerert with 15% $CaCO_3$ content and clayey soil texture (clay 48%, silt 37 %, and sand 15%). Three plots (0.8 ha each) were placed under different experimental treatments. The first one was non-irrigated plot and served as control, referred to as 0 YR. The second plot was irrigated with treated wastewater (TWW) for two years, referred to as 2 YR. The third plot was irrigated with TWW for five years, referred to as 5 YR. Flood basin irrigation was used to irrigate the alfalfa planted in the plot at 2 days' interval, each plot received a total amount of 172 m^3 per week (equivalent to

21.5 mm per week). Soil bulk density (BD) was determined using the core method. Soil alkalinity (pH) and electrical conductivity (EC) were measured in saturated paste extract, respectively. Cation exchange capacity (CEC) and sodium adsorption ratio (SAR) were determined by ammonium acetate method. Organic matter (OM) was measured using the loss on ignition. Particle size analysis was measured using the hydrometer method.

3.2. Irrigation Water Quality

The TWW used in this study was provided from a wastewater treatment plant located in the JUST campus. TWW samples were analyzed for various physicochemical characteristics following the standard methods described by the American Public Health Association (APHA) [27] and the average values are presented in Table 1. Overall, most of the measured parameters were below the recommended maximum concentrations and within guidelines for irrigation of agricultural crops (JS893/2006).

Table 1. Mean values of selected properties of the used TWW.

Parameter	Value	TWWS	Unit
pH	7.8	9	-
EC	1.6	2.5	dSm^{-1}
Na	359.4	230	
K	37.7		
Ca	97.0	230	
Mg	24.8	100	
Cl	297.8	400	mgL^{-1}
DOM	70		
TSS	30	300	
TDS	1050	1500	
COD	25	100	

TWWS: Treated wastewater standard (JS 893/2006); EC: electrical conductivity; DOM: Dissolved organic matter; TSS: Total suspended solid; TDS: Total dissolved solid.

3.3. Infiltration Measurements

In situ soil infiltration tests were performed following Schwärzel and Punzel [28] using the Hood infiltrometer (IL-2700, Umwelt-Gerate-Technik GmbH, Muncheberg, Germany). Using the Hood infiltrometer does not require any preparation of the soil prior to the measurements. The soil infiltration measurements were performed in five replicates per each treatment at a random location within each site. The infiltration measurements started by applying 0 mm tension and increased by 20 mm steps after that, until reaching the bubbling point of the soil. Each infiltration measurement took approximately 8 min to reach the steady rate, after which the tension was increased to the next level.

3.4. Model Parameter Estimation

In order to compare the infiltration models, parameter values for each model needed to be determined. The model parameters were estimated using iterative non-linear regression procedure of Solver tool, which is a component of the Excel software for Microsoft office for MAC version 14.5.5. This method can best fit the model to the experimental data while the sum of squared error (SSE) is minimized. Squared differences between measured and predicted data of cumulative infiltration can be calculated as:

$$SSE = \sum_{i=1}^{n}(I(m)_i - I(p)_i)^2 \qquad (6)$$

where $I(m)_i$ and $I(p)_i$ are measured and predicted cumulative infiltration for i-th measurement, respectively, and n is the number of cumulative infiltration measurements. Note that we used the same method to estimate the parameters of the physical model instead of determining those parameters by field measurements due to the violation of the assumptions such as uniform soil properties and uniform initial water content [29,30].

3.5. Model Performance Evaluation

The goodness of fit of each model was tested by several performance indices to evaluate how closely each model describes the measures and predicts infiltration. Models that best fit the data according to many goodness of fit indices were considered superior to the others. The commonly used, accepted, and recommended indices in published literature [29,31–36] are addressed below.

3.5.1. Mean Error (ME)

The ME statistic shows whether the selected model overestimates or underestimates the measured data of cumulative infiltration. ME can be calculated as:

$$ME = \sum_{i=1}^{n} \frac{I(p)_i - I(m)_i}{n} \qquad (7)$$

Its absolute value should be as small as possible, the closer to zero, the better the model.

3.5.2. Root Mean Square Error (RMSE)

RMSE is used as an index of absolute error, it is always positive. It can be calculated as:

$$RMSE = \sqrt{\frac{\sum_{i=1}^{n}(I(p)_i - I(m)_i)^2}{n}} \qquad (8)$$

The smaller the RMSE (close to zero), the better the agreement between the predicted and measured data.

3.5.3. Model Efficiency (EF)

EF proposed by Nash and Sutcliffe [37] is defined as one minus the sum of the absolute squared differences between the measured and predicted cumulative infiltration values normalized by the variance of the measured values during the period of the experiment. It can be calculated as:

$$EF\% = 1 - \frac{\sum_{i=1}^{n}(I(m)_i - I(p)_i)^2}{\sum_{i=1}^{n}(I(m)_i - \overline{I(m)})^2} \qquad (9)$$

where $\overline{I(m)}$ is the mean of the measured data. EF equal to 1 means the predicted cumulative infiltration values are in perfect agreement with the measured cumulative infiltration values.

3.5.4. Willmott's Index Agreement (W)

W index reflects the degree to which the data measured are accurately estimated by the data predicted [38]. It can be calculated as:

$$W = 1 - \frac{\sum_{i=1}^{n}(I(p)_i - I(m)_i)^2}{\sum_{i=1}^{n}\left(\left|I(p)_i - \overline{I(m)}\right| + \left|I(m)_i - \overline{I(m)}\right|\right)^2} \qquad (10)$$

W equal to 1 means the predicted cumulative infiltration values are in perfect agreement with the measured cumulative infiltration values.

Finally, the mean values of ME, RMSE, EF, and W have been calculated to compare the models' performances for all the different TWW irrigated soils tested. Lower values of mean ME and RMSE, and higher values of EF and W are expected for the better infiltration models of the TWW irrigated soils in this research.

4. Results and Discussion

4.1. Effect of TWW on Soil Properties

Table 2 shows that the soil irrigated with TWW had significantly higher EC values, OM, SAR, and CEC. However, soil bulk density decreased under TWW irrigation (Table 2). The basic infiltration rate and cumulative infiltration were 60.84, 14.04, and 8.42 mm hr^{-1} and were 140, 72, and 62 mm for the 0 YR, 2 YR, and 5 YR treatments after 90 min experiments' period, respectively. These results propose that using TWW in irrigation could lead to a significant decrease in soil infiltration rate and cumulative infiltration, which may lead to higher runoff and soil erosion vulnerability. The infiltration rate and cumulated infiltration rate were significantly lower in the case of 2 YR and 5 YR treatments compared to control. This reduction in infiltration could be explained by the changes in pore size distribution (higher micropore content and lower macrospore) (Table 2) as a result of the total suspended solid and salts existing in the TWW [29,31–36] (Table 2). The soil irrigated with TWW had relatively higher total porosity (56.7, 55.7, and 59% for 0 YR, 2 YR, and 5 YR, respectively) as a result of increasing the organic matter content in the treated soil (Table 2). However, it had significantly lower wide coarse pores (wCP), and narrow coarse pores (nCP) at 2 YR treatment than that of control (0 YR). Even though the total porosity (TP) increased significantly and insignificant difference of wCP and nCP at 5 YR treatment compared to control (0 YR) (Table 3), the decrease in infiltration rate and cumulative infiltration at 5 YR treatment can be related to soil surface sealing, and accumulating of organic material and grease on the upper thin layer of soil surface [39,40].

Table 2. Mean values of soil chemical properties, infiltration rate, and accumulated infiltration for the three treatments.

Treatment	pH	ECe (dS m^{-1})	OM (%)	CEC (cmole$_{(+)}$ kg^{-1})	SAR	i_r (mm hr^{-1})	I_{cu} (mm)	TP (%)	wCP (>50 μm) (%)	nCP (50 to 10 μm) (%)
0 YR	6.9 b	0.7 b	2.77 c	32.49 b	0.75 c	60.84 a	140 a	56.7 b	3.6 a	8.3 a
2 YR	7.7 a	1.68 a	4.37 b	31.16 b	3.85 b	14.04 b	72 b	55.7 b	1.8 b	6.5 b
5 YR	7.4 a	2.09 a	7.19 a	33.44 a	6.39 a	8.42 c	62 c	59.0 a	4.1 a	9.0 a

ECe: saturated paste extract electrical conductivity; OM: organic matter; CEC: cation-exchange capacity (CEC); i_r: basic infiltration (90 min experiment); I_{cu}: accumulated infiltration at the end of the 90-min; TP: total porosity; wCP: wide coarse pores; nCP: narrow coarse pores. Values followed by the same letter are not significantly different.

Table 3. The estimated parameters of the five evaluated soil infiltration models for TWW irrigated soils at JUST.

		KO		MK			PH		HO		SCS	
		k	A	k	a	i_f	S	A	i_f	i_o	β	k	a
0 YR	Mean	0.3016	0.4287	0.5372	0.2236	0.0020	0.1010	0.0012	0.0023	0.1581	0.0786	0.2106	0.4656
	SE	0.0192	0.0060	0.0950	0.0123	0.0001	0.0084	0.0000	0.0001	0.0293	0.0077	0.0263	0.0086
2 YR	Mean	0.1214	0.4796	0.5237	0.2169	0.0007	0.0903	0.0001	0.0010	0.1535	0.0775	0.3439	0.6103
	SE	0.0224	0.0210	0.0628	0.0024	0.0000	0.0069	0.0001	0.0000	0.0343	0.0091	0.0878	0.0375
5 YR	Mean	0.1032	0.4796	0.4961	0.2132	0.0006	0.0854	0.0000	0.0009	0.1085	0.0617	0.0755	0.5005
	SE	0.0136	0.0131	0.0449	0.0028	0.0001	0.0075	0.0000	0.0001	0.0124	0.0027	0.0121	0.0278
	Overall Max	0.3449	0.5482	0.8999	0.2458	0.0023	0.1299	0.0020	0.0026	0.2709	0.1067	0.6719	0.7050
	Overall Min	0.0571	0.4145	0.3431	0.1779	0.0005	0.0657	0.0000	0.0008	0.0710	0.0544	0.0324	0.4200

KO: Kostiakov; MK: Modified Kostiakov; PH: Philip; HO: Horton; NRCS: Natural Resources Conservation Service; k, A, a, are constants; if: final infiltration rate; S: soil sorptivity; io: initial infiltration rate; β: constant for soil and initial condition; SE: standard Error; 0 YR: non-irrigated plot (control); 2 YR: plots irrigated with treated wastewater (TWW) for two years; 5 YR: plots irrigated with TWW for five years; R: replicate number.

4.2. Model Parameter

The values of the estimated parameters are listed in Table 3 for the selected infiltration models in this study. In addition, Table 3 shows the mean and standard error for each treatment, and for each model used in this study.

In most publications, Modified Kostiakov model is basically the Kostiakov model plus i_f t term [41,42]. However, the results presented in Table 3 showed that these parameters of the above-mentioned models are completely different. Similar conclusion was drawn by Dashtaki et al. [29] who recommended to not use similar parameters when these two models are used. Regarding the discussed models, one may expect that the values of i_f in Modified Kostiakov and Horton models, and A in Philip model would be essentially equal to the measured final infiltration rates. However, based on the results presented in Table 2, they are not comparable and this could be explained by the fact that these parameters are empirical in nature [29].

The estimated parameter value for each model was different for each measurement in each treatment. This could be caused by the heterogeneity of soil properties (initial soil water content, soil and water temperature, etc.) under field condition. Moreover, there is a clear decreasing trend in the values of the estimated parameters for each model with increasing the TWW irrigation period. For example, the k parameter in the KO model decreased from 0.3016 for the 0 YR treatment to 0.1214 and 0.1032 for the 2 YR and 5 YR treatments, respectively. Similarly, the i_f parameter in the MK model decreased from 0.0020 for the 0 YR treatment to 0.0007 and 0.0006 for the 2 YR and 5 YR treatments, respectively. These decreases in the infiltration parameters suggest that long-term application of TWW results in a decrease in the cumulative infiltration which could be explained by clogging of the small pores by the suspended materials loaded in the TWW [6,22,43]. In general, the reported values of the estimated parameter are different from those reported in the literature because of the differences in soil and water quality. In a study conducted by Fan et al. [44] to analyze the influence of soil texture, initial water content, film hole diameter, and water depth on cumulative infiltration, numerical simulations carried out with HYDRUS-2D showed that cumulative infiltration from a film hole was affected by each of soil texture, film hole diameter, and water depth. Sorptivity, (S) parameter in Philip model, increases with the film hole diameter for the same soil texture, and the relationship is a power function [44]. Wang et al. [45] stated that the capillary suction sometimes can be approximated by sorption, can control the early stage of infiltration process in Philip model. However, gravity plays a bigger role as time goes, and the second term in Philip model will have greater values [45].

4.3. Model Performance

The performance of each infiltration model was evaluated based on the value of mean error (ME), root mean square error (RMSE), model efficiency (EF), and Willmott's index agreement (W). The results of the evaluation indices for the five selected soil infiltration models are presented in Table 4. Horton model had lowest ME followed by Philip model where NRCS model had the highest ME, and was found to be the worst in describing the infiltration in clayey soil irrigated by treated wastewater. Similar to the results presented by Li et al. [46], Horton model ranked number 1 with the lowest ME value compared to other models. Note that all models over-estimated the cumulative infiltration since the means of ME were positive (Table 4).

The theoretically-based Philip model was ranked number 1 and found to be the most predictive model considering RMSE followed by Modified Kostiakov model. As Machiwal et al. [11] concluded from a number of experiments conducted in a wasteland of Kharagpur, the best model to describe variability of infiltration process was Philip model [16]. NRCS model had the highest RMSE as the poorest model in predicting infiltration. In agreement to these predictions, Duan et al. [31] reported that NRCS model was the poorest in predicting the infiltration considering the RMSE. In contrast to our results, Zolfaghari et al. [35] reported that Modified Kostiakov model results in lowest RMSE and ranked as the best model to describe the soil infiltration. Additionally, Nie et al. [16] concluded that according to RMSE and R^2 results, Kostiakov–Lewis model was the best in predicting cumulative infiltration, and Philip model was the worst in prediction. These overall rankings are similar to within the treatments' ranking except that for Kostiakov model. It has an overall ranking of 4 considering both ME and RMSE but for the third treatment (soil irrigated with TWW for five years), it ranked number 2 and 1 considering ME and RMSE, respectively.

Table 4. Evaluation indices of the five evaluated soil infiltration models for TWW irrigated soils at JUST.

Model	Index	0 YR			2 YR			5 YR			Overall Mean	Overall Rank
		Mean	SE	Rank	Mean	SE	Rank	Mean	SE	Rank		
KO	ME	0.0866	0.0491	5	0.0305	0.0239	5	0.0034	0.0036	2	0.0402	4
	RMSE	1.1747	0.0455	4	0.1782	0.0356	2	0.1851	0.0186	1	0.5127	4
	EF	0.9011	0.0085	4	0.9879	0.0048	2	0.9850	0.0017	1	0.9959	2
	W	0.9656	0.0035	4	0.9967	0.0014	2	0.9960	0.0005	1	0.9990	2
MK	ME	0.0223	0.0132	3	0.0268	0.0028	3	0.0196	0.0066	4	0.0229	3
	RMSE	0.3108	0.0216	2	0.2906	0.0178	4	0.3236	0.0313	3	0.3083	2
	EF	0.9932	0.0004	2	0.9711	0.0030	4	0.9545	0.0031	3	0.9946	3
	W	0.9983	0.0001	2	0.9923	0.0008	4	0.9876	0.0009	3	0.9986	3
PH	ME	0.0206	0.0021	2	−0.0035	0.0142	2	0.0037	0.0079	3	0.0069	2
	RMSE	0.1438	0.0154	1	0.1737	0.0203	1	0.1925	0.0192	2	0.1700	1
	EF	0.9985	0.0003	1	0.9891	0.0026	1	0.9835	0.0025	2	0.9994	1
	W	0.9996	0.0001	1	0.9973	0.0006	1	0.9958	0.0006	2	0.9998	1
HO	ME	0.0004	0.0001	1	0.0006	0.0001	1	0.0013	0.0002	1	0.0008	1
	RMSE	0.4695	0.0307	3	0.4378	0.0385	5	0.4588	0.0434	5	0.4554	3
	EF	0.9846	0.0007	3	0.9334	0.0102	5	0.9083	0.0071	5	0.9869	5
	W	0.9961	0.0002	3	0.9824	0.0028	5	0.9753	0.0021	5	0.9967	5
NRCS	ME	0.0555	0.0403	4	0.0271	0.0022	4	0.0472	0.0277	5	0.0433	5
	RMSE	1.2261	0.0623	5	0.2099	0.0097	3	0.3334	0.0382	4	0.5898	5
	EF	0.8943	0.0032	5	0.9849	0.0016	3	0.9434	0.0188	4	0.9881	4
	W	0.9627	0.0014	5	0.9960	0.0004	3	0.9815	0.0074	4	0.9969	4

KO: Kostiakov; MK: Modified Kostiakov; PH: Philip; HO: Horton; NRCS: Natural Resources Conservation Service; ME: mean error; RMSE: root mean square error; EF: model efficiency; W: Willmott's index; SE: standard Error; 0 YR: non-irrigated plot (control); 2 YR: plots irrigated with treated wastewater (TWW) for two years; 5 YR: plots irrigated with TWW for five years; R: replicate number.

The results of model efficiency (EF) and Willmott's index agreement (W) evaluation for the five selected soil infiltration models are presented in Table 4. The overall ranking of both parameters (EF and W) were similar for each model of the evaluated models. Philip model had highest model efficiency (EF) as well as Willmott's index agreement (W), where the values was 0.9994 and 0.9998, respectively. Kostiakov model followed Philip model, and Horton model was the worst in prediction cumulative infiltration with lowest EF and W values (Table 4).

Similar to ME and RMSE evaluation indices, Kostiakov model had the highest EF and W values and ranked number 1 for the third treatment (soil irrigated with TWW for five years). In agreement with our results, Li et al. [46] had drawn a similar conclusion where both indices, EF and W, had similar ranking for all evaluated models. However, for the three models they evaluated, the superior model was Kostiakov model followed by Horton model, and the worst model in predicting soil infiltration was Philip model, whereas our ranking for the same models was Philip model followed by Kostiakov model and then Horton model. In another study, Jejurkar and Rajurkar [12] found that the models of Kostiakov and modified Kostiakov provide better fit to the measured cumulative infiltration in clayey soil with R^2 values 0.959 and 0.964 for Kostiakov and modified Kostiakov models, respectively.

In order to evaluate the overall performance of the five selected soil infiltration models considering all indices used in this study, the model ranking was summed up as final scores (Table 5).

Table 5. Overall ranking of the evaluated soil infiltration models for TWW irrigated soils at JUST.

Model	KO	MK	PH	HO	NRCS
Final Scores	12	11	5	14	18
Final Ranking	3	2	1	4	5

KO: Kostiakov; MK: Modified Kostiakov; PH: Philip; HO: Horton; NRCS: Natural Resources Conservation Service.

The final scores presented in Table 5 are the sum of the ranking number for each index (ME, RMSE, EF, and W) of each model. Philip model was superior over all models to describe and predict cumulative infiltration followed by Modified Kostiakov model, and the worst model was the NRCS model in predicting cumulative infiltration for the soil irrigated by treated wastewater. In contrast to these results, Dashtaki et al. [29] reported a superior performance for Horton model whereas Mirzaee et al. [36] reported a superior performance for Modified Kostiakov model. In another study, field experiments were conducted on a sandy soil to test the predictability of four infiltration models (Horton, Kostiakov, Philip, modified Kostiakov) for cumulative infiltration under local conditions by Ogbe et al. [24]. In terms of accuracy, the researchers found that cumulative infiltration was accurately predicted by Horton, followed by Kostiakov, then Philip and modified Kostiakov models. In agreement to our conclusion, Zolfaghari et al. [35] reported that NRCS model had the lowest ranking between all models to predict and describe soil infiltration. However, Al Maimuri [21] stated that Horton model has the least predictability and applicability among other three models tested in arid region. This result was attributed to the fact that Horton model is inefficient in predicting infiltration characteristics in regions where initial rainfall intensities are less than initial infiltration capacity of soils.

Based on the results of Table 5, Philip model was ranked number 1. Therefore, it has been used to plot the cumulative infiltration (Figure 1) based on the estimated parameters (Table 3). The results showed that at the end of the 90 min experiments, the cumulative infiltrations were 140 mm, 72 mm, and 63 mm for the 0 YR, 2 YR, and 5 YR treatments, respectively. These results suggest that long-term application of TWW results in significant decrease in the cumulative infiltration which may lead to susceptibility of more runoff and soil erosion [6,22,43]. This conclusion can be supported by the decrease in the sorptivity (S) parameter (Table 3), with increase in the TWW irrigation period which could be explained by clogging of the small pores.

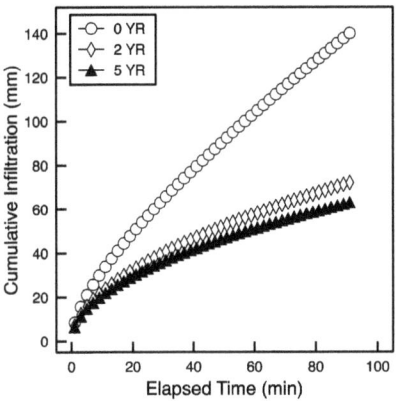

Figure 1. Cumulative infiltration curves fitted to Phillip's model parameters for the three experimental plots subjected to different irrigation treatments (0YR, 2YR, and 5YR).

5. Conclusions

Infiltration data of clayey soil irrigated with treated wastewater (TWW) were fitted to assess and evaluate Philip, Horton, Kostiakov, Modified Kostiakov, and NRCS models. Among the five selected infiltration models, Philip model and Modified Kostiakov models performed the best fitting results based on mean error, root mean square error, model efficiency, and Willmott's index of agreement. Philip model performed consistently well in all treatment, whereas the performance of Kostiakov model varied within the treatments, while NRCS model was the least predictable model. Moreover, using treated wastewater significantly decreased soil infiltration via accumulation of suspended solids and clogging of soil pores. Philip model captured best the adjustment to real field conditions and therefore, it could be recommended for the prediction of TWW infiltration in clayey soil.

Author Contributions: Conceptualization, investigation, methodology, writing—original draft preparation and writing—review and editing, A.A.A., M.A.G., M.Z.A. and O.M. All authors have read and agreed to the published version of the manuscript.

Funding: This research was funded by the Deanship of Research at the Jordan University of Science and Technology, grant number 76/2013.

Institutional Review Board Statement: Not applicable.

Informed Consent Statement: Not applicable.

Data Availability Statement: The authors confirm that the data supporting the findings of this study are available within the article.

Conflicts of Interest: The authors declare no conflict of interest.

References

1. Gharaibeh, M.A.; Eltaif, N.I.; Al-Abdullah, B. Impact of Field Application of Treated Wastewater on Hydraulic Properties of Vertisols. *Water Air Soil Pollut.* **2007**, *184*, 347–353. [CrossRef]
2. Makhadmeh, I.M.; Gharaiebeh, S.F.; Albalasmeh, A.A. Impact of Irrigation with Treated Domestic Wastewater on Squash (Cucurbita pepo L.) Fruit and Seed under Semi-Arid Conditions. *Horticulturae* **2021**, *7*, 226. [CrossRef]
3. Mohammad, M.J.; Mazahreh, N. Changes in Soil Fertility Parameters in Response to Irrigation of Forage Crops with Secondary Treated Wastewater. *Commun. Soil Sci. Plant Anal.* **2003**, *34*, 1281–1294. [CrossRef]
4. Ofori, S.; Puškáčová, A.; Růžičková, I.; Wanner, J. Treated wastewater reuse for irrigation: Pros and cons. *Sci. Total Environ.* **2021**, *760*, 144026. [CrossRef]
5. Lado, M.; Ben-Hur, M. Effects of Irrigation with Different Effluents on Saturated Hydraulic Conductivity of Arid and semiarid Soils. *Soil Sci. Soc. Am. J.* **2010**, *74*, 23–32. [CrossRef]
6. Gharaibeh, M.A.; Ghezzehei, T.A.; Albalasmeh, A.A.; Alghzawi, M.Z. Alteration of physical and chemical characteristics of clayey soils by irrigation with treated waste water. *Geoderma* **2016**, *276*, 33–40. [CrossRef]
7. Li, Y.; Peng, L.; Li, H.; Liu, D. Clogging in subsurface wastewater infiltration beds: Genesis, influencing factors, identification methods and remediation strategies. *Water Sci. Technol.* **2021**, *83*, 2309–2326. [CrossRef]
8. Levy, G.J.; Assouline, S. Physical aspects. In *Treated Wastewater in Agriculture Use and Impacts on the Soil Environment and Crops*; Levy, G.J., Fine, P., Bar-Tal, A., Eds.; Wiley-Blackwell: Hoboken, NJ, USA, 2011.
9. Ravi, V.; Williams, J.R. *Estimation of Infiltration Rate in the Vadose Zone: Compilation of Simple Mathematical Models—Volume I*; US Environmental Protection Agency: Washington, DC, USA, 1998.
10. Walker, W.R.; Prestwich, C.; Spofford, T. Development of the revised USDA—NRCS intake families for surface irrigation. *Agric. Water Manag.* **2006**, *85*, 157–164. [CrossRef]
11. Machiwal, D.; Jha, M.K.; Mal, B.C. Modelling Infiltration and quantifying Spatial Soil Variability in a Wasteland of Kharagpur, India. *Biosyst. Eng.* **2006**, *95*, 569–582. [CrossRef]
12. Jejurkar, C.L.; Rajurkar, M.P. An investigational approach for the modelling of infiltration process in a clay soil. *KSCE J. Civ. Eng.* **2015**, *19*, 1916–1921. [CrossRef]
13. Horton, R.E. An Approach Toward a Physical Interpretation of Infiltration-Capacity. *Soil Sci. Soc. Am. J.* **1940**, *5*, 399–417. [CrossRef]
14. Philip, J.R. The theory of infiltration: 2. the profile of infinity. *Soil Sci.* **1957**, *83*, 435–448. [CrossRef]
15. Kostiakov, A.N. On the Dynamics of the Coefficients of Water Percolation in Soils and on the Necessity of Studying It from a Dynamic Point of View for Purpose of Amelioration. *Trans. 6th Comm. Int. Soc. Soil Sci. Russ. Part A* **1932**, *1*, 17–21.
16. Nie, W.; Ma, X.; Fei, L. Evaluation of Infiltration Models and Variability of Soil Infiltration Properties at Multiple Scales. *Irrig. Drain.* **2017**, *66*, 589–599. [CrossRef]

17. Heber Green, W.; Ampt, G.A. Studies on Soil Phyics. *J. Agric. Sci.* **1911**, *4*, 1–24. [CrossRef]
18. Walter, R. Calibration of selected infiltration equations for the Georgia Coastal Plain. U.S. Agric. Res. Service. S. Region. ARS-S. 1976, 113, 14–15. *U.S. Agric. Res. Service. S. Region. ARS-S* **1976**, *113*, 14–15.
19. Deng, P.; Zhu, J. Analysis of effective Green–Ampt hydraulic parameters for vertically layered soils. *J. Hydrol.* **2016**, *538*, 705–712. [CrossRef]
20. Lei, G.; Fan, G.; Zeng, W.; Huang, J. Estimating parameters for the Kostiakov-Lewis infiltration model from soil physical properties. *J. Soils Sediments* **2020**, *20*, 166–180. [CrossRef]
21. Al Maimuri, N.M.L. Applicability of Horton model and recharge evaluation in irrigated arid Mesopotamian soils of Hashimiya, Iraq. *Arab. J. Geosci.* **2018**, *11*, 610. [CrossRef]
22. Bardhan, G.; Russo, D.; Goldstein, D.; Levy, G.J. Changes in the hydraulic properties of a clay soil under long-term irrigation with treated wastewater. *Geoderma* **2016**, *264*, 1–9. [CrossRef]
23. Parhi, P.K.; Mishra, S.K.; Singh, R. A Modification to Kostiakov and Modified Kostiakov Infiltration Models. *Water Resour. Manag.* **2007**, *21*, 1973–1989. [CrossRef]
24. Ogbe, V.B.; Jayeoba, O.J.; Ode, S.O. Comparison of Four Soil Infiltration Models on A Sandy Soil in Lafia, Southern Guinea Savanna Zone of Nigeria. *Prod. Agric. Technol. Nasarawa State Univ. J.* **2011**, *7*, 116–126.
25. Smith, R.E. The infiltration envelope: Results from a theoretical infiltrometer. *J. Hydrol.* **1972**, *17*, 1–22. [CrossRef]
26. Cuenca, R. *Irrigation System Design: An Engineering Approach*; Prentice Hall: Englewood Cliffs, NJ, USA, 1989.
27. APHA. *Standard Mehods for the Exmnination of Water and Wastewater*, 18th ed.; American Public Health Association: Washington DC, USA, 1992.
28. Schwärzel, K.; Punzel, J. Hood infiltrometer—A new type of tension infiltrometer. *Soil Sci. Soc. Am. J.* **2007**, *71*, 1438–1447. [CrossRef]
29. Dashtaki, S.G.; Homaee, M.; Mahdian, M.H.; Kouchakzadeh, M. Site-dependence performance of infiltration models. *Water Resour. Manag.* **2009**, *23*, 2777–2790. [CrossRef]
30. Dashtaki, S.G.; Homaee, M.; Loiskandl, W. Towards using pedotransfer functions for estimating infiltration parameters. *Hydrol. Sci. J.* **2016**, *61*, 1477–1488. [CrossRef]
31. Duan, R.; Fedler, C.B.; Borrelli, J. Field evaluation of infiltration models in lawn soils. *Irrig. Sci.* **2011**, *29*, 379–389. [CrossRef]
32. Mishra, S.K.; Tyagi, J.V.; Singh, V.P. Comparison of infiltration models. *Hydrol. Process.* **2003**, *17*, 2629–2652. [CrossRef]
33. Parhi, P.K. Another look at Kostiakov, modified Kostiakov and revised modified Kostiakov infiltration models in water resources applications. *Int. J. Agric. Sci.* **2014**, *4*, 138–142.
34. Razzaghi, S.; Khodaverdiloo, H.; Dashtaki, S.G. Effects of long-term wastewater irrigation on soil physical properties and performance of selected infiltration models in a semi-arid region. *Hydrol. Sci. J.* **2016**, *61*, 1778–1790. [CrossRef]
35. Zolfaghari, A.A.; Mirzaee, S.; Gorji, M. Comparison of different models for estimating cumulative infiltration. *Int. J. Soil Sci.* **2012**, *7*, 108–115. [CrossRef]
36. Mirzaee, S.; Zolfaghari, A.A.; Gorji, M.; Dyck, M.; Ghorbani Dashtaki, S. Evaluation of infiltration models with different numbers of fitting parameters in different soil texture classes. *Arch. Agron. Soil Sci.* **2014**, *60*, 681–693. [CrossRef]
37. Nash, J.E.; Sutcliffe, J. V River flow forecasting through conceptual models part I—A discussion of principles. *J. Hydrol.* **1970**, *10*, 282–290. [CrossRef]
38. Willmott, C.J. On the validation of models. *Phys. Geogr.* **1981**, *2*, 184–194. [CrossRef]
39. Mohawesh, O.; Mahmoud, M.; Janssen, M. Effect of irrigation with olive mill wastewater on soil hydraulic and solute transport properties. *Int. J. Environ. Sci. Technol.* **2014**, *11*, 927–934. [CrossRef]
40. Mohawesh, O.; Janssen, M.; Maaitah, O.; Lennartz, B. Assessment the effect of homogenized soil on soil hydraulic properties and soil water transport. *Eurasian Soil Sci.* **2017**, *50*, 1077–1085. [CrossRef]
41. Sepaskhah, A.R.; Afshar-Chamanabad, H. Determination of Infiltration Rate for Every-other Furrow Irrigation. *Biosyst. Eng.* **2002**, *82*, 479–484. [CrossRef]
42. Holzapfel, E.A.; Jara, J.; Zuñiga, C.; Mariño, M.A.; Paredes, J.; Billib, M. Infiltration parameters for furrow irrigation. *Agric. Water Manag.* **2004**, *68*, 19–32. [CrossRef]
43. Levy, G.J.; Fine, P.; Goldstein, D.; Azenkot, A.; Zilberman, A.; Chazan, A.; Grinhut, T. Long term irrigation with treated wastewater (TWW) and soil sodificatio. *Biosyst. Eng.* **2014**, *128*, 4–10. [CrossRef]
44. Fan, Y.; Gong, J.; Wang, Y.; Shao, X.; Zhao, T. Application of Philip infiltration model to film hole irrigation. *Water Supply* **2019**, *19*, 978–985. [CrossRef]
45. Wang, K.; Yang, X.; Li, Y.; Liu, C.; Guo, X. An Application of Chaos Gray-encoded Genetic Algorithm for Philip Infiltration Model. *Therm. Sci.* **2018**, *22*, 1581–1588. [CrossRef]
46. Li, G.; Feng, Q.; Zhang, F.; Cheng, A. Research on the infiltration processes of lawn soils of the Babao River in the Qilian Mountain. *Water Sci. Technol.* **2014**, *70*, 577–585. [CrossRef]

Article

Operation and Performance of Austrian Wastewater and Sewage Sludge Treatment as a Basis for Resource Optimization

Arabel Amann [1,*,†], Nikolaus Weber [1,†], Jörg Krampe [1], Helmut Rechberger [2], Ottavia Zoboli [1] and Matthias Zessner [1]

[1] Institute for Water Quality and Resource Management, Research Unit of Water Quality Management, TU Wien, Karlsplatz 13/E226-1, 1040 Vienna, Austria; nikolaus.weber@tuwien.ac.at (N.W.); joerg.krampe@tuwien.ac.at (J.K.); ozoboli@iwag.tuwien.ac.at (O.Z.); matthias.zessner-spitzenberg@tuwien.ac.at (M.Z.)

[2] Institute for Water Quality and Resource Management, Research Unit of Waste and Resource Management, TU Wien, Karlsplatz 13/E226-1, 1040 Vienna, Austria; helmut.rechberger@tuwien.ac.at

* Correspondence: arabel.amann@tuwien.ac.at; Tel.: +43-1-58801-22632

† These authors contributed equally to this work.

Abstract: Recent years came with a paradigm shift for wastewater treatment plants (WWTPs) to extend the sole purpose of contaminant removal to an additional function as resource recovery facilities. This shift is accompanied by the development of new European legislation towards better inclusion of resource recovery from wastewater. However, long operational lifespans and a multitude of treatment requirements demand thorough investigations into how resource recovery can be implemented sustainably. To aid the formulation of new legislation for phosphorus (P) recovery specifically, in 2017 we conducted a survey on Austrian WWTP-infrastructure, with a focus on P removal and sludge treatment, as well as disposal and sludge quality of all WWTPs above 2000 population equivalents (PE). Data were prepared for analysis, checked for completeness and cross-checked for plausibility. This study presents the major findings from this database and draws essential conclusions for the future recovery of P from wastewater. We see results from this study as useful to other countries, describing the current state of the art in Austria and potentially aiding in developing wastewater treatment and P recovery strategies.

Keywords: wastewater survey; phosphorus removal; phosphorus recovery; sludge stabilization; sludge disposal; sludge quality; sludge production

1. Introduction

While conventional wastewater treatment with the activated sludge system (mechanical, biological and chemical) was well established by the end of the 1990s [1], new issues have risen to the center of attention in the 21st century. The reduction of energy consumption [2,3], stricter effluent quality requirements for carbon and nutrients [4], the recovery of resources [5,6], wastewater reuse [7,8] and the removal of contaminants of emerging concern [9] and of antibiotic resistant genes [10,11] are only some of the new challenges to be addressed. This multitude of new tasks will in some cases not support themselves economically and might have negative side effects on the environment due to higher energy and/or material demands [12]. In addition, the primary function of wastewater treatment plants (WWTPs), the safe and cost-effective removal of major contaminants, should not be impeded by future task expansions or treatment changes. New legal requirements and plant configurations, therefore, need careful and thorough planning, best based on detailed information about the current status, infrastructure and performance of WWTPs.

The Federal Ministry of the Republic of Austria for Climate Action, Environment, Energy, Mobility, Innovation and Technology is currently underway in designing a new directive for the recovery of phosphorus (P) from municipal wastewater [13]. Eligibility of

WWTPs for P recovery is dependent on a multitude of design parameters and on sludge treatment infrastructure. Of high importance for the P recovery potential is the type of P removal (enhanced biological P removal (EBPR), or chemical with iron and aluminum) affecting its bio-availability in sludge [6] and sewage sludge ash [14]. For on-site recovery through struvite precipitation, higher P-concentrations in sludge will further positively affect the efficiency of the recovery units [15]. Recovery of P from sewage sludge ash is mainly competing with high concentrations in P raw materials (phosphate rock) of generally above 9% P [16]. Since incineration will leave only inorganic sludge components, sludge-P concentrations alone will not be a good predictor for sludge ash-P concentrations in these cases. Information on the inorganic sludge content is, therefore, a necessity. Next to P, heavy metal concentrations will also determine the suitability of sludge ash for P-recovery [17,18].

For WWTPs that are better suited for the extraction of P from sewage sludge ash (e.g., with no EBPR), recovery will in most cases require prior mono-incineration of their sewage sludge, a practice that is yet uncommon in Austria [13] and generally more expensive than co-incineration [19]. Therefore, additional costs due to legal changes towards P recovery requirements will depend largely on the current costs for sludge disposal, on sludge transport distances as well as on sludge production amounts.

To aid in the formulation of new legislation, in 2017 we commenced an extensive survey of the Austrian wastewater and sludge treatment system. We then established a comprehensive database through the help of a multitude of WWTP operators and federal state authorities. In accordance with the mentioned factors affecting P recovery, it covers information on wastewater treatment design, P removal and sludge stabilization methods, sewage sludge production, water content and quality, as well as on sludge disposal. This study presents the major findings from this database and draws essential conclusions for the future recovery of P from wastewater. We see results from this study as useful to other countries, describing the current state of the art in Austria and potentially aiding in developing wastewater treatment and P recovery strategies.

2. Materials and Methods

In 2017, a survey (see Supplementary Material 1) on Austrian WWTP-infrastructure and P-removal, as well as on sludge treatment, disposal and quality was sent out through the Austrian Water and Waste Management Association to all WWTPs with a design capacity above 2000 population equivalents (PE, 1 PE equals cumulative oxygen demand (COD) load of 120 g per day). WWTPs below 2000 PE were excluded from the analysis, as information on these small-scale plants is generally scarce due to less stringent or non-existent effluent criteria [20]. On the other hand, as WWTP above 20,000 PE treat approximately 95% of the COD load and 86% of the P-load in Austria (see Table 2), obtaining results from these plants was set as a priority. In addition, all nine federal states authorities of Austria were contacted for auxiliary data to advance the completeness of the data basis.

Data on WWTP loads were obtained from the national Austrian emission inventory on surface water bodies [21] based on the data collection requirements of the Austrian water act [22]. At the time of survey preparation in 2017 the general year of interest was set for 2016, because it was the year for which the most updated and complete information would be likely available. However, in some cases, data from 2015 and 2017 were included as well if no data were available for 2016. All data were grouped based on WWTP design capacity PE, although for comparison of resource use and production, actual yearly PE loads were used.

Data were then prepared for analysis, checked for completeness and cross-checked for plausibility by comparing values of each parameter within its group and with other available parameters (e.g., sludge stabilization method). For data analysis, the GUI *R Studio* and programming language *R* was used (Version 4.1.1). Data distributions are plotted using R package *ggplot*. The middle horizontal line gives the median, the upper and lower "hinges" correspond to the 25th and 75th percentiles. Upper and lower whiskers reach to the highest and lowest values within a distance of 1.5 times the interquartile range (distance of 25th and 75th percentile) starting from the upper and lower hinges, respectively.

Pearson correlation method was used to check for linear correlation between two normally distributed variables. A two-way ANOVA was applied to test for significant differences in means of a quantitative dependent variable according to the levels of a related categorical independent variable. Statistical tests were considered significant at $p < 0.05$.

3. Results and Discussion
3.1. Data Availability

Data availability for WWTPs with a design capacity bigger than 20,000 PE was high and generally exceeded 50% availability for most parameters (see Table 1). Only sludge quality data (loss on ignition (LOI) and P content) was available for less than 50% of these plants. As WWTPs between 2000 and 20,000 PE were not the primary focus of this study and plant count is generally much higher, data availability for these plants was much lower. For all parameters, only 35% of plants or lower provided any data. However, since most wastewater is treated in WWTPs above 20,000 PE (Table 2) and a decent number of observations was still achieved for WWTPs below 20,000 PE, this dataset is seen as highly representative of the current Austrian status quo.

Table 1. Recovered data and data availability after data curation for wastewater treatment plants (WWTPs) bigger than 20,000 and smaller than 20,000 PE.

Data Availability in % of N	<20,000 PE (N = 439)	>20,000 PE (N = 194)
EBPR yes/no	11	78
Chemical P-removal yes/no	32	86
Flocculating agent demand	7	52
Primary clarifier yes/no	14	86
Sludge stabilisation method	35	99
Sludge production (dry matter)	35	100
Sludge production (wet)	32	96
Loss on ignition (LOI) of sewage sludge	4	42
Sludge dewatering yes/no	23	92
Type of sludge dewatering unit	18	86
Sludge dry matter content	3	76
Sludge drying yes/no	15	84
Type of sludge dryer	1	84
P content of sewage sludge	3	38
Heavy metal content of sewage sludge	30	60

Table 2. Share of treated COD and P, and distribution of precipitating agents according to the size class of WWTPs in percent of population equivalent (PE) loads.

	2000 to 20,000 PE	>20,000 to 50,000 PE	>50,000 PE 100,000 PE	>100,000 PE
Total number of WWTPs	439	125	33	36
Share of treated COD in %	5	19	10	66
Share of treated P in %	14	19	9	58
Precipitating agent; n = in % of PE load	114	97	25	28
Aluminium	22	20	9	6
Iron	49	47	59	89
Aluminium/Iron Mix	19	28	27	4
Others	10	5	6	0

3.2. Phosphorus Removal

Table 2 provides an overview on the obtained data for P and COD removal. Information is given separately for four size groups of WWTP design capacities: (1) smaller than 20,000 PE, (2) between 20,000 and 50,000 PE, (3) between 50,000 and 100,000 PE, and (4) larger than 100,000 PE. The 36 largest plants of more than 100,000 PE treat the major share of COD (66%) and P (58%).

Austrian WWTPs remove 90% of all P from wastewater influent [23]. All surveyed WWTPs (n = 308) have chemical P removal in place. In addition, around 30% state that they also have an anaerobic tank for EBPR installed. Iron is the primarily used precipitating agent (77% of total PE load) in all size groups (Table 2), and its probability of use increases with larger WWTP sizes. Aluminum is applied to treat around 10% of PE load. It is more commonly used in WWTPs below 50,000 PE (~21%). Aluminum–iron mixes treat 11% of PE load and are used below 100,000 PE (19–27%). Other agents like lime are rarely applied (2% of PE load), and if, only at plants <100,000 PE.

For P-recovery or direct use of sludge in agriculture, EBPR would be the preferred method of choice, enhancing the bio-availability of P in sewage sludge [24]. It is, however, questionable if iron-P-removal can be fully replaced, especially in larger WWTPs with anaerobic digestion, as it additionally functions in sulfide (odor) control in anaerobic digestors [25]. This is additionally supported by higher use of iron in WWTPs with anaerobic sludge treatment (92%) than aerobic treatment (80% of PE load; see Table S1).

The fact that all Austrian WWTPs with EBPR use at least little additional chemical dosing shows that EBPR-performance is limited in Austrian treatment plants. Therefore, chemical precipitant use might be reduced with the application of EBPR, but never fully redundant. A switch to EBPR will also come with additional operational tasks for plant operators, most importantly the prevention of scaling of pipes through uncontrolled P precipitation as struvite [26]. Controlling this struvite precipitation would be beneficial, as it would create more easily accessible P forms for recovery both on-site as well as from ash [27], but it is unclear if plant operators can be persuaded to take on these additional challenges.

For a better understanding of precipitating agent use, the declared demand by WWTP operators calculated as mol per year was plotted against the theoretical demand (Figure S1). Theoretical demand was estimated from P inflow minus P effluent loads, subtracting P demand for biomass production (1% of biological oxygen demand loads) and P removal if a primary clarifier is present. Assumed β values (mol of agent dosed per mol of P) for precipitation were set at 1.2 for effluent limits of 2 mg L^{-1}, at 1.5 for a limit of 1 mg L^{-1} and at 2.5 for precipitation after the secondary clarifier if stringent limits of 0.5 mg L^{-1} are set. In general, the derived theoretical precipitating agent demand from P and BOD WWTP loads is a good predictor for actual agent use (r = 0.83, p-value < 0.005). We further analyzed the calculated ß-values from plants with a limit p value of 1 mg L^{-1} (n = 120) and with and without EBPR (Figure 1). Median demand per mol of P precipitated was significantly different for those two groups (t (113) = −2.6589; p = 0.009), with plants with EBPR having an mean reduction in demand of 18%.

3.3. Primary Clarification, Sludge Production and Stabilization

Sewage treatment in Austria occurs mainly by the activated sludge system. Conventional aerobic sludge stabilization is mostly used in WWTPs smaller than 20,000 PE, while anaerobic sludge stabilization via digestion is the more frequently applied method in WWTPs larger than 20,000 PE (Table 3). Other designs are rare, with most cases being sequencing batch reactors in smaller WWTPs (largest SBR plant with 60,000 PE). Out of the plants with aerobic sludge stabilization, simultaneous stabilization (sludge age of more than 25 days) is most common, comprising ~73% of aerobically stabilized sludge. Separate aerobic stabilization has its highest share (28% of aerobically stabilized sludge) in plants between 20,000 and 100,000 PE.

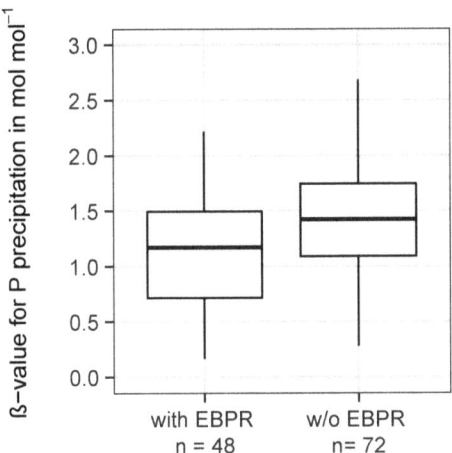

Figure 1. Comparison of ß-values for P-precipitation with and without additional enhanced biological phosphorus removal (EBPR) in mol precipitation agent per mol P. Statistical values are provided in Table S2.

Primary clarifiers are mostly installed in larger plants (89% for >100,000 PE) and less abundant in WWTPs smaller than 20,000 PE (24%). This size dependency is generally derived from the combined use of primary clarifiers with anaerobic sludge stabilization. Approximately 95% of clarifier capacity is installed on sites with anaerobic digesters. The remaining 5% capacity are installed in aerobic plants, mostly in combination with simultaneous sludge stabilization (80%).

Table 3. Primary clarifier abundance and use of various sludge stabilization methods in different size groups of Austrian wastewater treatment plants.

Shares Given in % of PE Treated	2000 to 20,000 PE	>20,000 to 50,000 PE	>50,000 PE 100,000 PE	>100,000 PE	Total
Primary clarifiers					
n	61	104	29	33	227
Occurrence	24	68	82	89	82
Sludge stabilisation					
n	155	124	33	36	358
anaerobic	11	66	79	97	85
aerobic, of which...	89	34	21	3	15
... simultaneous	74	70	59	100	73
... separated	18	28	28	0	21
... unknown	8	2	13	0	5

A thorough knowledge on sludge production is the basis for developing new sewage sludge mono-incineration plant concepts in Austria. Through the ongoing transformation of Austrian WWTPs towards anaerobic digestion and better stabilization, sludge amounts have been decreasing for years [23]. WWTPs with sludge utilization in agriculture often add legitimate amounts of lime for hygienization. If these plants switch to mono-incineration instead, a further decrease in sludge production is expected, since adding inorganic material to sludge will only further reduce P ash concentrations and thereby hinder recycling. To estimate sludge amounts for incineration, data on total sludge yield per PE (organic and inorganic) were analyzed according to their primary sludge treatment method (aerobic/anaerobic) and lime addition (Figure 2). Anaerobic treatment achieves the lowest total sludge yield of 37 g PE^{-1} d^{-1} (Table S3). Simultaneous aerobic treatment produces more sludge (52 g PE^{-1} d^{-1}) than separated aerobic treatment (45 g PE^{-1} d^{-1}). Addition of lime resulted in around 54 to 57% higher yield with median values of 57 g for anaerobic and 77 g PE^{-1} d^{-1} for aerobic treatment.

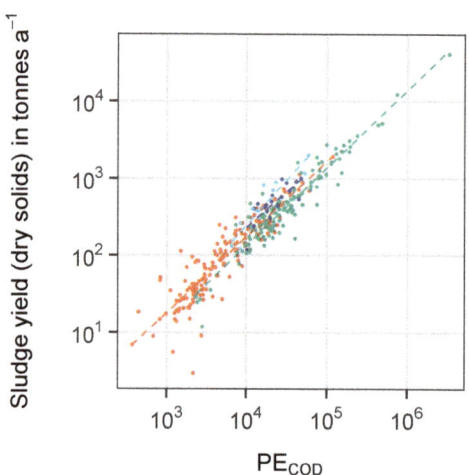

Figure 2. Total sludge yield (as dry matter) in tons per year as a function of treated population equivalents (PE; derived from COD with 120 g COD PE^{-1} d^{-1}) and dependent on the primary sludge stabilization method as well as on potential lime addition. Statistical values are provided in Table S3.

Derived values are in the range of observed and modeled values from literature, however, large variations were found for total sludge yield (27–82 g PE^{-1} d^{-1} [28,29]). Data on loss of ignition (LOI; Figure 3) can put sludge yield into context with the degree of sludge stabilization. Anaerobic treatment generally achieves the best stabilization and lowest LOI (59%), followed by separated aerobic treatment (64%). Simultaneous stabilization shows highest LOI of 71%. With the addition of lime (inorganic matter) LOI in sludge decreases considerably to 34–35%.

Previous detailed analysis on Austrian sludge production found a volatile suspended solids (VSS) yield of 16 to 20 g PE^{-1} d^{-1} for anaerobic and separated aerobic stabilization [30]. Non-sufficiently stabilized sludge from simultaneous aerobic stabilization showed a VSS yield of 20 to 35 g PE^{-1} d^{-1}. If LOI values are taken into account, total sludge yield can be estimated from VSS production according to Equation (1).

$$\text{Total sludge yield [g PE}^{-1}\text{ d}^{-1}\text{]} = \text{VSS [g PE}^{-1}\text{ d}^{-1}\text{]}/\text{LOI [\%]} \quad (1)$$

Assuming a sludge LOI of 60% for anaerobic or separated aerobic treatment and 71% for simultaneous stabilization (Table S4), the corresponding total sludge yields would be in the area of 26–33 g (anaerobic/separated) and 35–50 g PE^{-1} d^{-1} (simultaneous), respectively. In comparison, the higher median total sludge yields derived by this study (>37 g PE^{-1} d^{-1}) suggest a slightly incomplete stabilization and a VSS yield after stabilization of commonly above 20 g PE^{-1} d^{-1}.

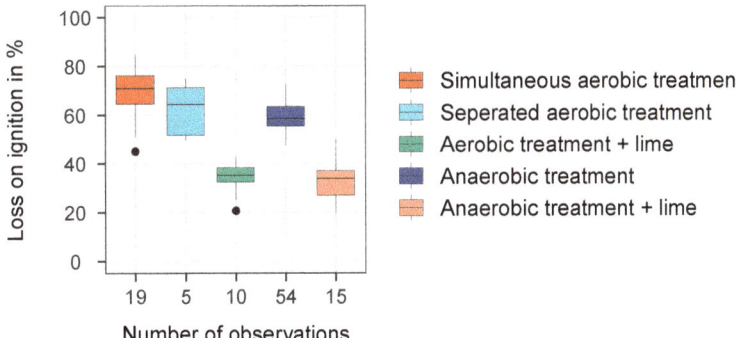

Figure 3. Loss on ignition (LOI) in percent as a function of the primary sludge stabilization method. Statistical values are provided in Table S4.

3.4. Sludge Processing: Dewatering, Drying and Hygienization

Types of dewatering units are unequally distributed between the different size groups (Table 4). Filter presses are most common in smaller WWTPs (39%), while above 20,000 PE, centrifuges are the standard method of choice (37–76%). Screw presses are more abundant in WWTP below 50,000 PE and have occupied the market only in recent years. Data from a German inquiry on sewage sludge treatment in 2003 showed no occurrence of screw presses for dewatering at that time [31]. Belt presses are rarely in use. Some WWTPs use mobile dewatering, however this is mostly common for smaller plants. Addition of lime for hygienization is done in 10–38% of the plants, correlated to the primary sludge disposal method (agricultural and composting). Sludge drying on-site is rarely implemented in Austria, with around 19 known installations, namely, 11 solar dryers, two belt dryers and four convection dryers.

Table 4. Summary table of dewatering units and use of lime according to the size group of the respective wastewater treatment plants.

	2000 to 20,000 PE	>20,000 to 50,000 PE	>50,000 PE 100,000 PE	>100,000 PE
Dewatering units				
n	74	103	28	33
Centrifuge in %	26	37	45	76
Belt press in %	6	8	10	2
Filter press in %	39	25	25	10
Screw press in %	24	26	19	12
Others in %	5	4	1	0
Hygienisation with lime				
n	5	76	21	21
Share in %	21	24	38	10

Dewatering with the addition of lime achieved the highest dry matter content in sewage sludge (Figure 4). Arguably, this is partially from an increase of the related solid mass to water ratio, but calcium addition is also known to increase floc strength and dewaterability [32]. Out of the different types of dewatering units, filter presses showed the highest median dry matter content (28%) followed by centrifuges (25%), screw presses (24%) and finally belt presses with the lowest median value of 22%. Mobile units showed a high range from 21 to 34%. Values correspond well with data from DWA guideline M 366 for sludge dewatering (filter press 22–28%, with lime 30–40%, centrifuges 22–30%, screw presses 20–28% and belt presses 20–28% [33]).

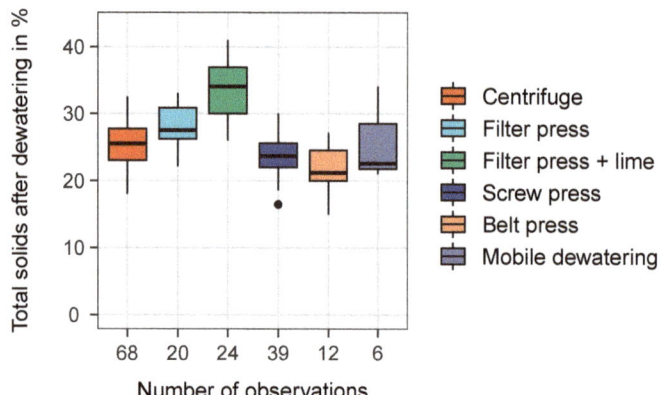

Figure 4. Dry matter content of dewatered sludge in percent according to the type of dewatering unit used and the (non-)addition of lime. Statistical values are provided in Table S5.

Further, a WWTP-size dependency was observed, with median dry matter content (dewatered and without lime addition) increasing from 23 to 25 to 26 to 26.7%, for size groups 1, 2, 3 and 4, respectively. While reasons for this cannot be derived from the data itself, it is assumed that the improved dewatering performance is achieved both by better aggregates as well as by better supervision and operation of larger plants.

Out of the installed dryers, the two convection dryers achieved the highest dry matter content of around 84%, with belt dryers and solar dryers at approximately 71–73% (Table S6).

3.5. Sludge Quality

For the development of sustainable P recovery strategies, information on sludge P concentrations and accompanying heavy metals is required. Not all treatment plants monitor their sludge quality; therefore, it will be necessary to resort to easily available data—e.g., recorded WWTP-P inflow and outflow loads—to determine P loads and concentration in sewage sludge. We plotted measured P concentrations from sludge analysis against theoretical P concentrations derived from yearly inflow/outflow loads and sludge quantities (Figure 5). As is depicted, theoretical P concentration is a good predictor for actual P concentrations ($t(103) = 9.9897$, $p < 0.001$), with a general deviation of smaller than 25%. Theoretical analysis might also give a better understanding of mean yearly concentrations, as measurements only represent a moment in time.

Observed P concentrations range from 9 to 63 g kg^{-1} (Figure 6a) and are in part connected to the degree of sludge stabilization, which is commonly well represented by LOI (Figure 3). Due to a lower LOI and a reduced sludge mass, P concentrations are higher in anaerobically treated sludge with 34 g kg^{-1}. As inorganic matter is added with lime addition, LOI is also lower. However, P is diluted by this treatment, leading to median P concentrations in the area of 22 g kg^{-1} only.

For recovery from sewage sludge ash, P concentrations in ash are of interest. We estimated P ash concentrations from LOI, sludge mass and P concentrations. As depicted (Figure 6b), Austrian ash P concentrations without added lime would rest around the 9% mark, which P rock is rarely falling short of [16]. For a better cost effectiveness of P recovery from ash, operators should try to reduce inorganic additives on-site, without compromising the effectiveness of wastewater treatment. Observed median levels of other nutrients in sludge were 31 g nitrogen, 7 g magnesium, 1.4 g potassium and 0.9 g sodium per kg of sludge (Figure S2 and Table S11).

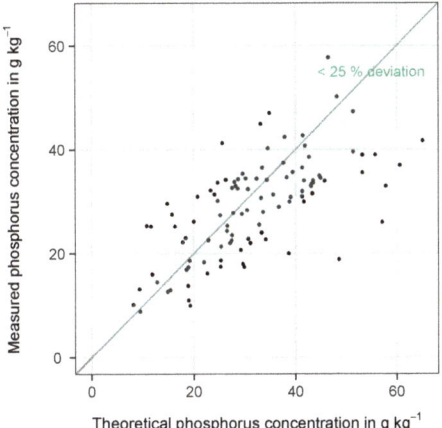

Figure 5. Comparison of measured sludge P concentrations from sludge monitoring data to theoretical sludge P concentrations derived from WWTP-sewage P-inflow minus P-outflow.

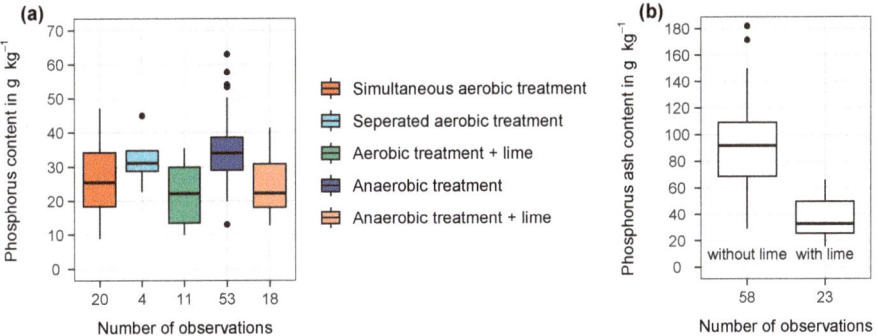

Figure 6. (a) P content in grams per kilogram as a function of the primary sludge stabilization method, and (b) derived P ash content as a function of lime addition. Statistical values are provided in Table S10.

While it is known that WWTPs partially remove heavy metals from the liquid stream, information on the impact of different operating conditions on heavy metal removal is scarce and sometimes contradictory [34,35]. Sludge heavy metal concentrations will further relate to the abundance of pollutant sources (municipal, industrial) in the WWTP drainage area [36]. As current German legislation obliges only certain WWTP size groups to recover P, it is of interest if there exists a correlation of heavy metal concentrations in sludge with WWTP sizes. Figure 7 shows Austrian heavy metal concentrations in sewage sludge as a function of the size group of the WWTP. It can be seen, that most heavy metals are similarly distributed across all WWTP groups. Exceptions are chromium and nickel, which show comparatively high values for WWTPs larger than 100,000 PE. Accordingly, sludge quality in Austria shows for the most part no size-dependency. The majority of WWTPs can fulfill even the more stringent Austrian heavy metal limits in sludge for use in agriculture (limits shown as dashed lines, taken from the work in [37]). Further analysis of the data, considering the influence of different treatment schemes and of the type of drainage area, should be performed for a better understanding of final metal concentrations in sewage sludge.

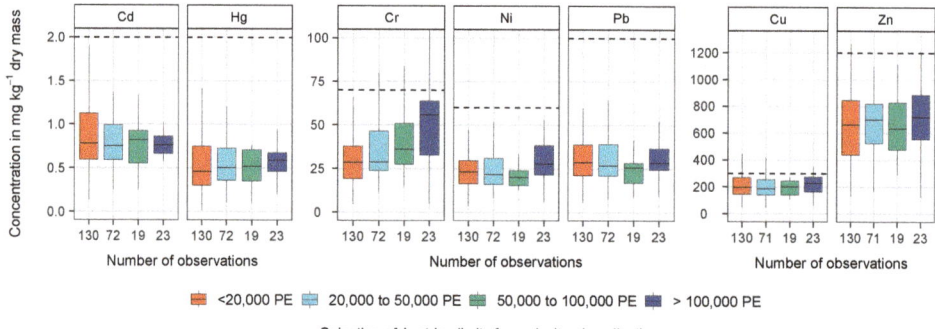

Figure 7. Heavy metal (cadmium, mercury, nickel, lead, chromium, copper and zinc) sewage sludge concentrations in milligrams per kilogram. Dashed lines give the most common Austrian heavy metal limits for the application of sludge in agriculture [37]. Statistical values are provided in Table S12.

3.6. Sludge Disposal

For 91% of Austrian sewage sludge, disposal routes could be successfully tracked (Figure 8). The largest amounts are treated via mono-incineration or external composting. Sludge disposal routes are diverse, with some states (Vienna = W) with 100% thermal treatment and others (Burgenland = B) with close to a 100% of soil-based sludge use. Therefore, changes to sludge disposal due to potential P recycling and increased incineration will have very different degrees of effect in different states. Currently, only direct agricultural disposal (wet or dewatered = 18%) reliably brings P to arable land. Informal talks with WWTP stakeholders confirmed that compost from sludge is often not used in agricultural land with high P demand, but for recultivation of landfills or for landscaping. Specific estimates could not be derived, since tracking of composted sludge proved highly time consuming or impossible.

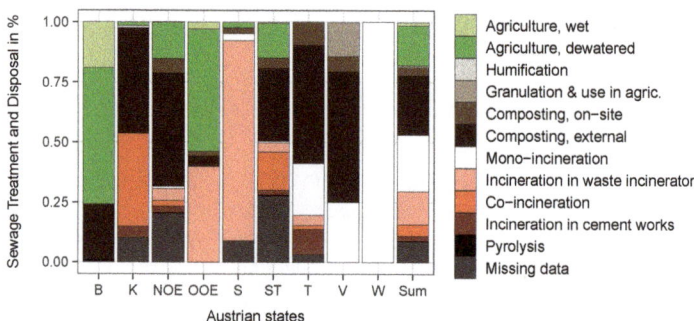

Figure 8. Sludge treatment and disposal routes in Austria in 2016 in % and for each federal state.

Sludge disposal costs in 2016 ranged from 3.5 to 100 EUR t^{-1} (wet mass) or 21 up to 560 EUR t^{-1} (dry mass) (Figure 9). These are well in line with published sludge disposal costs in Germany (160 to 480 EUR t^{-1} dry mass [38]). Disposal in agriculture and composting on-site was comparatively cheap with median values between 6 to 40 EUR and 94 to 180 EUR t^{-1} for wet and dry mass, respectively. External composting through 3rd party contractors and incineration was correlated to higher costs with median values of 59 to 75 EUR t^{-1} (wet) and 230 to 290 EUR t^{-1} (dry mass). Compared to the cost of mono-incineration (280 to 480 EUR t^{-1} dry mass [38]), an increase in costs for sludge disposal is likely if P recovery from ash is pursued.

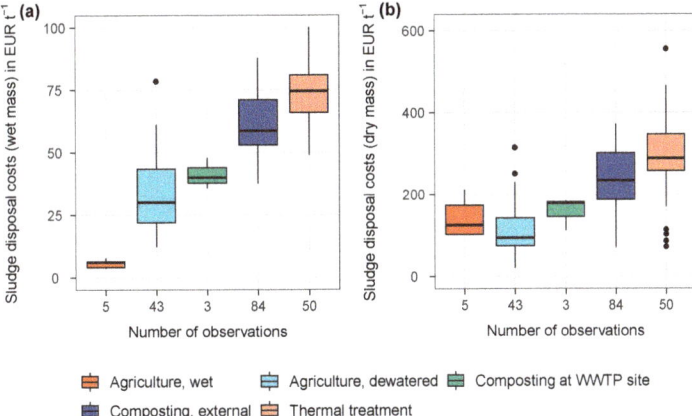

Figure 9. Cost of sludge disposal in Euro per ton as a function of their disposal route (**a**) based on sludge wet mass and (**b**) based on sludge dry mass. Statistical values are provided in Tables S7 and S8.

Similarly to cost, transport distances were highest for external composting and thermal treatment, with some plants transporting over 530 km (one-way; Figure 10). Median values for thermal treatment were more than twice as high (120 km) than external composting (50 km). Agricultural disposal and composting at the WWTP site rarely exceeded 20 km with median values of 15 and 0.25 km, respectively. A move towards more incineration might result in longer transport distances. Careful evaluation of strategic locations for mono-incineration sites will, consequently, be decisive to limit future emissions from sludge transport.

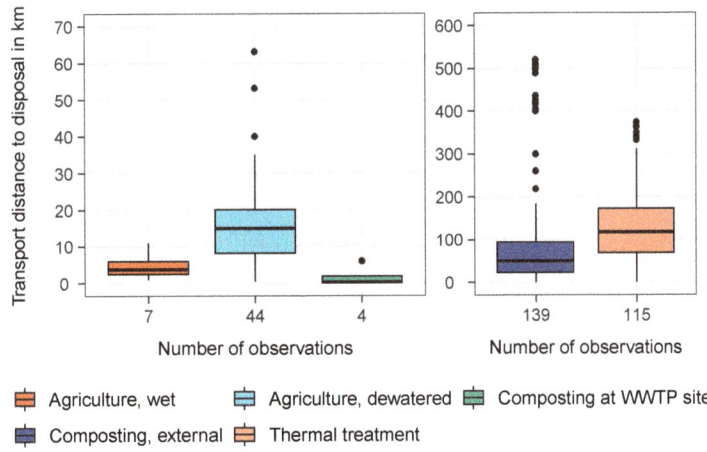

Figure 10. Distance of sewage sludge transport to disposal as a function of their disposal route. Statistical values are provided in Table S9.

4. Conclusions

This study presented the main findings from a carefully developed database on wastewater treatment and sludge disposal in Austria for the years 2015–2017. Data availability was high for most analyzed parameters, and the database posed a good basis for further deliberations on changes in sludge P management. Austrian wastewater treatment plants perform well, and have seen proper updates to state of the art technology, with a high share in anaerobic stabilization and well functioning dewatering units. Though EBPR would be the preferred P removal method of choice for P recovery, chemical precipitants are still vital for the secured removal of P from wastewater. Even WWTPs with EPBR installed use valid amounts of iron and aluminum, which will likely remain so in the years to come.

Sludge quantities are expected to decrease further with a shift from agricultural valorization to incineration of sludge. In turn, it is estimated that sludge disposal costs and transport distances will increase, from lower costs for agricultural disposal to higher costs for mono-incineration. To keep P concentrations high for an efficient recovery of P, efforts should focus on reducing inorganic additives to sludge as much as possible without inhibiting the treatment process. Heavy metals in sludge generally do not exceed the more stringent limits posed by Austrian legislators. Nevertheless, caution should be exercised, as with an increased recycling of sludge, sludge ash or of sludge derived products, total heavy metal loads to agriculture will increase if metals in sewage sludge ash for P recovery are not (partially) removed.

Before legislation for P recovery is implemented, care should be taken to further analyze different options of P recovery for their cost and environmental efficiency in the Austrian context. As mono-incineration of sludge is still rare in Austria, additional planning should focus on finding sustainable and strategically well-placed locations for new plants, in order to reduce impacts from sewage sludge transport.

Supplementary Materials: The following are available online at https://www.mdpi.com/article/10.3390/w13212998/s1, Table S1: Summary data table of use of iron and aluminum in WWTPs with aerobic and anaerobic sludge treatment. Figure S1: Comparison of theoretical precipitating agent demand derived from precipitable phosphorus amounts and declared demand from WWTP operators in mol per year. Table S2: Summary data table of observed β values (mol of added precipitating agent divided by mol of phosphorus that need to precipitated) with or without enhanced biological phosphorus removal (EBPR). Table S3: Summary data table of sewage sludge yield in g PE^{-1} and d^{-1} according to their sludge treatment process and with or without lime addition for hygienization. Table S4: Summary data table of sewage sludge loss on ignition in % according to their sludge treatment process and with or without lime addition for hygienization. Table S5: Summary data table of sewage sludge dry solid concentrations in % after dewatering with various devices. Table S6: Type and number of installed sludge dryers in Austria as well as mean dry matter content after drying in percent. Table S7: Summary data table of sewage sludge disposal costs (wet substance) in € t^{-1} according to the applied sludge disposal method Table S8: Summary data table of sewage sludge disposal costs (dry substance) in € t^{-1} according to the applied sludge disposal method. Table S9: Summary data table of sewage sludge transport distances (one-way) in km according to the applied sludge disposal method. Table S10: Summary data table of sewage sludge phosphorus content in g kg^{-1} according to their sludge treatment process and with or without lime addition for hygienization Figure S2: Nutrient (nitrogen, magnesium, potassium and sodium) sewage sludge concentrations in grams per kilogram. Table S11: Summary data table of sewage sludge nutrient content in g kg^{-1}. Table S12: Summary data table of heavy metal concentrations in sewage sludge in mg kg^{-1} according to the size group of the respective wastewater treatment plants.

Author Contributions: Conceptualization, A.A., H.R., O.Z. and M.Z.; Data curation, A.A.; Formal analysis, A.A. and N.W.; Funding acquisition, A.A., H.R., O.Z. and M.Z.; Investigation, A.A. and N.W.; Methodology, A.A., N.W. and M.Z.; Project administration, J.K., H.R., O.Z. and M.Z.; Resources, J.K. and H.R.; Software, A.A. and N.W.; Supervision, J.K., H.R. and M.Z.; Validation, A.A., N.W., J.K. and M.Z.; Visualization, A.A. and N.W.; Writing—original draft, A.A. and O.Z.; Writing—review and editing, A.A., J.K., H.R., O.Z. and M.Z. All authors have read and agreed to the published version of the manuscript.

Funding: This research was funded by the Federal Ministry of the Republic of Austria for Climate Action, Environment, Energy, Mobility, Innovation and Technology. Open Access Funding by TU Wien.

Data Availability Statement: Restrictions apply to the availability of these data. The raw data are not publicly available due to privacy issues. However, data presented in this study are available as summarized statistical data in the Supplementary Material.

Acknowledgments: We want to thank all treatment plant operators and the federal state authorities of Austria for their help with data curation in this project. We further acknowledge TU Wien Bibliothek for their financial support through its Open Access Funding Programme.

Conflicts of Interest: The authors declare no conflict of interest. The funders had no role in the design of the study; in the collection, analyses or interpretation of data; in the writing of the manuscript, or in the decision to publish the results.

Abbreviations

The following abbreviations are used in this manuscript:

BOD	Biological oxygen demand
COD	Chemical oxygen demand
EBPR	Enhanced biological phosphorus removal
LOI	Loss on ignition
P	Phosphorus
PE	Population equivalents based on COD (1 PE = 120 g COD d^{-1})
VSS	Volatile suspended solids
WWTP	Wastewater treatment plant

References

1. Jenkins, D.; Wanner, J. *Activated Sludge-100 Years and Counting*; IWA Publishing: London, UK, 2014.
2. Longo, S.; d'Antoni, B.M.; Bongards, M.; Chaparro, A.; Cronrath, A.; Fatone, F.; Lema, J.M.; Mauricio-Iglesias, M.; Soares, A.; Hospido, A. Monitoring and Diagnosis of Energy Consumption in Wastewater Treatment Plants. A State of the Art and Proposals for Improvement. *Appl. Energy* **2016**, *179*, 1251–1268. [CrossRef]
3. Ganora, D.; Hospido, A.; Husemann, J.; Krampe, J.; Loderer, C.; Longo, S.; Bouyat, L.M.; Obermaier, N.; Piraccini, E.; Stanev, S.; et al. Opportunities to Improve Energy Use in Urban Wastewater Treatment: A European-Scale Analysis. *Environ. Res. Lett.* **2019**, *14*, 044028. [CrossRef]
4. Charlton, M.B.; Bowes, M.J.; Hutchins, M.G.; Orr, H.G.; Soley, R.; Davison, P. Mapping Eutrophication Risk from Climate Change: Future Phosphorus Concentrations in English Rivers. *Sci. Total Environ.* **2018**, *613-614*, 1510–1526. [CrossRef] [PubMed]
5. Egle, L.; Rechberger, H.; Krampe, J.; Zessner, M. Phosphorus Recovery from Municipal Wastewater: An Integrated Comparative Technological, Environmental and Economic Assessment of P Recovery Technologies. *Sci. Total Environ.* **2016**, *571*, 522–542. [CrossRef]
6. Melia, P.M.; Cundy, A.B.; Sohi, S.P.; Hooda, P.S.; Busquets, R. Trends in the Recovery of Phosphorus in Bioavailable Forms from Wastewater. *Chemosphere* **2017**, *186*, 381–395. [CrossRef]
7. Fatta-Kassinos, D.; Manaia, C.; Berendonk, T.U.; Cytryn, E.; Bayona, J.; Chefetz, B.; Slobodnik, J.; Kreuzinger, N.; Rizzo, L.; Malato, S.; et al. COST Action ES1403: New and Emerging Challenges and Opportunities in Wastewater REUSe (NEREUS). *Environ. Sci. Pollut. Res.* **2015**, *22*, 7183–7186. [CrossRef] [PubMed]
8. Jaramillo, M.F.; Restrepo, I. Wastewater Reuse in Agriculture: A Review about Its Limitations and Benefits. *Sustainability*, **2017**, *9*, 1734. [CrossRef]
9. Schaar, H.; Clara, M.; Gans, O.; Kreuzinger, N. Micropollutant Removal during Biological Wastewater Treatment and a Subsequent Ozonation Step. *Environ. Pollut.* **2010**, *158*, 1399–1404. [CrossRef]

10. Berendonk, T.U.; Manaia, C.M.; Merlin, C.; Fatta-Kassinos, D.; Cytryn, E.; Walsh, F.; Bürgmann, H.; Sørum, H.; Norström, M.; Pons, M.N.; et al. Tackling Antibiotic Resistance: The Environmental Framework. *Nat. Rev. Microbiol.* **2015**, *13*, 310–317. [CrossRef]
11. Slipko, K.; Reif, D.; Wögerbauer, M.; Hufnagl, P.; Krampe, J.; Kreuzinger, N. Removal of Extracellular Free DNA and Antibiotic Resistance Genes from Water and Wastewater by Membranes Ranging from Microfiltration to Reverse Osmosis. *Water Res.* **2019**, *164*, 114916. [CrossRef]
12. Amann, A.; Zoboli, O.; Krampe, J.; Rechberger, H.; Zessner, M.; Egle, L. Environmental Impacts of Phosphorus Recovery from Municipal Wastewater. *Resour. Conserv. Recycl.* **2018**, *130*, 127–139. [CrossRef]
13. BMK. *Federal Waste Management Plan 2017. Part 1*; Technical Report; Federal Ministry of the Republic of Austria for Climate Action, Environment, Energy, Mobility, Innovation and Technology (BMK): Vienna, Austria, 2018.
14. Steckenmesser, D.; Vogel, C.; Adam, C.; Steffens, D. Effect of Various Types of Thermochemical Processing of Sewage Sludges on Phosphorus Speciation, Solubility, and Fertilization Performance. *Waste Manag.* **2017**, *62*, 194–203. [CrossRef] [PubMed]
15. Tansel, B.; Lunn, G.; Monje, O. Struvite Formation and Decomposition Characteristics for Ammonia and Phosphorus Recovery: A Review of Magnesium-Ammonia-Phosphate Interactions. *Chemosphere* **2018**, *194*, 504–514. [CrossRef]
16. Kratz, S.; Schnug, E. Trace Elements in Rock Phosphates and P Containing Mineral and Organo-Mineral Fertilizers Sold in Germany. *Sci. Total Environ.* **2016**, *542*, 1013–1019. [CrossRef] [PubMed]
17. Krüger, O.; Grabner, A.; Adam, C. Complete Survey of German Sewage Sludge Ash. *Environ. Sci. Technol.* **2014**, *48*, 11811–11818. [CrossRef] [PubMed]
18. Smol, M.; Adam, C.; Anton Kugler, S. Inventory of Polish Municipal Sewage Sludge Ash (SSA)—Mass Flows, Chemical Composition, and Phosphorus Recovery Potential. *Waste Manag.* **2020**, *116*, 31–39. [CrossRef]
19. Kacprzak, M.; Neczaj, E.; Fijałkowski, K.; Grobelak, A.; Grosser, A.; Worwag, M.; Rorat, A.; Brattebo, H.; Almås, A.; Singh, B.R. Sewage Sludge Disposal Strategies for Sustainable Development. *Environ. Res.* **2017**, *156*, 39–46. [CrossRef]
20. Verordnung des Bundesministers für Land- und Forstwirtschaft über die Begrenzung von Abwasseremissionen aus Abwasserreinigungsanlagen Für Siedlungsgebiete (1. AEV Für Kommunales Abwasser; 1st Ordinance on Municipal Wastewater Emissions). BGBl. Nr. 210/1996. Available online: https://www.ris.bka.gv.at/GeltendeFassung.wxe?Abfrage=Bundesnormen&Gesetzesnummer=10010980 (accessed on 29 September 2021).
21. EmRegV-OW. Verordnung des Bundesministers für Land- und Forstwirtschaft, Umwelt und Wasserwirtschaft über ein Elektronisches Register zur Erfassung aller Wesentlichen Belastungen von Oberflächenwasserkörpern Durch Emissionen von Stoffen aus Punktquellen 2017 (Emissionsregisterverordnung 2017—EmRegV-OW 2017) (Emission Inventory Ordinance). BGBl. II Nr. 207/2017. Available online: https://www.ris.bka.gv.at/GeltendeFassung.wxe?Abfrage=Bundesnormen&Gesetzesnummer=20009954 (accessed on 29 September 2021).
22. WRG. Wasserrechtsgesetz 1959—WRG. 1959. (Water Rights Act). BGBl. Nr. 215/1959. Available online: https://www.ris.bka.gv.at/GeltendeFassung.wxe?Abfrage=Bundesnormen&Gesetzesnummer=10010290 (accessed on 29 September 2021).
23. Überreiter, E.; Lenz, K.; Zieritz, I. *Kommunales Abwasser. Österreichischer Bericht 2018*; Technical Report; Bundesministerium für Nachhaltigkeit und Tourismus: Vienna, Austria, 2018.
24. Kratz, S.; Vogel, C.; Adam, C. Agronomic Performance of P Recycling Fertilizers and Methods to Predict It: A Review. *Nutr. Cycl. Agroecosyst* **2019**, *115*, 1–39. [CrossRef]
25. Park, C.M.; Novak, J.T. The Effect of Direct Addition of Iron(III) on Anaerobic Digestion Efficiency and Odor Causing Compounds. *Water Sci. Technol.* **2013**, *68*, 2391–2396. [CrossRef] [PubMed]
26. Krishnamoorthy, N.; Dey, B.; Unpaprom, Y.; Ramaraj, R.; Maniam, G.P.; Govindan, N.; Jayaraman, S.; Arunachalam, T.; Paramasivan, B. Engineering Principles and Process Designs for Phosphorus Recovery as Struvite: A Comprehensive Review. *J. Environ. Chem. Eng.* **2021**, *9*, 105579. [CrossRef]
27. Egle, L.; Rechberger, H.; Zessner, M. Overview and Description of Technologies for Recovering Phosphorus from Municipal Wastewater. *Resour. Conserv. Recycl.* **2015**, *105*, 325–346. [CrossRef]
28. Kelessidis, A.; Stasinakis, A.S. Comparative Study of the Methods Used for Treatment and Final Disposal of Sewage Sludge in European Countries. *Waste Manag.* **2012**, *32*, 1186–1195. [CrossRef] [PubMed]
29. Mininni, G.; Laera, G.; Bertanza, G.; Canato, M.; Sbrilli, A. Mass and Energy Balances of Sludge Processing in Reference and Upgraded Wastewater Treatment Plants. *Environ. Sci. Pollut. Res.* **2015**, *22*, 7203–7215. [CrossRef]
30. Nowak, O.; Franz, A.; Svardal, K.; Müller, V. Specific Organic and Nutrient Loads in Stabilized Sludge from Municipal Treatment Plants. *Water Sci. Technol.* **1996**, *33*, 243–250. [CrossRef]
31. Durth, A.; Schaum, C.; Meda, A.; Wagner, M.; Hartmann, K.H.; Jardin, N.; Kopp, J.; Otte-Witte, R. Ergebnisse Der DWA-Klärschlammerhebung 2003 (Results from the DWA-Sewage Sludge Inquiry 2003). *KA Abwasser Abfall* **2003**, *52*, 1099–1107.
32. Christensen, M.L.; Keiding, K.; Nielsen, P.H.; Jørgensen, M.K. Dewatering in Biological Wastewater Treatment: A Review. *Water Res.* **2015**, *82*, 14–24. [CrossRef]
33. DWA-M 366. *Merkblatt DWA-M 366 Schlammentwässerung (Sewage Sludge Dewatering)*; Technical Report; German Association for Water, Wastewater and Waste: Hennef, Germany, 2013.
34. Cantinho, P.; Matos, M.; Trancoso, M.A.; dos Santos, M.M.C. Behaviour and Fate of Metals in Urban Wastewater Treatment Plants: A Review. *Int. J. Environ. Sci. Technol.* **2016**, *13*, 359–386. [CrossRef]

35. Mailler, R.; Gasperi, J.; Chebbo, G.; Rocher, V. Priority and Emerging Pollutants in Sewage Sludge and Fate during Sludge Treatment. *Waste Manag.* **2014**, *34*, 1217–1226. [CrossRef]
36. Clara, M.; Windhofer, G.; Weilgony, P.; Gans, O.; Denner, M.; Chovanec, A.; Zessner, M. Identification of Relevant Micropollutants in Austrian Municipal Wastewater and Their Behaviour during Wastewater Treatment. *Chemosphere* **2012**, *87*, 1265–1272. [CrossRef]
37. ÖWAV. *ÖWAV ExpertInnenpapier Kritische Ressource Phosphor. Erstellt Durch Die AG 1 "Klärschlamm Und Tierische Nebenprodukte in Einem Optimierten P-Management" Des ÖWAV Arbeitsausschusses "Klärschlammplattform"*; Technical Report; Austrian Association for Water, Wastewater and Waste (ÖWAV): Vienna, Austria, 2018.
38. German Environment Agency. *Klärschlammentsorgung in der Bundesrepublik Deutschland (Sewage Sludge Disposal in the Republic of Germany)*; Technical Report; German Environment Agency (Umweltbundesamt): Dessau, Germany, 2018.

Article

Feasibility Study of Water Reclamation Projects in Industrial Parks Incorporating Environmental Benefits: A Case Study in Chonburi, Thailand

Weeraya Intaraburt [1], Jatuwat Sangsanont [2], Tawan Limpiyakorn [1], Piyatida Ruangrassamee [3], Pongsak Suttinon [3] and Benjaporn Boonchayaanant Suwannasilp [1,*]

1. Department of Environmental Engineering, Faculty of Engineering, Chulalongkorn University, Phayathai Rd., Patumwan, Bangkok 10330, Thailand; 6170276121@student.chula.ac.th (W.I.); tawan.l@chula.ac.th (T.L.)
2. Department of Environmental Science, Faculty of Science, Chulalongkorn University, Bangkok 10330, Thailand; jatuwat.s@chula.ac.th
3. Department of Water Resources Engineering, Faculty of Engineering, Chulalongkorn University, Bangkok 10330, Thailand; piyatida.h@chula.ac.th (P.R.); pongsak.su@chula.ac.th (P.S.)
* Correspondence: benjaporn.bo@chula.ac.th

Citation: Intaraburt, W.; Sangsanont, J.; Limpiyakorn, T.; Ruangrassamee, P.; Suttinon, P.; Suwannasilp, B.B. Feasibility Study of Water Reclamation Projects in Industrial Parks Incorporating Environmental Benefits: A Case Study in Chonburi, Thailand. *Water* **2022**, *14*, 1172. https://doi.org/10.3390/w14071172

Academic Editors: Martin Wagner, Sonja Bauer and Elias Dimitriou

Received: 7 February 2022
Accepted: 4 April 2022
Published: 6 April 2022

Publisher's Note: MDPI stays neutral with regard to jurisdictional claims in published maps and institutional affiliations.

Copyright: © 2022 by the authors. Licensee MDPI, Basel, Switzerland. This article is an open access article distributed under the terms and conditions of the Creative Commons Attribution (CC BY) license (https://creativecommons.org/licenses/by/4.0/).

Abstract: Financial feasibility is usually a concern in water reclamation projects. Aside from internal benefits, water reclamation in industrial parks delivers health and environmental benefits not normally considered in cost–benefit analyses (CBA). This study investigated the influence of environmental benefits on the feasibility of water reclamation projects with flow rate scenarios in accordance with industrial parks in Chonburi, Thailand. CBAs of water reclamation plants for industrial water supply, consisting of ultrafiltration (UF) and reverse osmosis (RO), with flow rates of 5200, 10,000, 15,000, and 25,000 m^3/day and discount rates of 3%, 5%, 7%, 9% and 11% were conducted. Considering only the direct costs and benefits, none of the projects were financially feasible. However, when the environmental benefits were included, the projects became profitable in all cases except those with a flow rate of 5200 m^3/day and discount rates of 5%, 7%, 9%, and 11% and those with flow rates of 10,000 and 25,000 m^3/day and an 11% discount rate. Further, CBAs of water reclamation projects in industrial parks for irrigation were conducted with post-treatment processes consisting of sand filtration and chlorine disinfection for flow rates of 240, 480, 2400, 3600, and 4800 m^3/day. The projects are profitable, regardless of environmental benefits.

Keywords: cost–benefit analysis; industrial water reuse; environmental benefit; irrigation

1. Introduction

Industrialization, urbanization, and population growth have increased the demand for water worldwide, leading to water scarcity in many countries [1]. Interannual and intra-annual climate variability are becoming greater, resulting in higher fluctuations in the water supply. The observed changes in heavy precipitation, droughts, and tropical cyclones have strengthened since 2014 [2]. The provision of clean water and water use efficiency are part of the Sustainable Development Goals (SDGs) adopted by the United Nations in 2015 [3]. A circular economy is an economic framework that maximizes service using cyclical material flows, renewable energy sources, and cascading-type energy flows [4], and it is also reflected in SDG 12—responsible consumption and production [3]. "Wastewater" is an important component of the circular economy [5], and water reclamation and reuse are increasingly being integrated into water resource management, as they are expected to mitigate water scarcity worldwide and provide flexibility to respond to both short- and long-term water supply needs [6,7].

Industrial parks, particularly those located in water-stress areas, have high water demands that may compete with municipal and agricultural water demands, necessitating appropriate water reuse to conserve natural water resources and assure industrial park sustainable development [8]. Water reclamation has been shown to provide various benefits, including economic, environmental, health, and social benefits [9–17]. Apart from the direct benefits of reclaimed water, water reclamation can effectively lessen pollution emissions, provide nutrients as fertilizers, conserve freshwater resources, and offer recreational benefits [9,11,17,18]. According to a life cycle analysis (LCA) of water reuse in an industrial park is environmentally beneficial compared to the no-reuse scenario in most aspects, that is, marine aquatic toxicity, abiotic depletion, acidification, eutrophication, freshwater aquatic ecotoxicity, global warming, human toxicity, terrestrial ecotoxicity, and photochemical oxidation, with eutrophication and freshwater aquatic ecotoxicity potentials greatly depending on the pollutants discharged into the environment [19]. Moreover, water reclamation can potentially provide ecosystem service values, although research in this area is still limited [20,21].

However, water reclamation can be challenging due to social, technological, economic, and regulatory constraints [22]. Economic feasibility, along with public acceptance, is considered a key factor in the implementation of water reclamation projects [12,22], since high-cost advanced treatment technologies are usually required to ensure that reclaimed water is of high quality [23–25]. Despite its importance, public information on the economic feasibility of water reclamation projects in industrial parks is still limited [10]. Few studies have compared the costs of different water reuse options with those of the water supply options available in the areas under study [26,27]. Furthermore, there are still no studies that consider externalities, such as environmental and social costs/benefits, of water reclamation projects in industrial parks, although these issues have been suggested for inclusion in an economic analysis of water reclamation projects [28,29] and have been addressed in previous studies on the water reclamation of effluents from municipal wastewater treatment plants (WWTPs) [12–17]. The inclusion of externalities in cost–benefit analysis (CBA) can disclose the genuine value of water reclamation projects [13], which can be used to obtain governmental support for such initiatives. Further, although LCA can provide information on all environmental aspects of water reclamation projects, it cannot be coherently combined with an economic feasibility analysis in the decision-making process. The economic and environmental dimensions are still considered separately [27]. CBA, with the inclusion of externalities, is an emerging tool for merging the two dimensions in the decision-making process, which can ensure economic, environmental, and social sustainability [12].

Chonburi is a province in Thailand located in the Eastern Economic Corridor (EEC), initiated by the Thai government to promote industrial growth. Water scarcity in the EEC is predicted to affect all sectors in the next 20 years [30]. Thus, there is an urgent need to place water reuse in the EEC on the policy agenda. To increase internal water reuse in industrial parks, the Industrial Estate Authority, Thailand, issued regulations requiring the reuse of at least 15% of the water supply in industrial parks established after 2015. Based on our 2018 survey, Chonburi has a high percentage of internal water reuse in industrial parks, up to 70.5%, with most industrial parks in Chonburi mainly reclaiming water for two activities: (1) industrial water supply and (2) irrigation of plants and green areas [31]. Thus, Chonburi can serve as a model for internal water reuse in industrial parks, and its data are readily available for economic feasibility analyses. Such analyses will provide valuable information for other industrial parks in Thailand and other developing countries with similar industrial park types and socioeconomic statuses.

Therefore, the objectives of this study are (1) to evaluate the feasibility of water reclamation projects in industrial parks using CBA that incorporates both economic and environmental aspects, and (2) to examine the impact of environmental benefits on the feasibility of water reclamation projects in industrial parks under different scenarios (water reclamation activities, flow rates, and discount rates). This article is divided into four sections: introduction, methods, results and discussion, and conclusions. In the methods section, the background of Chonburi province is described, and the methodology of cost and benefit calculations and CBA are presented with an explanation of various water reclamation scenarios. The CBA of two water reclamation activities—that is, (1) for industrial water supply and (2) for irrigation of plants and green areas—are examined. We selected different flow rate scenarios in accordance with industrial parks in Chonburi, Thailand, which fall within the typical range for small-to-large industrial parks (Table S1). The discount rates used varied from 3% to 11%, covering those generally used for this type of project. The environmental benefits of water reclamation from not discharging pollutants into the environment were estimated using shadow prices for pollutants, which directly reflect the cost of damages if discharged to the environment. Net present values (NPVs), internal rates of return (IRRs), benefit–cost ratios (BCRs), and payback periods (PPs) were estimated and compared in cases with the inclusion and exclusion of environmental benefits. The results of the CBA are then presented in the results and discussion section, along with practical implications, recommendations for governmental support, and limitations and suggestions for future research. Lastly, all results and discussion are summarized in the conclusions.

Although the inclusion of externalities in the economic analysis of water reclamation has long been proposed [28,29], research in this area has been incomplete, particularly for water reclamation in industrial parks in developing countries. This study used the CBA framework with the inclusion of environmental benefits in real applications to examine the economic feasibility of water reclamation projects in industrial parks across a wide range of scenarios, utilizing data from industrial parks in Chonburi, Thailand, as a case study. Limitations and research gaps required for practical applications are also addressed. Moreover, shadow prices of pollutants were first applied to estimate the environmental benefits of water reclamation in industrial parks for not discharging them to the environment. The findings of this study can assist us to appreciate the true worth of water reclamation projects in industrial parks and to fully comprehend how environmental benefits affect the feasibility of water reclamation projects. Furthermore, they can serve as guidelines for the necessary governmental support that will substantially assist industrial parks in completing their initiatives. The CBA of different water reclamation scenarios can also help identify the most appropriate options under different circumstances.

2. Methods

2.1. Background of Water Reuse in Industrial Parks in Chonburi, Thailand

Chonburi Province is located in the eastern region of Thailand in the EEC, along with nearby provinces, that is, Rayong and Chachoengsao. Under the EEC initiated by the Thai government, the province is now set to become a major industrial center for Thailand's eastern region, with the port of Laem Chabang serving as a major commercial port. It is also a tourist attraction with a diverse and beautiful natural landscape. Chonburi's gross provincial product (GPP) was THB 1,030,949 million, equivalent to 6.3% of Thailand's gross domestic product (GDP) in 2018, making it the third-highest GPP in the country [32]. The economic structure of Chonburi Province in 2018 was divided into three sectors: industry (64.78%), agriculture (2.44%), and tourism and services (32.78%) [33].

Based on a study conducted by the Office of the National Water Resources [30], the water demand in Chonburi around 2017 was 469 million cubic meters (MCM), while in Rayong it was 494 MCM, and Chachoengsao recorded 1456 MCM. The major water use in Chonburi and Rayong is industrial water use, while Chachoengsao mainly uses water for irrigation. The study also projected that in the next 20 years, around 2037, the domestic and industrial water demand in the three provinces would increase from approximately 857 MCM to 1258 MCM.

Eight industrial parks are located in Chonburi, as shown in Figure 1, with 11 central WWTPs. Most factories in the industrial parks in Chonburi produce auto parts and electronic parts. According to our 2018 survey [31], which obtained secondary data from industrial parks in Chonburi, the flow rates of wastewater in the central WWTPs in the industrial parks in Chonburi varied from 226 to 12,704 m^3/day (Table S1), with a total flow rate of 51,119 m^3/day or 18,658,501 m^3 per year. The concentrations of pollutants in WWTP effluents, that is, biochemical oxygen demand (BOD), chemical oxygen demand

(COD), total Kjeldahl nitrogen (TKN), total phosphorus (TP), suspended solids (SS), total dissolved solids (TDS), and heavy metals, are shown in Table S2. A total of 70.5% of the wastewater was reused for various activities, that is, industrial water supply (49.7% of water reuse), irrigation of plants and green areas (21.1% of water reuse), power plant cooling water (28.7% of water reuse), and sale of reused water without post-treatment (0.5% of water reuse).

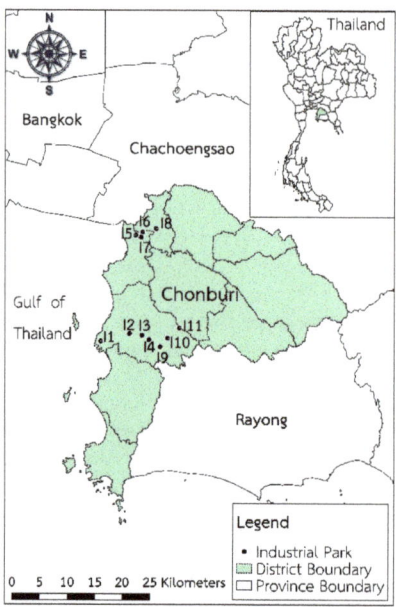

Figure 1. Locations of industrial parks in Chonburi Province, Thailand.

2.2. Scenarios of Water Reclamation Projects in Industrial Parks

In this study, we evaluate the economic feasibility of two types of water reclamation projects in industrial parks: (1) water reclamation projects for industrial water supply and (2) water reclamation projects for the irrigation of plants and green areas. These are the primary water reclamation activities in Chonburi's industrial parks, accounting for 70.8% of water reuse. Although reclaimed water used for power plant cooling accounted for 28.7% of total water reuse, Chonburi has only one industrial park with a power plant. The majority of industrial parks lack their own power plants. As a result, water reclamation for power plant cooling is uncommon in industrial parks.

For water reclamation projects for industrial water supply, we assumed that the system consisted of ultrafiltration (UF) and reverse osmosis (RO) units. This assumption is based on the actual systems that have been successfully constructed and operated in several industrial parks in Thailand. CBAs were conducted for different wastewater flow rates—5200, 10,000, 15,000, and 25,000 m^3/day—, covering typical flow rates in industrial parks in Chonburi that reclaim water for industrial water supply. RO permeate was assumed to be 75% of the influent. The assessment was conducted for cases that included and excluded externalities.

For water reclamation plants for the irrigation of plants and green areas in industrial parks, we assumed that the system consisted of sand filtration and chlorine disinfection. This system is considered appropriate and highly affordable in developing countries. CBAs were conducted for wastewater flow rates of 240, 480, 2400, 3600, and 4800 m^3/day. These flow rates were selected based on the reclaimed water for the irrigation of plants and green areas in industrial parks in Chonburi, which were typically less than 20% of wastewater flow rates, according to our 2018 survey [31], together with the highest maximum capacity of the central WWTPs in the industrial parks in Chonburi (24,000 m^3/d). Thus, the maximum flow rate selected was 4800 m^3/day, and the lower flow rates varied. This irrigation activity is assumed to have no negative environmental impact on the soil, and the volume is not too large to infiltrate groundwater and affect groundwater quality.

Notably, the flow rates chosen for these two water reclamation activities (industrial water supply and irrigation of plants and green areas) were not the same. The irrigation of plants and green areas was usually limited by the green areas available in industrial parks; thus, the flow rates were typically less than 20% of the WWTP effluent. By contrast, water reclamation for industrial water supply is not limited in this way.

2.3. Estimation of the Costs and Benefits of Water Reclamation Projects

For water reclamation projects for industrial water supply, the direct costs and benefits of these projects included construction costs, operation and maintenance costs, and revenues from the sale of reclaimed water. The information used for the estimation of direct costs and benefits is shown in Table 1.

Table 1. Information for estimating the direct costs and benefits of water reclamation projects for industrial water supply, consisting of ultrafiltration and reverse osmosis.

Direct Costs and Benefits	Wastewater Flow Rates (m^3/d)			
	5200	10,000	15,000	25,000
Volume of reclaimed water produced (m^3/d) [1]	3900	7500	11,250	18,750
Direct costs				
Construction costs (THB million) [2]	108.2	145	187	332
Operating and maintenance costs (million THB/year) [3]	37.96	73	109.5	182.5
Direct benefits				
Water prices (THB/m^3) [4]	26	26	26	26

[1] We assumed that the filtrate flow rates were 75% of the wastewater flow rates. [2] These are the actual construction costs of water reclamation plants in Thailand disclosed by industrial estates that used ultrafiltration and reverse osmosis systems, which included the costs of pipeline systems. [3] Ratanathamsakul et al. (2020) [34]. [4] This is the water price in an industrial estate (Industrial Estate Authority of Thailand).

For cases that included externalities, the environmental benefits of not discharging WWTP effluent into the river were included as positive externalities. The costs of environmental damage from pollutants in wastewater, such as nitrogen, phosphorus, SS, and COD, were estimated based on their average concentrations in the WWTP effluents of industrial parks in Chonburi in 2018 (Table S3). In the absence of recent information on the costs of environmental damage from water pollutants discharged into water bodies in Thailand, these values were derived from the shadow prices of pollutants discharged into a river in Spain (−16.353 EURO/kg N; −30.944 EURO/kg P; −0.005 EURO/kg SS; and −0.098 EURO/kg COD) in [35]. The shadow prices of pollutants were estimated from the associated costs of pollutant removal in the treatment process. When a water reclamation project exists, no pollutants are released. The shadow prices of pollutants, therefore, reflect environmental benefits that are positive externalities in the CBA. Equation (1) was then used to account for the exchange rates of the currencies and the difference in gross domestic product based on purchasing power parity (GDP (PPP)) per capita in Spain and Thailand, reflecting the difference in the ability to purchase goods and services per capita in the two countries [36]. The values of the money in 2020 were estimated using Equation (2).

$$\text{CFR}_{\text{TH2010}} = \frac{\text{GDP (PPP) per capita}_{\text{TH2010}}}{\text{GDP (PPP) per capita}_{\text{SP2010}}} \times E_{2010} \quad (1)$$

where

$\text{CFR}_{\text{TH2010}}$ = Thailand's PPP conversion in 2010 (THB/EURO)

GDP (PPP) per capita $_{TH2010}$ = GDP (PPP) per capita of Thailand in 2010
= 13,195.36 USD [37]
GDP (PPP) per capita $_{SP2010}$ = GDP (PPP) per capita of Spain in 2010
= 31,593.85 USD [37]
E_{2010} = average exchange rate in 2010 = 42.4 THB/EURO [38]

$$FV = PV(1+r)^n \qquad (2)$$

where

FV = future value of the investment of present value (PV) (the values in 2020)
PV = present value of an investment (the values in 2010)
r = 10-year average inflation rate (2010–2020), which is 1.42% [39]
n = number of compounding periods (10 years)

Regarding water reclamation plants for the irrigation of plants and green areas in industrial parks, the direct costs and benefits included construction costs, operation and maintenance costs, revenues from the sale of reclaimed water, and savings on fertilizer costs due to the nitrogen and phosphorus contents in reclaimed water. The information used for the estimation of direct costs and benefits is shown in Table 2. Indirect environmental benefits from not discharging WWTP effluents into rivers were estimated from the average pollutant concentrations in WWTP effluents together with their shadow prices [35] using Equations (1) and (2) in the same manner as that used for water reclamation projects for industrial water reuse.

Table 2. Information for estimating the direct costs and benefits of water reclamation projects for the irrigation of plants and green areas, consisting of sand filtration and chlorination.

Direct Costs and Benefits	Wastewater Flow Rates (m³/d)				
	240	480	2400	3600	4800
Volume of reclaimed water produced (m³/d) [1]	240	480	2400	3600	4800
Direct costs					
Construction costs (THB million) [2]	1.4	2	12	16	24
Pipeline system costs (THB million) [3]	0.505	0.87	2.795	2.795	5.21
Operating and maintenance costs (million THB/year) [4]	0.1971	0.3942	1.971	2.9565	3.942
Direct benefits					
Reclaimed water prices (THB/m³) [5]	12	12	12	12	12
Nitrogen fertilizer price [6] (THB/kg)	11.5	11.5	11.5	11.5	11.5
Phosphorus fertilizer price [6] (THB/kg)	17.1	17.1	17.1	17.1	17.1

[1] We assumed no loss in the water reclamation systems. [2] The construction costs of sand filtration and chlorine contact tank units in water treatment plants are obtained from the Provincial Waterworks Authority of Thailand. [3] The costs of pipeline installation in Thailand are obtained from contractor companies under the assumption that the water distribution distance was 5 km. [4] Ratanathamsakul et al. (2020) [34]. [5] These are the reclaimed water (second grade water) prices disclosed by an industrial estate. [6] The fertilizer prices are from the Bureau of Agricultural Economic Research, Thailand.

2.4. Cost–Benefit Analysis of Water Reclamation Projects

CBA was used to analyze the economic feasibility associated with reclaimed wastewater projects in industrial parks with different flow rate scenarios. The analysis was completed by assessing various indicators: NPVs, IRRs, BCRs, and PPs [15,40]. The methodology flowchart of the CBA of water reclamation projects in industrial parks, including all of the scenarios conducted in this study, is illustrated in Figure 2.

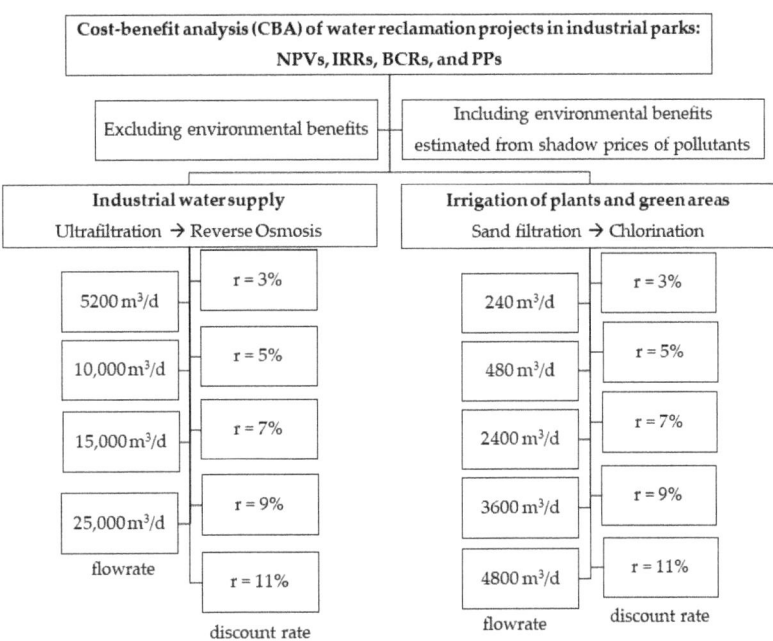

Figure 2. Methodology flowchart of the CBA of water reclamation projects in industrial parks.

NPV is the difference between a project's benefits and its cost over the project's lifetime, as shown in Equation (3). A 20-year lifespan was assumed for all water reclamation plants [13]. Discount rates of 3%, 5%, 7%, 9%, and 11% were used in the analysis, covering the rates commonly used for water projects in Thailand and internationally [13,41,42]. In 2020, the Comptroller General's Department, Thailand, recommended the use of a 5% lending interest rate to estimate the reference prices of construction projects. The social discount rate recommended by the Asian Development Bank (ADB) is 9% [43]. For the private sector, discount rates can be considered through weighted average cost of capital analysis to reflect the opportunity cost of private investment, which largely depends on the capital structure of companies and can be as high as 11%. Alternatively, lending interest rates can be adopted as discount rates. In Thailand, for loans from four major commercial banks, the minimum retail rate is 6.75% [44]. However, if projects are eligible to receive loans from the Environment Fund in Thailand, the interest rates are as low as 2–3% [45]. Nevertheless, large private companies can often afford to undertake projects with environmental benefits as their corporate social responsibility projects, regardless of whether they yield no or low profits. Water reclamation projects may fall into this category, where the discount rates used are usually low. Therefore, the discount rates of 3% to 11% were chosen in this study to cover typical ranges of discount rates.

$$\text{NPV} = \sum_{t=0}^{n} \frac{-C_t + B_t}{(1+r)^t} \qquad (3)$$

where

C_t = costs in year t
B_t = benefits in year t
r = discount rate
n = project's lifetime (20 years)

Costs (C_t) and benefits (B_t) of different scenarios are summarized in Table S4. If the result of the calculation is an NPV that is ≥ 0, then a project is economically acceptable;

however, if the result of the calculation is an NPV that is <0, then a project is not acceptable from an economic perspective.

The IRR is used to estimate the profitability of potential investments. The IRR is a discount rate that makes the NPV of all cash flows equal to zero, as shown in Equation (4). A project is economically feasible if the IRR is $\geq$ r.

$$\sum_{t=0}^{n} \frac{-C_t + B_t}{(1 + IRR)^t} = 0 \qquad (4)$$

where

C_t = costs in year t
B_t = benefits in year t
n = project's lifetime (20 years)

In CBA, the BCR is another useful indicator. It is defined as the ratio of project benefits to project costs, as shown in Equation (5). Project implementation is acceptable if the BCR is ≥ 1.

$$BCR\ ratio\ =\ \frac{\sum_{t=0}^{n-1} \frac{B_t}{(1+r)^t}}{\sum_{t=0}^{n-1} \frac{C_t}{(1+r)^t}} \qquad (5)$$

where

C_t = costs in year t
B_t = benefits in year t
r = discount rate
n = project's lifetime (20 years)

The PP is the time at which the benefits of a project surpass its costs, as shown in Equation (6). In other words, in the year after the PP of a project, the net gains or benefits of the project become visible.

$$PP = number\ of\ years\ before\ payback + \frac{unrecovered\ present\ value}{present\ value\ of\ cash\ flows\ in\ payback\ year} \qquad (6)$$

3. Results and Discussion

3.1. Cost–Benefit Analysis of Water Reclamation Projects for Industrial Water Supply

CBAs of water reclamation projects for industrial water supply were conducted based on wastewater flow rates of 5200, 10,000, 15,000, and 25,000 m^3/day. The NPVs of water reclamation projects for industrial water supply for cases that included and excluded environmental benefits are shown in Figure 3 (Table S5). The details of the costs and benefits in the NPVs are summarized in Table S6. Considering only the direct costs and benefits resulted in NPVs less than zero in all cases, this indicates that the projects were not economically feasible. The total direct benefits of a project were less than the investment and operation and maintenance costs. However, when the environmental benefits of not discharging WWTP effluents into public waters were considered along with the direct costs and benefits, the NPVs were greater than 0 in all cases except for those with a flow rate of 5200 m^3/day and discount rates of 5%, 7%, 9% and 11% and those with flow rates of 10,000 and 25,000 m^3/day and an 11% discount rate.

In all cases, the IRRs were smaller than the fixed discount rates when only the direct costs and benefits were considered, as shown in Table 3. Given that the NPVs of the projects were negative in these cases, it was impossible to calculate the IRRs such that the NPVs became zero. By contrast, when the indirect benefits to the environment were considered together with the direct costs and benefits, in most cases, the IRRs were larger than the chosen discount rates, which ranged from 4.6% to 11.5%.

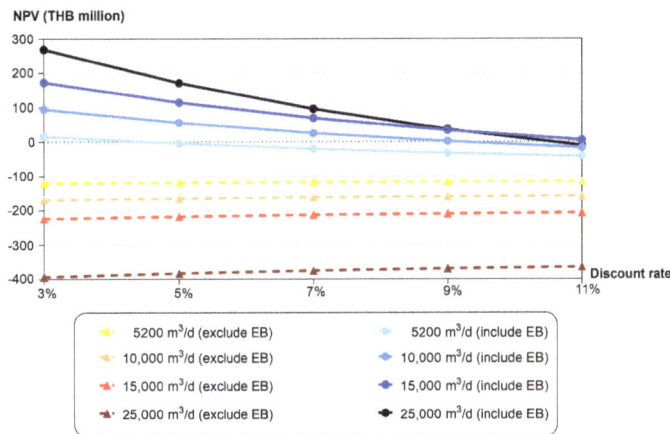

Figure 3. Net present values (NPVs) for water reclamation projects for industrial water supply, with the inclusion or exclusion of environmental benefits (EB).

Table 3. Internal rates of return for water reclamation projects for industrial water supply when including environmental benefits.

Flow Rate (m^3/d)	Including Environmental Benefits
5200	4.6%
10,000	9.2%
15,000	11.5%
25,000	10.5%

Similarly, the BCR results in Figure 4 (Table S5) show that when only the direct benefits were considered, the BCRs were smaller than 1 and ranged from 0.72 to 0.88. However, when the indirect environmental benefits were also considered, the BCRs were 0.90–1.10. These results are consistent with the PPs shown in Figure 5 (Table S5). The PPs were more than 20 years when only direct costs and benefits were considered. When indirect environmental benefits were included, the PPs ranged from 8.9 to more than 20 years. However, if the PP exceeds 20 years, the projects are not considered feasible.

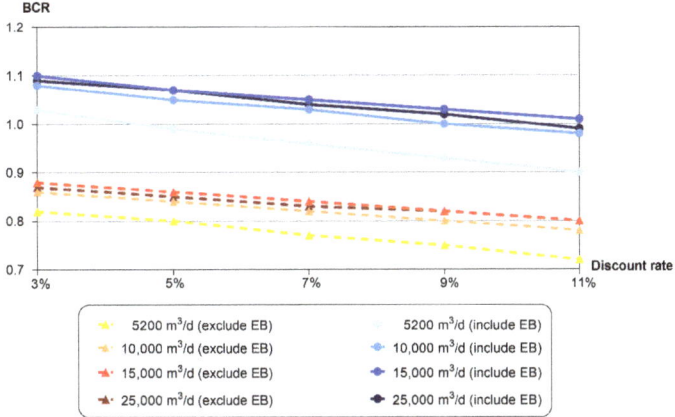

Figure 4. Benefit–cost ratios (BCRs) for water reclamation projects for industrial water supply when including and excluding environmental benefits (EB).

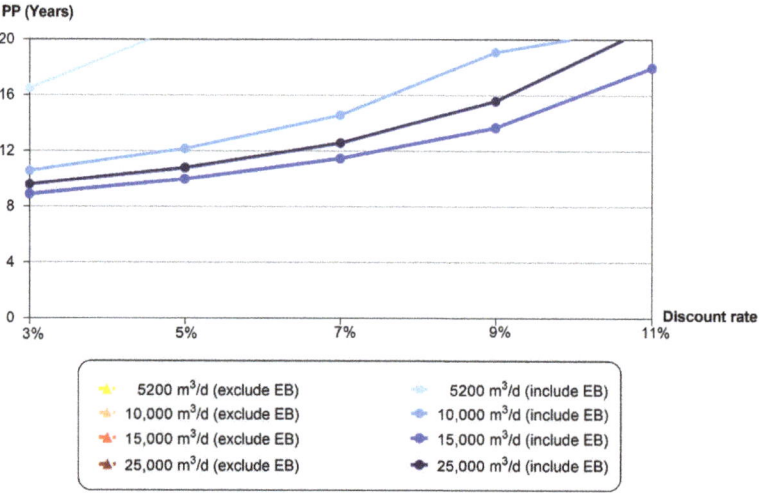

Figure 5. Payback periods (PPs) for water reclamation projects for industrial water supply when including and excluding environmental benefits (EB).

Moreover, the selling prices of reclaimed water appeared to have a major impact on project feasibility. If the selling prices of reclaimed water were too low, the projects were not feasible. Therefore, we further determined the minimum selling prices of reclaimed water that would still make the projects feasible. The results are shown in Table 4. In all cases, when only the direct costs and benefits were considered, the minimum selling prices of reclaimed water ranged from 29.73 to 36.24 THB/m^3. By contrast, when environmental benefits were included, the minimum selling prices were 23.23–29.73 THB/m^3, which were lower than the prices without environmental benefits. At these prices, the NPVs became positive, the BCRs were equal to 1, and the IRRs were equal to the selected discount rates. However, these prices were associated with a PP of 20 years. Therefore, selling prices should be set higher to make the investment more profitable and to reduce the PP.

Table 4. Minimum selling price of reclaimed water (THB) for water reclamation projects for industrial water supply when including and excluding environmental benefits (EB).

Discount Rate	5200 m^3/d		10,000 m^3/d		15,000 m^3/d		25,000 m^3/d	
	Exclude EB	Include EB	Exclude EB	Include EB	Exclude EB	Include EB	Exclude EB	Include EB
3%	31.78	25.28	30.23	23.73	29.73	23.23	29.93	23.43
5%	32.77	26.27	30.93	24.42	30.33	23.83	30.56	24.06
7%	33.85	27.35	31.67	25.17	30.97	24.47	31.25	24.75
9%	35.00	28.50	32.49	25.97	31.66	25.16	31.99	25.49
11%	36.24	29.73	33.33	26.82	32.39	25.88	32.76	26.26

These results clearly show that water reclamation projects for industrial water supply within industrial parks are more economically feasible when environmental benefits are included. However, such projects are not economically feasible under all scenarios when only the direct costs and benefits are considered. This consideration aligns with the perspective of industrial parks, where externalities are generally not considered. In other words, water reclamation for industrial water supply is infeasible from the perspective of industrial parks, and thus, it does not provide financial incentives for project implementation. However, from the public's perspective, the benefits outweigh the costs when all aspects, including environmental benefits, are considered. Therefore, the government may have to intervene to support such projects. Recommendations for governmental support are discussed in Section 3.4.

Similarly, the cost-effectiveness analysis conducted by Giurco et al. (2011) [27] found that financial support in the form of a capital grant and a 0% interest loan was required for industrial water reuse projects to be financially feasible. Water reuse options were found to be more expensive than other water supply options available in the region under study (Port Melbourne, Australia). Therefore, financial assistance should be offered to businesses as an incentive to implement water reuse projects [26].

The CBA of water reclamation projects using effluents from municipal WWTPs in Beijing, China, was conducted by Fan et al. (2015) [16]. In their study, reclaimed water (680 million m^3 in 2010) was used for industrial reuse (20%), agricultural irrigation (47%), environmental reuse (30%), and miscellaneous urban reuse (3%). Both internal costs and benefits as well as positive and negative externalities, including environmental benefits, public health impacts, and groundwater recharge and pollution, were considered in the CBA [16]. The CBA revealed a relatively high BCR of 1.7 when externalities were included, providing further incentives for project implementation.

The environmental benefits of water reuse in a model industrial park were previously analyzed using LCA for several environmental categories, including climate change, freshwater eutrophication, marine eutrophication, and resource depletion (minerals, fossils, renewables, as well as water) [27]. All water reuse options in the industrial park, which was assumed to be located in Germany, provided environmental benefits for all environmental categories, except resource depletion. Similarly, using LCA, Tong et al. (2013) [19] evaluated the environmental impacts of water reuse in an industrial park in China. Their results revealed the environmental benefits of water reuse. The LCA conducted in previous studies clearly demonstrated the environmental benefits of water reuse in industrial parks. The results of our study reemphasize the importance of the environmental benefits associated with water reclamation projects, which, when factored into the CBA, can strongly influence economic feasibility.

3.2. Cost–Benefit Analysis of Water Reclamation Projects for the Irrigation of Plants and Green Areas

Water reclamation for the irrigation of plants and green areas provides several direct and indirect benefits, including revenues from the sale of reclaimed water, savings on fertilizer costs due to the nitrogen and phosphorus contents in reclaimed water, and environmental benefits from not discharging WWTP effluent into public waters. Feasibility studies of water reclamation projects for the irrigation of plants and green areas in industrial parks with post-treatment processes consisting of sand filtration and chlorine disinfection were conducted under wastewater flow rates of 240, 480, 2400, 3600, and 4800 m^3/day. The NPVs were greater than 0 in all cases, regardless of whether the indirect benefits to the environment were included or not (Figure 6, Table S7). The details of the costs and benefits in the NPVs are summarized in Table S8. Notably, the investment and operation and maintenance costs were much lower than those for water reclamation projects for industrial water supply, as membrane units were not included in the treatment system. The selling prices of reclaimed water (12 THB/m^3) appeared to be high. As a result, the benefits exceed the investment and operation and maintenance costs. Nevertheless, when the indirect benefits to the environment were included, the NPVs were greater than the analyses that considered only the direct costs and benefits.

In all cases, the IRRs were higher than the fixed discount rates (Table 5). When only the direct costs and benefits were considered, the IRRs ranged from 45.5% to 69.2%. When indirect environmental benefits were included, the IRRs ranged from 68.0% to 103.3%. Further, the BCRs were greater than 1 in all cases, as shown in Figure 7 (Table S7), ranging from 2.44 to 3.78 when only the direct costs and benefits were considered, and from 3.42 to 5.30 when environmental benefits were included.

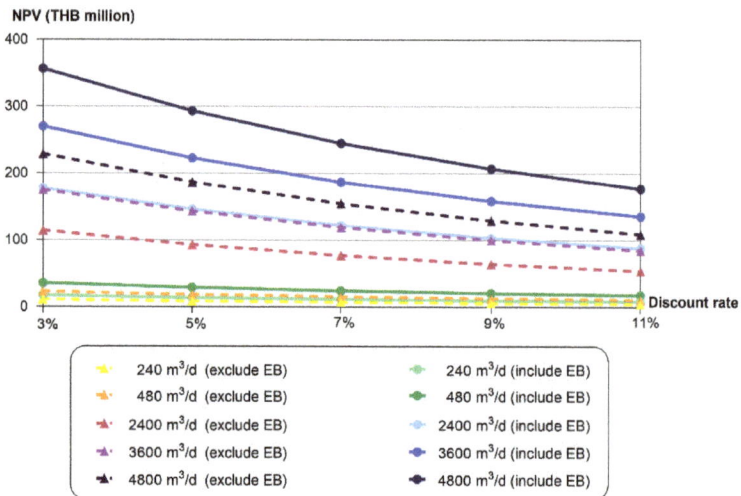

Figure 6. NPVs for water reclamation projects for the irrigation of plants and green areas when including and excluding environmental benefits (EB).

Table 5. Internal rates of return for water reclamation projects for the irrigation of plants and green areas when including and excluding environmental benefits (EB).

Flow Rate (m³/d)	Exclude EB	Include EB
240	45.5%	68.0%
480	60.5%	90.2%
2400	58.6%	87.5%
3600	69.2%	103.3%
4800	59.4%	88.6%

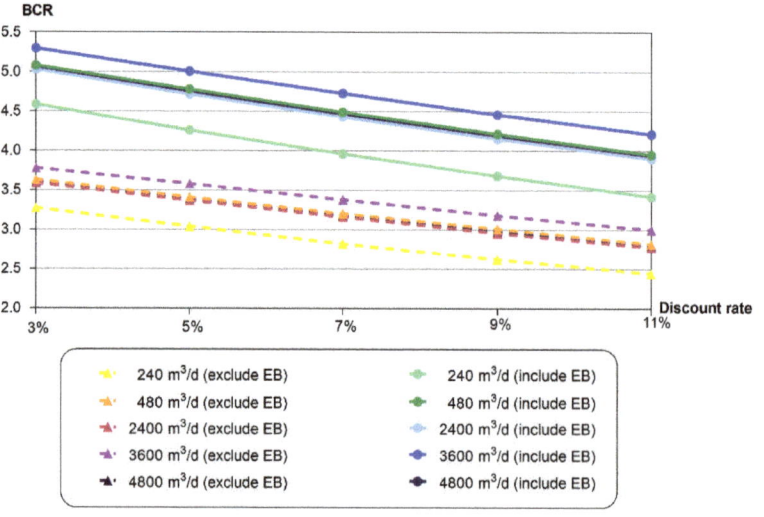

Figure 7. Benefit–cost ratios (BCRs) for water reclamation projects for the irrigation of plants and green areas when including and excluding environmental benefits (EB).

As shown in Figure 8 (Table S7), the PPs ranged from 1.5 to 2.7 years when only the direct costs and benefits were considered, and they were between 1.0 and 1.7 years when the indirect environmental benefits were considered together with the direct benefits. In summary, the PP was always less than 3 years, indicating a good investment.

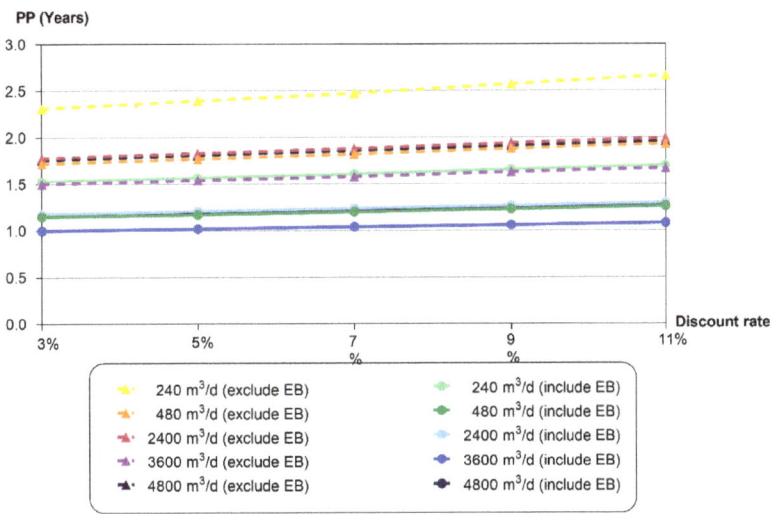

Figure 8. Payback periods (PPs) for water reclamation projects for the irrigation of plants and green areas when including and excluding environmental benefits (EB).

According to the feasibility analyses, all cases were highly cost-effective. Table 6 shows that the minimum selling prices were well below 12 THB/m^3 in all cases. For cases where only the direct costs and benefits were evaluated, the minimum selling price of reclaimed water ranged from 3.06 to 4.83 THB/m^3. At these prices, the NPVs became positive, the BCRs were equal to 1, and the IRRs were equal to the selected discount rates. Moreover, the minimum selling price decreased to zero when environmental benefits were included. In other words, reclaimed water used for the irrigation of plants and green areas can be offered free of charge. The NPVs were positive at these prices, and the BCRs were 1.01–1.57. The IRRs were 11.3–18.8%, well above the discount rates used, while the PPs were 5.7–18.9 years.

Table 6. Minimum selling price of reclaimed water (THB) for water reclamation projects for the irrigation of plants and green areas when including and excluding environmental benefits (EB).

Discount Rate	240 m^3/d		480 m^3/d		2400 m^3/d		3600 m^3/d		4800 m^3/d	
	Exclude EB	Include EB	Exclude EB	Include EB	Exclude EB	Include EB	Exclude EB	Include EB	Exclude EB	Include EB
3%	3.56	0.00	3.20	0.00	3.24	0.00	3.06	0.00	3.22	0.00
5%	3.85	0.00	3.42	0.00	3.46	0.00	3.25	0.00	3.44	0.00
7%	4.15	0.00	3.65	0.00	3.70	0.00	3.67	0.00	3.67	0.00
9%	4.48	0.00	3.90	0.00	3.95	0.00	3.67	0.00	3.93	0.00
11%	4.83	0.00	4.16	0.00	4.22	0.00	3.90	0.00	4.19	0.00

From the CBA of water reclamation projects for the irrigation of plants and green areas within industrial parks, projects are profitable for all scenarios, regardless of whether environmental benefits are included. Therefore, from the perspective of industrial parks, there are financial incentives for project implementation. Nevertheless, when including environmental benefits, projects had higher NPVs, IRRs, and BCRs and lower PPs compared with cases excluding environmental benefits.

The findings of this study are consistent with those of the CBAs of projects for the reclamation of effluents from municipal WWTPs for irrigation, recreation, and environmental purposes, which found that projects were more economically feasible when externalities such as environmental and social benefits were included [13,15,17]. Verlicchi et al. (2012) [15] evaluated the feasibility of using reclaimed wastewater from a central WWTP in Po Valley, Italy, for irrigation and environmental purposes. The reclamation treatment train consisted of rapid sand filtration, a horizontal subsurface flow bed, and a lagoon. The results suggested that the agricultural, environmental, financial, and recreational benefits offset the high construction costs, thereby making the project economically feasible [15]. Similarly, Arena et al. (2020) [17] conducted a CBA on a reclamation project in Puglia, Italy, which reused effluent from a municipal WWTP for irrigation, and the CBA considered environmental and recreational benefits. The water reclamation units included clariflocculation, filtration, and ultraviolet (UV) disinfection. In almost all cases, environmental benefits must be included to make a project economically feasible.

Molinos-Senante et al. (2011) [13] conducted CBA on 13 WWTP in the Valencia region of Spain that reused effluent for environmental purposes. Their results suggested that some projects (4 out of 13) were not economically feasible when only the internal costs and benefits were assessed, whereas all projects were feasible when externalities, that is, environmental benefits, were included. However, the water reclamation system in the Molinos-Senante et al. study [13] was unclear and might include secondary treatment and/or membrane units, differing from this study and the studies conducted by Verlicchi et al. (2012) [15] and Arena et al. (2020) [17]. Despite the differences in water reclamation facilities, the inclusion of the environmental benefits similarly improved the economic feasibility of projects.

Similarly, LCA and eco-efficiency assessment have pointed to a similar direction that water reclamation for irrigation delivered economic and environmental benefits [10,46], although the results are likely site-specific [47]. An LCA study by Meneses et al. (2010) [10] demonstrated that the agricultural use of reclaimed water from a WWTP located on the Mediterranean coast offers environmental and economic benefits, especially when compared to desalinated water, and that water reclamation should be encouraged when freshwater is scarce. Based on an eco-efficiency assessment, Canaj et al. (2021) [46] suggested that reclaimed water could be used to generate an economically profitable yield of vineyard cultivation in Acquaviva Delle Fonti, Italy while also offering net environmental benefits.

3.3. Practical Implications

Comparing the two options, we observed that water reclamation projects for the irrigation of plants and green areas are considerably more cost-effective than those for industrial water supply. Therefore, the irrigation of plants and green areas is considered a recommended entry point for water reclamation in industrial parks due to its high NPVs, IRRs, and BCRs and short PPs (<3 years). However, the water demand for irrigation in industrial parks is generally low, not more than 23% of the WWTP effluent, based on the actual amount of irrigation water in industrial parks in Chonburi (Table S9). Therefore, another water reclamation option, such as reclaimed water for industrial water supply, can be combined to achieve a higher percentages of water reuse.

In this study, a wide range of discount rates was applied in the feasibility studies to provide an overall picture of how environmental benefits affect the feasibility of water reclamation projects. It was not our intent to perform feasibility studies for specific projects. Investments in water reclamation projects could be made by different companies with different discount rates. Given that each company may have vastly different financial costs, lending interest rates, and/or shareholder returns, discount rates are likely to vary widely. It is recommended that the results of this study be used with regard to appropriately selected discount rates.

3.4. Recommendations for Governmental Support

The economic feasibility analyses conducted in this study indicate that water reclamation projects for industrial water supply are infeasible without accounting for environmental benefits. Given that pollution discharge fees do not exist in Thailand or in many developing countries, environmental benefits currently do not directly benefit the industrial sector. Their incentives for pursuing water reclamation projects are likely to be limited. Environmental benefits often benefit the general public. As a result, the government should establish supportive policies to ensure the success of such water reclamation efforts.

Governmental support for water reclamation projects can be in the form of funding support for investment in water reclamation projects [48,49], the addition of adders to the prices of reclaimed water, or tax reductions for businesses that invest in water reclamation projects [50]. These measures can be chosen in a variety of ways, depending on the government and local context. Economic feasibility analyses, such as those undertaken in this study, may also aid in determining the appropriate level of support. For instance, funding support for water reclamation projects could be assessed using the difference between the NPVs with and without environmental benefits. The amount of the adders added to reclaimed water prices can be estimated by comparing the minimum selling prices of reclaimed water with and without consideration of the environmental benefits. Tax savings should be proportional to the environmental benefits created by such projects. Notably, giving low-interest loans did not appear to make water reclamation projects economically feasible, as the IRRs were negative for all flow rates when environmental benefits were excluded.

Although water reclamation projects for the irrigation of plants and green areas were found to be economically feasible, even without considering environmental benefits, governmental support is still considered beneficial because it can increase the incentives for investing in such projects. Governmental support for water reclamation projects for the irrigation of plants and green areas could take the form of tax reductions or low-interest loans.

The government can also play an important role in initiating technology adoption and in developing a regulatory framework, indicators, and monitoring procedures to ensure transparency and to guarantee health and safety, ultimately leading to public acceptance of water reclamation.

3.5. Limitations and Suggestions for Future Research

The CBA of water reclamation projects in industrial parks conducted in this study shows the importance of environmental benefits of water reclamation projects, which increase the economic feasibility of the projects when they are taken into account. However, it should be noted that the CBA is based on many assumptions, and the applicability of the results critically depends on the validity of the assumptions. For example, the water treatment technology used in a targeted industrial park could be different from the ones chosen in this study, namely, UF and RO for water reclamation for industrial water supply and sand filtration, followed by chlorination for irrigation of plants and green areas, resulting in different costs of construction, operation, and maintenance. Further, the flow rate of irrigation water for each industrial park is quite specific. The amount of water supply for irrigation is uncontrollable and depends on the vegetation area, rainfall, evapotranspiration, season, and other factors related to water use by vegetation. Rough estimates of irrigation water in industrial parks might be obtained based on their green areas using the value suggested for calculating irrigation water demand in public parks, which is 1.7 mm/d in Thailand [51]. Taking into consideration local meteorological conditions and grass evapotranspiration, the irrigation water demands should be calculated as shown in Text S1, Tables S10 and S11. The calculation was made for grass as a representative crop, which resulted in an irrigation water demand of 1.9 mm/d in Chonburi. With this irrigation water demand, the irrigated green areas based on the flow rate scenarios would be in the range of 9–177 ha (Table S12). Although we conducted economic feasibility studies for a wide range of irrigation water flow rates (240–4800 m^3/d), it is possible that the irrigation water flow rate for a specific industrial park would be out of this range, particularly at low flow rates, such as industrial parks I2–I4 (43–56 m^3/d) in Chonburi (Table S9). In these cases, the results of this study would not be applicable. Furthermore, it was assumed in the CBA that irrigation water would not be large enough to infiltrate groundwater and that there would be no adverse environmental impacts on soil and groundwater. Indeed, these assumptions can be considered limitations of our study, as they may not be true; thus, the environmental impacts of irrigation by reclaimed water should be considered in future studies.

Further, the environmental benefits of not releasing heavy metals into water bodies when wastewater treatment plant effluent is reclaimed were not considered in this study, as their shadow prices are not currently available. Research on the environmental costs of pollutants other than COD, SS, N, and P, such as heavy metals, chemicals, and emerging pollutants, is still needed. Further research is also required on the ecosystem service values resulting from water reclamation projects in industrial parks. The socioeconomic impacts of water reclamation projects in industrial parks on other stakeholders, such as communities and the public, were also not considered in this study. For example, water reclamation in industrial parks can reduce overall water demand, making more water available to other sectors, such as agriculture, especially during drought periods. Additionally, there are health benefits to not releasing pollutants into the environment, which may result in less human exposure to these pollutants. Future research could translate and incorporate all aspects of environmental impacts analyzed in detail in LCA into CBA. The costs and benefits associated with these impacts could demonstrate the full value of water reclamation projects, potentially making them more feasible and appealing.

4. Conclusions

Water reclamation projects for the irrigation of plants and green areas with post-treatment consisting of sand filtration and chlorination in industrial parks were feasible regardless of the environmental benefits. Moreover, reclaimed water for irrigation could be given for free if environmental benefits are considered. However, in many cases, water reclamation projects for industrial water supply consisting of UF and RO, which were otherwise financially infeasible, became feasible when environmental benefits were considered. Comparing the two options, we observed that water reclamation for irrigation was substantially more cost-effective than the industrial water supply. As water reclamation projects often benefit the general public, the government should provide supportive measures to encourage their implementation and ensure their success. Nevertheless, the CBA in this study has certain limitations, as the environmental costs associated with using reclaimed water for irrigation to soil and groundwater if provided in excess were not considered. The entire benefits of water reclamation projects, including the environmental benefits of not releasing heavy metals, chemicals, or emerging pollutants, and socioeconomic and health benefits to communities and the general public, still require more investigation in the future.

Supplementary Materials: The following supporting information can be downloaded at: https://www.mdpi.com/article/10.3390/w14071172/s1, Table S1: Flow rates of wastewater to wastewater treatment plants in industrial parks in Chonburi and Rayong, which are among the most important industrial provinces in the Eastern Economic Corridor (EEC), Thailand. Table S2: Concentrations of pollutants in the wastewater treatment plant effluents of industrial estates in Chonburi, according to our 2018 survey. Table S3: Average concentrations of pollutants and their environmental damage costs. Table S4: Costs and benefits of different scenarios. Table S5: Net present values (NPVs), internal rates of return (IRRs), benefit–cost ratios (BCRs), and payback periods (PPs) for water reclamation projects for industrial water supply when including and excluding environmental benefits. Table S6: Costs and benefits in the net present values (NPVs) of water reclamation projects for industrial water supply. Table S7: Net present values (NPVs), internal rates of return (IRRs), benefit–cost ratios (BCRs), and payback periods (PPs) for water reclamation projects for irrigation of plants and green areas when including and excluding environmental benefits. Table S8: Costs and benefits in the net present values (NPVs) of water reclamation projects for irrigation of plants and green areas. Table S9: Actual amount of irrigation water in industrial parks in Chonburi in 2018. Table S10: Effective monthly rainfall in Chonburi. Table S11: Irrigation water demand of grass in Chonburi. Table S12: The irrigated green areas based on flow rate scenarios. Text S1: Calculation of irrigation water demand.

Author Contributions: Conceptualization, B.B.S. and P.S.; methodology, B.B.S., P.S. and J.S.; investigation, W.I.; data curation, W.I.; writing—original draft preparation, W.I., B.B.S. and J.S.; writing—review and editing, B.B.S., P.S., J.S., P.R. and T.L.; supervision, B.B.S. and P.S.; funding acquisition, B.B.S., P.R., J.S. and T.L. All authors have read and agreed to the published version of the manuscript.

Funding: This research was funded by Thailand Science Research and Innovation (TSRI) under grant numbers SRI6230304, ORG6310016, and CU_FRB640001_01_21_6.

Institutional Review Board Statement: Not applicable.

Informed Consent Statement: Not applicable.

Data Availability Statement: Not applicable.

Acknowledgments: The authors would like to express their gratitude to all of the industrial parks that provided the data for this study. We would also like to thank the Industrial Estate Authority of Thailand for their tremendous support.

Conflicts of Interest: The authors declare no conflict of interest.

References

1. UNESCO World Water Assessment Programme. *The United Nations World Water Development Report 2021: Valuing Water*; UNESCO: Paris, France, 2021; ISBN 978-92-3-100434-6.
2. IPCC. Summary for Policymakers. In *Climate Change. The Physical Science Basis*; Contribution of Working Group I to the Sixth Assessment Report of the Intergovernmental Panel on Climate Change; Masson-Delmotte, V., Zhai, P., Pirani, A., Connors, S.L., Péan, C., Berger, S., Caud, N., Chen, Y., Goldfarb, L., Gomis, M.I., et al., Eds.; IPCC: Geneva, Switzerland, 2021; *in press*.
3. United Nations Development Programme. Sustainable Development Goals (SDGs). 2021. Available online: https://www.undp.org/sustainable-development-goals (accessed on 27 August 2021).
4. Korhonen, J.; Honkasalo, A.; Seppala, J. Circular Economy: The concept and its limitations. *Ecol. Econ.* **2018**, *143*, 37–46. [CrossRef]
5. Sgroi, M.; Vagliasindi, F.G.; Roccaro, P. Feasibility, sustainability and circular economy concepts in water reuse. *Curr. Opin. Environ. Sci.* **2018**, *2*, 20–25. [CrossRef]
6. Tian, Y.; Hu, H.; Zhang, J. Solution to water resource scarcity, water reclamation and reuse. *Environ. Sci. Pollut. Res.* **2017**, *24*, 5095–5097. [CrossRef] [PubMed]
7. Leverenze, H.L.; Asano, T. Wastewater reclamation and reuse system. *Treatise Water Sci.* **2011**, *4*, 63–71.
8. Bauer, S.; Dell, A.; Behnisch, J.; Chen, H.; Bi, X.; Nguyen, V.A.; Linke, H.J.; Wagner, M. Water-reuse concepts for industrial parks in waterstressed regions in South East Asia S. *Water Sci. Technol. Water Supply.* **2020**, *20*, 296–306. [CrossRef]
9. Lyu, S.; Chen, W.; Zhang, W.; Fan, Y.; Jiao, W. Wastewater reclamation and reuse in China: Opportunities and challenges. *J. Environ. Sci.* **2016**, *39*, 86–96. [CrossRef] [PubMed]
10. Meneses, M.; Pasqualino, J.C.; Castells, F. Environmental assessment of urban wastewater reuse: Treatment alternatives and applications. *Chemosphere* **2010**, *81*, 266–272. [CrossRef]
11. Munoz, I.; Rodriguez, A.; Rosal, R.; Fernandez-Alba, A.R. Life cycle assessment of urban wastewater reuse with ozonation as tertiary treatment: A focus on toxicity-related impacts. *Sci. Total Environ.* **2009**, *407*, 1245–1256. [CrossRef]
12. Arborea, S.; Giannoccaro, G.; De Gennaro, B.C.; Iacobellis, V.; Piccinni, A.F. Cost-benefit analysis of wastewater reuse in Puglia, Sourthern Italy. *Water* **2017**, *9*, 175. [CrossRef]
13. Molinos-Senante, M.; Hernández-Sancho, F.; Sala-Garrido, R. Cost-benefit analysis of water-reuse projects for environmental purposes: A case study for Spanish wastewater treatment plants. *J. Environ. Manag.* **2011**, *92*, 3091–3097. [CrossRef]
14. Liang, X.; van Dijk, M.P. Cost Benefit Analysis of Centralized Wastewater Reuse Systems. *J. Benefit-Cost Anal.* **2012**, *3*, 1–30. [CrossRef]
15. Verlicchi, P.; Al Aukidy, M.; Galletti, A.; Zambello, E.; Zanni, G.; Masotti, L.A. Project of reuse of reclaimed wastewater in the Po Valley, Italy: Polishing sequence and cost benefit analysis. *J. Hydrol.* **2012**, *432*, 127–136. [CrossRef]
16. Fan, Y.; Chen, W.; Jiao, W.; Chang, A.C. Cost-benefit analysis of reclaimed wastewater reuses in Beijing. *Desalination Water Treat.* **2015**, *53*, 1224–1233. [CrossRef]
17. Arena, C.; Genco, M.; Mazzola, M.R. Environmental Benefits and Economical Sustainability of Urban Wastewater Reuse for Irrigation—A Cost-Benefit Analysis of an Existing Reuse Project in Puglia, Italy. *Water* **2020**, *12*, 2926. [CrossRef]
18. Pasqualino, J.C.; Meneses, M.; Castells, F. Life cycle assessment of urban wastewater reclamation and reuse alternatives. *J. Ind. Ecol.* **2010**, *15*, 49–63. [CrossRef]

19. Tong, L.; Liu, X.; Yuan, Z.; Zhan, Q. Life cycle assessments of water reuse systems in an industrial park. *J. Environ. Manag.* **2013**, *129*, 471–478. [CrossRef] [PubMed]
20. Vouloulis, N. The potential of water reuse as a management option for water security under the ecosystem services approach. *Desalin. Water Treat.* **2015**, *53*, 3263–3271. [CrossRef]
21. Bellver-Domingo, A.; Hernandez-Sancho, F. Circular economy and payment for ecosystem services: A framework proposal based on water reuse. *J. Environ. Manag.* **2022**, *305*, 114416. [CrossRef]
22. Mannina, G.; Badalucco, L.; Barbara, L.; Cosenza, A.; Trapani, D.D.; Gallo, G.; Laudicina, V.A.; Marino, G.; Muscarella, S.M.; Presti, D.; et al. Enhancing a Transition to a Circular Economy in the Water Sector: The EU Project WIDER UPTAKE. *Water* **2021**, *13*, 946. [CrossRef]
23. Gerrity, D.; Pecson, B.; Trussell, R.S.; Trussell, R.R. Potable reuse treatment trains throughout the world. *J. Water Supply Res. Technol.* **2013**, *62*, 321–338. [CrossRef]
24. Baresel, C.; Dahlgren, L.; Almemark, M.; Lazic, A. Municipal wastewater reclamation for non-potable reuse-environmental assessments based on pilot-plant studies and system modelling. *Water Sci. Technol.* **2015**, *72*, 1635–1643. [CrossRef] [PubMed]
25. Chen, Z.; Wu, Q.; Wu, G.; Hu, H.-Y. Centralized water reuse system with multiple applications in urban areas: Lessons from China's experience. *Resour. Conserv. Recycl.* **2017**, *117*, 125–136. [CrossRef]
26. Giurco, D.; Bossilkov, A.; Patterson, J.; Kazaglis, A. Developing industrial water reuse synergies in Port Melbourne: Cost effectiveness, barriers and opportunities. *J. Clean. Prod.* **2011**, *19*, 867–876. [CrossRef]
27. Boysen, B.; Cristóbal, J.; Hilbig, J.; Güldemund, A.; Liselotte, S.; Rudolph, K.-U. Economic and environmental assessment of water reuse in industrial parks: Case study based on a model industrial park. *J. Water Reuse Desalin.* **2011**, *10*, 475–489. [CrossRef]
28. Raucher, R. *An Economic Framework for Evaluating the Benefits and Costs of Water Reuse: Final Project Report and User Guidance*; WateReuse Foundation: Alexandria, VA, USA, 2006.
29. Souza, S.D.; Medellin-Azuara, J.; Burley, N.; Lund, J.R.; Howitt, R.E. *Guidelines for Preparing Economic Analysis for Water Recycling Projects*; Center for Watershed Sciences, University of California: Davis, CA, USA, 2011.
30. Office of The National Water Resources (ONWR). *Final Report on Master Plan on Development and Management of Water Resources in Eastern Region*; ONWR: Bangkok, Thailand, 2019.
31. Suwannasilp, B.B.; Sangsanont, J.; Intaraburt, W.; Sanohwong, K. *Survey of Current Situation of Wastewater in the Circular Economy in Chonburi Province and Prediction of Its Potential*; Research Report; Bangkok, Thailand, 2020. Available online: https://scholar.google.com.hk/scholar?q=Survey+of+Current+Situation+of+Wastewater+in+the+Circular+Economy+in+Chonburi+Province+and+Prediction+of+Its+Potential&hl=zh-CN&as_sdt=0&as_vis=1&oi=scholart (accessed on 1 March 2022).
32. Parliamentary Budget Office. *Budget Expenditure Analysis Report for Fiscal Year 2021: Provinces and Regions*; Parliamentary Budget Office: Bankok, Thailand, 2021.
33. Chonburi Provincial Labour Office. *Thailand. Labour Situation Report*; Chonburi Provincial Labour Office: Chonburi, Thailand, 2020.
34. Ratanathamsakul, C.; Sirikulvadhana, S.; Thongthammachat, C.; Buathet, C.; Thanaphongphan, K.; Yamamoto, K. *Development of Industrial and Urban Areas by Wastewater Reclamation in Eastern Economic Corridor (EEC) Area*; Research Report; Bangkok, Thailand, 2020. Available online: https://thaiembdc.org/eastern-economic-corridor-eec/ (accessed on 1 March 2022).
35. Hernández-Sancho, F.; Molinos-Senante, M.; Sala-Garrido, R. Economic valuation of environmental benefits from wastewater treatment processes: An empirical approach for Spain. *Sci. Total Environ.* **2010**, *408*, 953–957. [CrossRef]
36. Haputta, P.; Puttanapong, N.; Silalertruksa, T.; Bangviwat, A.; Prapaspongsa, T.; Gheewala, S.H. Sustainability analysis of bioethanol promotion in Thailand using a cost-benefit approach. *J. Clean. Prod.* **2020**, *251*, 119756. [CrossRef]
37. International Monetary Fund. GDP per Capita, Current Prices. 2021. Available online: https://www.imf.org/external/datamapper/PPPPC@WEO/ESP/THA (accessed on 31 January 2021).
38. Bank of Thailand. Average Exchange Rate of Commercial Banks in Bangkok (2002-Present). 2021. Available online: https://www.bot.or.th/App/BTWS_STAT/statistics/ReportPage.aspx?reportID=123&language=th (accessed on 31 January 2021).
39. International Monetary Fund. Inflation Rate, Average Consumer Prices (Annual Percent Change). 2021. Available online: https://www.imf.org/external/datamapper/PCPIPCH@WEO/THA?zoom=THA&highlight=THA (accessed on 31 January 2021).
40. Feangthee, A.; Mankeb, P.; Suwanmaneepong, S. The financial feasibility analysis of rice transplanting machine business service in Uttaradit province. *King Mongkut's Agric. J.* **2019**, *37*, 559–569.
41. Djukic, M.; Jovanoski, I.; Ivanovic, O.M.; Lazic, M.; Bodroza, D. Cost-benefit analysis of an infrastructure project and a cost-reflective tariff: A case study for investment in wastewater treatment plant in Serbia. *Renew. Sustain. Energy Rev.* **2016**, *59*, 1419–1425. [CrossRef]
42. Thailand Development Research Institute Foundation. Agency Budget Reform (Environment). 2004. Available online: http://tdri.or.th/wp-content/uploads/2013/04/budget_20_2545.pdf (accessed on 21 May 2020).
43. Asian Development Bank (ADB). *Guidelines for the Economic Analysis of Projects*; ADB: Mandaluyong, Philippines, 2017; ISBN 978-92-9257-763-6.
44. Wasinsombat, W.; Chantuk, T. Feasibility study of investment in prestressed concrete product business, Kamphaeng Saen district, Nakhon Pathom province. *Veridian E-J. Silpakorn Univ. Humanit. Soc. Sci. Arts* **2016**, *9*, 669–684.

45. Environment Fund. Criteria/Conditions of Low Interest Private Loans. 2021. Available online: http://envfund.onep.go.th/home/detailView/20/0 (accessed on 18 August 2021).
46. Canaj, K.; Morrone, D.; Roma, R.; Boari, F.; Cantore, V.; Todorovic, M. Reclaimed water for vineyard irrigation in a mediterranean context: Life cycle environmental impacts, life cycle costs, and eco-efficiency. *Water* **2021**, *13*, 2242. [CrossRef]
47. Maeseele, C.; Roux, P. An LCA framework to assess environmental efficiency of water reuse: Application to contrasted locations for wastewater reuse in agriculture. *J. Clean. Prod.* **2021**, *316*, 128151. [CrossRef]
48. U.S. Government Accountability Office. *Bureau of Reclamation: Water Reuse Grant Program Supports Diverse Projects and Is Managed Consistently with Federal Regulations*; A Report to Congressional Requesters; U.S. Government Accountability Office: Washington, DC, USA, 2018.
49. De Paoli, G.; Mattheiss, V. *Deliverable 4.7 Cost, Pricing and Financing of Water Reuse Against Natural Water Resources*; Demoware research project FP7-ENV-2013-WATER-INNO-DEMO; DemoWare: Santa Monica, CA, USA, 2016.
50. Freedman, J.; Enssle, C. *Addressing Water Scarcity through Recycling and Reuse: A Menu for Policy Makers*; GE Ecomagination: Boston, MA, USA, 2015.
51. Udomsinroj, K. *Water Supply Engineering*; Mitr Nara Karnpim Ltd., Part: Bangkok, Thailand, 1993.

Article

Long-Term Toxicological Monitoring of a Multibarrier Advanced Wastewater Treatment Plant Comprising Ozonation and Granular Activated Carbon with In Vitro Bioassays

Lam T. Phan [1,2,3], Heidemarie Schaar [1,*], Daniela Reif [1], Sascha Weilguni [1], Ernis Saracevic [1], Jörg Krampe [1], Peter A. Behnisch [4] and Norbert Kreuzinger [1]

[1] Institute for Water Quality and Resource Management, TU Wien, Karlsplatz 13/226-1, 1040 Vienna, Austria; lam.thanh@tuwien.ac.at (L.T.P.); daniela.reif@tuwien.ac.at (D.R.); sascha.weilguni@gmail.com (S.W.); ernis.saracevic@tuwien.ac.at (E.S.); joerg.krampe@tuwien.ac.at (J.K.); norbkreu@iwag.tuwien.ac.at (N.K.)
[2] Faculty of Environment and Natural Resources, Ho Chi Minh City University of Technology (HCMUT), 268 Ly Thuong Kiet Street, District 10, Ho Chi Minh City 70000, Vietnam
[3] Viet Nam National University Ho Chi Minh City, Linh Trung Ward, Ho Chi Minh City 70000, Vietnam
[4] BioDetection Systems bv, Science Park 406, 1098 XH Amsterdam, The Netherlands; Peter.Behnisch@bds.nl
* Correspondence: heidemarie.schaar@tuwien.ac.at

Abstract: A set of CALUX in vitro bioassays was applied for long-term toxicity monitoring at an advanced wastewater treatment plant comprising ozonation and granular activated carbon filtration for the abatement of contaminants of emerging concern (CEC). During the 13-month monitoring, eight reporter gene assays targeting different modes of action along the cellular toxicity pathway were accessed to evaluate the suitability and robustness of the technologies. Two approaches were followed: on the one hand, signal reduction during advanced treatment was monitored; on the other hand, results were compared to currently available effect-based trigger values (EBTs). A decrease of the corresponding biological equivalent concentrations after the multibarrier system could be observed for all modes of action; while the estrogenic activity decreased below the EBT already during ozonation, the potencies of oxidative stress-like and toxic PAH-like compounds still exceeded the discussed EBT after advanced treatment. Overall, the long-term monitoring confirmed the positive effect of the multibarrier system, commonly evaluated only by CEC abatement based on chemical analysis. It could be demonstrated that advanced WWTPs designed for CEC abatement are suitable to significantly decrease toxicity responses not only in the frame of pilot studies but under real-world conditions as well.

Keywords: ozonation; granular activated carbon; CALUX reporter gene bioassays; effect-based trigger value; urban wastewater

1. Introduction

The amount and diversity of chemicals in use by our modern society are constantly increasing, reaching 156 million chemicals in 2019, according to the statistics of the chemical abstracts service of the American Chemical Society [1]. A wide variety of these substances enter the water cycle via sewerage and wastewater treatment plants after application and use [2], where they are considered contaminants of emerging concern (CEC). CEC may have been present in the aquatic environment in the past, but only recently have concerns been raised about their potential ecological or human health impacts. In this paper, CEC refers to organic trace substances present in the low microgram to nanogram per liter concentration range. Currently, applied best available technologies (BAT) for conventional biological wastewater treatment cover carbon and nutrient removal. However, these wastewater treatment plants (WWTPs) are not designed to target organic trace compounds, thus resulting in a release into the environment in line with diverse substance-specific removal patterns [3].

The current European Union legislative framework to assess water quality for wastewater and surface water goes back to 1991 and 2000, respectively [4,5]. Then, impacts on aquatic systems were still dominated by carbon and nutrient pollution, while CEC were of no immediate concern. Despite the implementation of follow up amendments, e.g., by directives on environmental quality standards (EQS) and the EU watch list [6–9], an approach to adequately deal with the increasing number, diversity, and change of chemical substances in use did not exist until now [10].

During the last decade, upgrading conventional biological WWTPs for CEC abatement by ozonation or activated carbon proved feasible and financially affordable [11]. Thus, these advanced treatment steps can be implemented on a short- to mid-term scale. Despite existing technical solutions, the issue of required removal efficiencies and linked treatment goals is still pending, since current criteria and approaches based on single chemical analyses are not fit to tackle the chemical diversity of known and yet unknown CEC in water. However, defining robust treatment targets is of paramount significance for developing technical solutions and their resource-efficient operation, especially since single substance analysis does not cover the biological effects of unknown substances (including metabolites) and mixtures [1].

In contrast to single chemical analysis, effect-based methods (EBMs) with bioanalytical tools can account for mixture effects of known and unknown compounds showing common modes of action (MOA) [1,12,13]. Thus, bioassays can be applied to quantify a specific biological effect without knowing its chemical composition. Nowadays, a wide variety of cell-based in vitro bioassays is available, and there are no ethical issues compared to in vivo testing [10,14–16]. Methodological synergies for chemical analysis and sample preparation for EBMs such as enrichment and extraction steps and the possibility of high throughput analysis further helped to increase acceptance and applicability. Due to these advances, EBMs are increasingly applied for water quality assessment, ranging from drinking water to environmental samples such as surface water and wastewater [17–19]. Thus, in vitro bioassays are suitable analytical tools to quantify mixture effects since they are, per definition, detecting the impact of all chemicals inducing the same toxicity endpoint in a given MOA-specific bioassay.

EBMs yield quantitative effect measures that can be translated into biological equivalent concentrations (BEQ). As an example for EBMs, estrogenic activity as one of the most relevant MOA for endocrine disruption [19] can be given. Estrogenic effects in water can be attributed to the occurrence of steroidal estrogens, e.g., 17β-estradiol (E2) or 17α-ethinylestradiol, and industrial chemicals, e.g., bisphenol A (BPA, cf. [20]) or nonylphenols. Results for EBMs are given in BEQ, in the case of estrogenicity, calibrated to E2 equivalents (ng/L). By definition, 1 ng of E2 has a relative effect potency (REP) of one. Less active substances such as BPA have a molar REP of 1.95×10^{-5} [10], meaning the effect is approximately five orders of magnitude less potent than E2. Therefore, the bioassay result in E2 equivalents gives an integrated view of the summary effect for all estrogenic chemicals in the water as if evoked only by E2. As an advantage, even unknown estrogenic compounds are assessed, and chemical multi-target analysis can be avoided.

Only recently, discussions on the implementation of EBMs in the EU Water Framework Directive have started [18,21,22], together with the development of linked environmental quality standards (EQS). In that regard, several authors [10,22–24] developed and proposed effect-based trigger values (EBT), which help differentiate acceptable and unacceptable effect levels for different MOA. These EBT form an essential base for the implementation of EBMs into the regulatory framework. Reviewing the current status and immediate applicability of effect-based assays for assessing risks associated with the reuse of treated wastewater, the COST action NEREUS (EU-COST Action ES 1403) proposed including bioassays in monitoring programs for WWTP effluents. A suggested bioassay battery and EBT from the literature were published in the joint NORMAN and Water Europe Position paper "New and emerging challenges and opportunities in wastewater reuse" [25].

Implementing EBMs and advanced treatment steps address organic CEC and their mixture but with a different focus. Whereas EBMs quantify effects of mixture toxicity, advanced treatment tackles these effects from the causative substance-specific side by decreasing reactive known and yet unknown chemical agents. This makes it obvious to link both approaches by using in vitro bioassays with their linked EBTs as treatment goals and quality criteria for designing, operating, and evaluating advanced wastewater treatment. So far, most investigations on advanced treatment technology assessment with bioassays [26–28] were conducted on lab- or pilot-scale plants operated by scientific staff over a comparably short period with high control efforts. Despite the successful combination of EBMs and advanced treatment technologies, a routine application under real-life conditions with fluctuations of the wastewater, operational, technical and maintenance failures, resulting in suboptimal operation of technologies and therefore out-of-target efficiency is still not considered.

The overall objective of the present study was the long-term toxicological monitoring of multibarrier advanced wastewater treatment under actual conditions, applying a MOA-based in vitro bioassay battery to target relevant toxicological endpoints. After installation, setup of a proper and robust operation, and training, the WWTP operators were committed to integrating the plant operation into their daily routine. Monthly routine monitoring samplings over one year formed the basis to assess the performance and suitability of the applied technologies for broader implementation.

2. Materials and Methods

2.1. Sampling Site and Advanced Wastewater Treatment Plant

The sampling site was located at a municipal wastewater treatment plant (7250 p. e.) with a conventional activated sludge process, including full nitrification/denitrification, and phosphorus removal. This corresponds to the best available technologies for biological wastewater treatment in Austria and to the requirements for eutrophication-sensitive areas according to the EU urban wastewater treatment directive [5]. Over the period investigated, the mean chemical oxygen demand (COD) and dissolved organic carbon (DOC) concentrations in the WWWTP effluent were 14.85 ± 2.67 mg/L and 4.26 ± 0.49 mg/L, respectively. The nutrient concentrations averaged 0.62 ± 1.25 mg/L for ammonia (NH_4-N), 0.75 ± 0.55 mg/L for nitrate (NO_3-N), and 0.08 ± 0.08 mg/L for nitrite nitrogen (NO_2-N). The mean daily wastewater flow was 1540 ± 612 m^3/d. The WWTP treats wastewater from a combined sewer system. In case of a stormwater event, wastewater exceeding the hydraulic design capacity of the plant is pumped to rainwater basins and afterward continuously treated according to the hydraulic capacity. The treatment plant's effluent is buffered in a basin, which serves as a feed tank for the advanced treatment plant.

Figure 1 shows the flow scheme of the multibarrier system comprising ozonation and granular activated carbon filtration. The advanced treatment demonstrator plant is on transition of the technical readiness level (TRL) 7 to 8. The three ozone reactors operated in series had a total volume of 12 m^3 and the hydraulic retention time varied between 9 and 40 min, depending on the inflow dynamics of wastewater. The activated carbon filter was filled with 1.8 m^3 of granular activated carbon (GAC), type Epibon A (Donau Carbon, Frankfurt, Germany) and treated a side stream of 8 m^3/h, which resulted in an empty bed contact time of 13.5 min. During routine operation a specific nitrite compensated ozone dose (D_{spec}) of 0.55 g O_3/g DOC was targeted in the automated process control system that was based on a UV-DOC-correlation model and continuous UV absorption measurement. A posteriori it ranged between 0.4 and 0.7 g O_3/g DOC. For specific research campaigns the specific ozone dose was varied between 0.2 and 0.9 g O_3/g DOC. The sampled bed volumes (BV) of the granular activated carbon filter ranged from approx. 1000 (start of monitoring) to 33,100 (final sampling campaign).

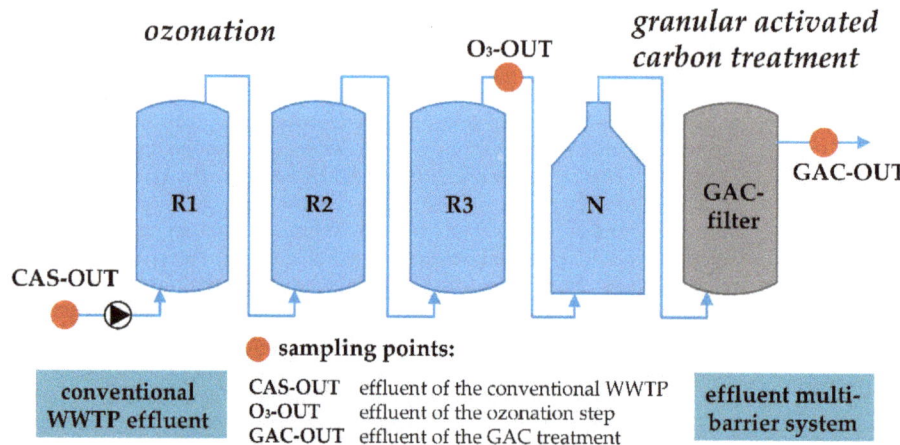

Figure 1. Flow scheme of the advanced treatment demonstrator plant with sampling points (CAS: conventional activated sludge, R: ozone reactor, N: feed tank for GAC-filter, GAC: granular activated carbon).

2.2. Sampling Campaigns

A monthly routine monitoring was performed between May 2018 and May 2019, including all three sampling points indicated in Figure 1. Additionally, two scientific research campaigns were conducted to evaluate the impact of a broader dose range on the toxicity by comparing the inlet and outlet of the ozonation step. After assessing the sampling type, it was decided to take all samples as grab samples in 1.5 L aluminum bottles, according to the recommendations of BioDetection Systems BV (Amsterdam, The Netherlands). Over the sampling period of 13 months, 16 samples were taken, see Table S1.

2.3. Sample Extraction for Bioanalysis

All wastewater samples were filtered through a glass fiber filter (pore size 3 μm), and the maximum sample volume after filtration was 1000 mL.

The samples were concentrated by solid-phase-extraction (SPE) with Oasis HLB cartridges (500 mg, 6cc, Waters 186000115; Waters Corporation, Taunton, MA, USA) according to the described protocol of BioDetection Systems bv (Amsterdam, The Netherlands) with slight modifications regarding the final resuspension of the sample that had been evaporated to dryness. The cartridges were pre-conditioned with 6 mL acetonitrile and 6 mL deionized water, both of which were drawn through the cartridges under a low vacuum with a vacuum manifold to remove residual bonding agents. The filtered samples were loaded onto the cartridge under a slight vacuum; the flow over the cartridge was adjusted to a few drops per second to avoid exceeding 10 mL/min. After loading, the cartridges were washed with 6 mL 5% methanol (*w/w*) and then dried for 30 min under vacuum to remove excess water remaining on the cartridge. Subsequently, the adsorbed analytes were eluted from the cartridges to a 20 mL culture tube with 10 mL methanol and 10 mL acetonitrile at a flow rate of approx. 5 mL/min. Afterward, the samples were evaporated to dryness ($\pm$ 0.5 mL) under a stream of nitrogen at room temperature. This volume was transferred from the culture tube to the vial and rinsed with 0.5 mL methanol and 0.5 mL acetonitrile. The final volume of 1.5 mL extracted sample was kept in the fridge at 7 °C before analysis.

2.4. Bioassays

The wastewater extracts were analyzed by BioDetection Systems bv (Amsterdam, The Netherlands) with nine CALUX (Chemical Activated Luciferase eXpression) reporter gene bioassays. The principle of the bioassay is described in [29], and the corresponding key references are given in Table S2.

The in vitro bioassay test battery was designed to target MOA based on well-defined toxic mechanisms that cover relevant steps along the cellular toxicity pathway as recommended in the literature [10,13,14], see Figure 2. Even though positive signal responses cannot be directly translated into higher-order effects, every adverse outcome begins with a molecular initiating event. It demonstrates the link between biological response at the cellular level with higher-order effects on the organ, followed by the organism and eventually the population level, which is summarized under the concept of adverse outcome pathways [30].

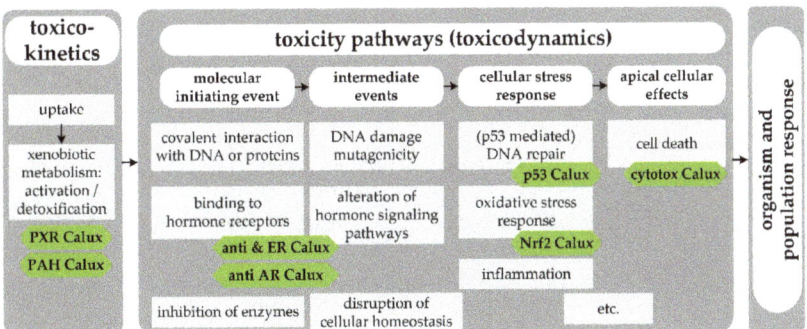

Figure 2. In vitro CALUX bioassay panel allocated to the corresponding events on the toxicity pathway (according to [14], modified).

Five of the nine modes of action investigated in this long-term monitoring were suggested for WWTP effluent monitoring in the joint NORMAN and Water Europe Position paper [25] by the NEREUS COST Action ES 1403. Additionally, genotoxicity and cytotoxicity were included, being amongst the first endpoints applied for water quality assessment [31]. Anti-estrogenicity was investigated as an additional hormone-mediated assay.

A summary of the applied CALUX bioassays covering the measured endpoints, reference compounds, and EBTs is given in Table 1. For quantification of the analyzed effect, the results of the CALUX bioassays and corresponding EBTs are provided as biological equivalents (BEQ) per liter sample related to reference compounds given in Table 1. BEQ were expressed as tributyltin acetate equivalent concentration (TBT-EQ) for cytotoxicity, 17β-estradiol equivalent concentration (EEQ) for estrogenicity, tamoxifen equivalent concentration (Tam-EQ) for anti-estrogenicity, flutamide equivalent concentration (Flu-EQ) for anti-androgenicity, curcumin equivalent concentration (Cur-EQ) for oxidative stress response, cyclo-phosphamide equivalent concentration (CPA-EQ) for genotoxicity with metabolic activation S9, actinomycin equivalent concentration (ACT-EQ) for genotoxicity without metabolic activation S9, benzo[a]pyrene equivalent concentration (B[a]P-EQ) for toxic PAH and nicardipine equivalent concentration (Nic-EQ) for xenobiotic sensing with PXR CALUX.

An individual limit of quantification (LOQ) is determined for every single analysis. Genotoxicity was analyzed with and without the addition of S9 for metabolic activation. Different results with S9 addition elucidate a metabolization or detoxification of ingredients [13] and helps differentiate between directly and indirectly acting genotoxic compounds.

Not each endpoint was targeted in every sample. While the hormone-mediated MOA ERα (in short, ER) and anti-AR CALUX were measured in all samples, the remaining six endpoints were analyzed alternately according to the frequency depicted in Table S2.

Table 1. Information on the CALUX in vitro bioassay panel and frequency of analysis.

Bioassay	Effect	Measured Endpoint or Molecular Target	Reference Compound	EBT [)](**) [BEQ]
Cytotox	Cytotoxicity	Repression of constitutive transcriptional activation	Tributyltin acetate	-
ERα [)](*)	Estrogenicity	Estrogen receptor α-mediated signalling	17β-Estradiol	0.1 ng/L
anti-ERα	Anti-estrogenicity	Repression of estrogen receptor α-mediated signalling	Tamoxifen	not available
anti-AR [)](*)	Anti-androgenicity	Repression of androgen receptor activation	Flutamide	14 µg/L
Nrf2 [)](*)	Oxidative stress response	Activation of the Nrf2 pathway	Curcumin	10 µg/L
p53 + S9	Genotoxicity response +S9	p53-dependent pathway activation with S9	Cyclo-phosphamide	-
p53 − S9	Genotoxicity response -S9	p53-dependent pathway activation without S9	Actinomycin	-
PAH [)](*)	Toxic PAH-xenobiotics metabolism	Aryl-hydrocarbon receptor activation	Benzo[a]pyrene	6.2 ng/L
PXR [)](*)	Xenobiotic metabolism and sensing	Activation of pregnane X receptor	Nicardipine	3 µg/L

[)](*) bioassays suggested in the joint position paper by NORMAN and Water Europe [25]; for endpoints given in this paper, the lower EBT suggested was applied. [)](**) EBTs linked to the MOA were retrieved from the literature [10,24].

2.5. Data Interpretation

If the BEQ was below the LOQ, half the LOQ was used as a result. This approach was applied in order to not exclude results < LOQ from statistical analysis. Due to the sample-specific LOQs, the BEQ derived from results < LOQ can slightly deviate and, in some cases, give the impression of an increased signal along with the treatment steps.

3. Results

This paper presents long-term toxicological monitoring of a multibarrier treatment system with ozonation and granular activated carbon treatment under realistic conditions. The applied technologies were investigated with a panel of in vitro bioassays involving selected water-relevant MOA along the cellular toxicity pathway. Two approaches were employed to assess the suitability and the performance of the treatment technologies:

1. The BEQ decrease was determined for the various steps of the multibarrier system (CAS-OUT, O_3-OUT, and GAC-OUT).
2. The BEQ were compared to currently discussed MOA-specific EBTs to identify the impact of advanced treatment.

Figure 3a gives an overview of the cytotoxic effects after conventional and advanced treatment. Figure 3b demonstrates the TBT-equivalents for each sampling campaign. After conventional treatment, TBT-EQ ranged between 0.19 and 3.3 µg/L, with a median of 0.64 µg TBT-EQ/L. Cytotoxicity was below LOQ in seven out of sixteen samples (Table S3). In the rest of the samples, ozonation decreased the effect below LOQ, where it remained after the subsequent GAC treatment (Figure 3b and Table S3). The median decline in cytotoxicity achieved 83% (Figure 4) already after the first step of the multibarrier treatment system.

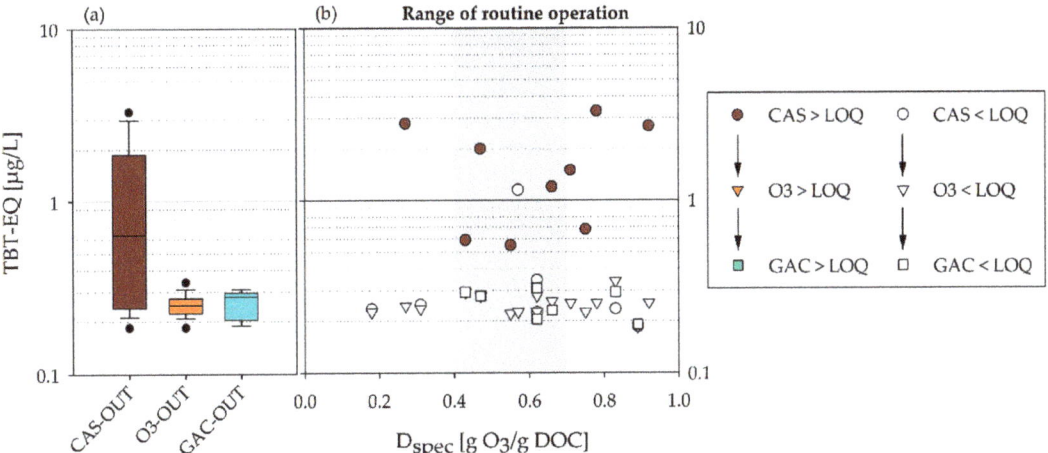

Figure 3. (a) Boxplot showing the range of tributyltin acetate equivalents over all campaigns along the multibarrier system; the box indicates the 25th to 75th percentile and the line within the box indicates the median. The whiskers show the 10th and the 90th percentiles and data points that lie outside these percentiles are plotted as dots. (b) TBT-EQ along the multibarrier system for each sampling campaign.

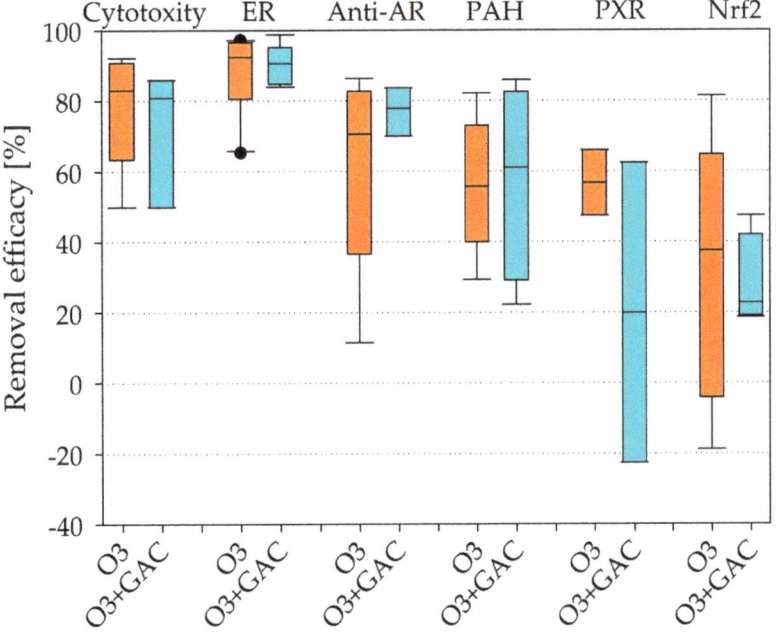

Figure 4. Boxplots showing the range of removal for the investigated MOA along the multibarrier treatment system over the one-year monitoring.

In the effluent of the conventional WWTP, all but one sample were above the LOQ and EBT for estrogenic activity, respectively (Figure 5a,b, and Table S3). EEQ ranged from 0.15 (0.09)–1.2 ng/L with a median at 0.42 ng/L. Ozonation resulted in a substantial EEQ decline; half of the samples decreased < LOQ, and only one out of 14 samples exceeded the EBT of 0.1 ng EEQ/L by 0.02 ng/L (0.12 ng EEQ/L). Similar results were obtained after GAC filtration with a single exceedance of the EBT (0.16 ng EEQ/L). Except for this sampling campaign, GAC-OUT results were in the same order of magnitude as in O3-OUT, impeding the determination of a significant further reduction.

A median decrease of 92% was achieved for the ozonation process (n = 13, Figure 4), only for the two lowest D_{spec} < 0.3 g O_3/g DOC the EEQ reduction was below 70%. Taking the smaller sample size for the GAC filter (n = 7) into account, the removal of the multibarrier system (O3 + GAC) was in the same range.

Anti-androgenicity was <LOQ in ten out of sixteen effluent samples of the conventional WWTP (Figure S1, Table S4). The EBT of 14 µg Flu-EQ/L was exceeded twice (16 and 18 µg/L). Even though ozonation resulted in a decrease below LOQ for the remaining six samples, the further removal is not evident in the median (Figure S1a); this is a result of the sample-specific LOQ, which were partly lower in CAS-OUT than in O3-OUT. Considering the single campaigns where anti-androgenic effects were detected in CAS-OUT (Figure S1b), however, the removal potential of the ozonation step becomes apparent. In GAC-OUT the effect remained < LOQ. Thus, none of the multibarrier effluent samples exceeded the EBT of 14 µg Flu-EQ/L. The median anti-androgenicity removal in the ozonation and the multibarrier system was 72 (n = 6) and 78% (n = 3, Figure 4), calculated with $\frac{1}{2}$ LOQ.

Anti-estrogenic effects were analyzed only twice, and it was < LOQ during the first screening. The second screening gave 0.51 µg Tam-EQ/L in the effluent of the conventional treatment, and after ozonation (0.7 g O_3/g DOC) and GAC-treatment the effect decreased < LOQ.

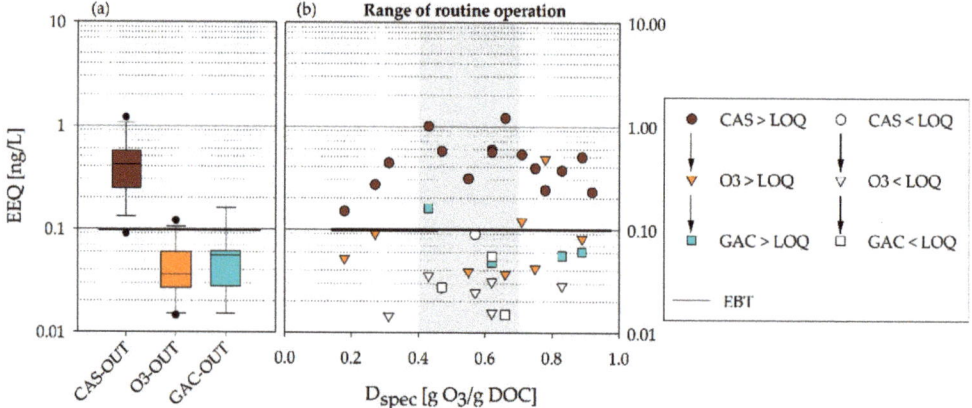

Figure 5. (a) Boxplot showing the range of 17β estradiol equivalents over all campaigns along the multibarrier system; (b) EEQ along the multibarrier system for each sampling campaign.

In the effluent of the conventional WWTP, the benzo[a]pyrene equivalents ranged between 100 and 270 ng/L, with a median at 145 ng/L. The ozonation resulted in a decrease of the B[a]P-EQ (Figure 6). Despite a median decline of 56–61% (cf. Figure 4), Figure 6 demonstrates the constant exceedance of the proposed and here applied EBT. This EBT is mainly based of B[a]P, while many kinds of different PAHs can occur in such complex mixtures.

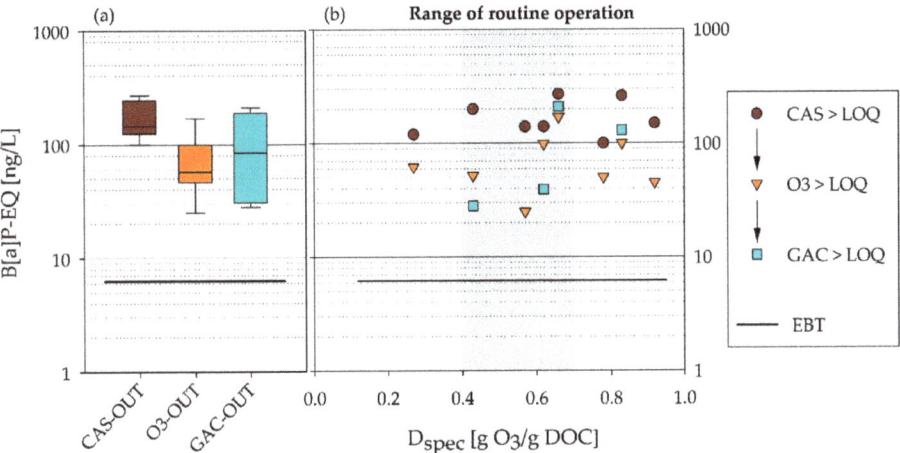

Figure 6. (a) Boxplot showing the range of benzo[a]pyrene equivalents over all campaigns along the multibarrier system; (b) B[a]P-EQ along the multibarrier system for each sampling campaign.

Results on xenobiotic sensing are depicted in Figure S2. For one of the three analyzed campaigns, BEQ were below the LOQ for all three sampling sites. The remaining two campaigns delivered a removal of 48 and 66% during ozonation. For the GAC filter, however, contradictory results were obtained. While an increase of the PXR activity was observed for the first sampling that was done at approx. 1000 bed volumes (BV) treated, a further removal after ozonation occurred in a sampling campaign after one year of operation at approx. 33,100 BV.

Four out of 13 analyzed samples of CAS-OUT were below the LOQ (Table S4). Despite a reduction during advanced treatment, the samples still had the potency to trigger oxidative stress response mechanisms. Due to the varying LOQ for the different treatment stages, the signal after ozonation and after GAC was reduced below LOQ only once (Table S4). An increase was determined for the ozonation in two sampling campaigns, but GAC resulted in an overall decrease. All analyzed samples were above the EBT of 10 ng Cur-EQ/L (Figure S3).

The genotoxic activity was below the LOQ both with (n = 5) and without the addition of the metabolic activation mix S9 (n = 3).

4. Discussion

The long-term toxicological monitoring offered a valuable opportunity to encounter realistic operating conditions, including fluctuations in wastewater quantity and quality, which can impact conventional and advanced treatment.

Nevertheless, most ozonation plants apply flow-proportional dose control due to enhanced efforts required for water quality-related operation. Concerning the process stability of conventional biological treatment, insufficient nitratation within the two-step nitrification process can lead to nitrite occurrence. Nitrite has a stochiometric ozone consumption of 3.43 mg O_3/mg NO_2-N and decreases the effective D_{spec} if nitrite compensation is not implemented in the process control strategy. A model calculation with a setpoint of 0.55 g O_3/g DOC and a DOC effluent concentration of 4.5 mg/L demonstrates that 0.2 mg NO_2-N/L decreases the effective D_{spec} to 0.4 g O_3/g DOC, which equals a decrease of 28%. Taking fluctuations in the effluent quality (nitrite or DOC) into account, the effective D_{spec} can readily vary between 0.4 and 0.7 g O_3/g DOC with flow-proportional control.

Due to the potential variations, Figures 3, 5 and 6 in the results chapter indicate a D_{spec} of 0.4 to 0.7 g O_3/g DOC as the range of routine operation. This range coincides quite well with the recommended D_{spec} of 0.4 to 0.6 g O_3/g DOC for CEC abatement from tertiary treated wastewater [32], and with the D_{spec} of 0.55 g O_3/g DOC applied at the first full-scale WWTP upgraded with ozonation at Neugut, Switzerland [33].

The sampling campaigns with lower specific ozone doses simulate nitrite occurrence and the lack of adequate process control in order to assess the impact of low ozone doses on the investigated MOA. Apart from the estrogenic activity that revealed a signal reduction <70% for the two campaigns with a D_{spec} < 0.3 g O_3/g DOC, no clear correlation with the tested doses (0.18–0.92 g O_3/g DOC) and the range of routine operation, respectively, could be determined.

After the conventional treatment, cytotoxicity was in the range representative for other Austrian CAS-plants with full nitrification and denitrification [34]. The advanced treatment proved beneficial for baseline toxicity removal, confirming the suitability of the multibarrier system. Since transformation products formed during ozonation are more hydrophilic, they are less cytotoxic, but still contribute to mixture effects [35]. Biodegradation during biologically activated GAC theoretically offers the potential to reduce these effects, but in the present study it was not possible to prove this due to the non-detects after ozonation. A significant reduction after ozonation was also determined in a study on three German WWTPs [36]. In addition, they also revealed the effect reduction potential of biological posttreatment with a fluidized bed reactor. Even though GAC, applied in the present study, differs from the fluidized bed reactor, both systems represent biological posttreatment processes. Thus, it is a strong indication for an additional benefit of GAC and the strength of the multibarrier approach.

As cytotoxicity is a non-specific toxicity endpoint that provides an estimate of the overall toxic burden in a mixture, it is considered important to be investigated [37].

The range of 17β estradiol equivalents (EEQs) observed in the effluent of the conventional treatment was in accordance with nine Austrian WWTPs [34] and can be considered as representative for CAS-plants operated according to the EU requirements for eutrophication sensitive areas [5] applying biological nitrogen removal (tertiary treatment). Biological nitrogen removal can only be achieved at low-loaded WWTPs with high solids retention time, a parameter that correlates well with estrogenicity removal [38]. This correlation is partly reflected by data for ERα CALUX determined in the effluent of 12 European WWTPs along the Danube River [29]. High-loaded WWTP with only secondary treatment (carbonaceous biological oxygen demand removal, but no biological nitrogen removal) are mostly characterized by higher effluent EEQ compared to tertiary treatment.

According to NEREUS [39], an average decrease of estrogenic activity by approx. one order of magnitude was observed during conventional treatment. The results of this paper showed that an average decrease by another order of magnitude could be accomplished with advanced treatment. A significant EEQ decrease by ozonation was also observed during other full-scale studies [36,40,41]. The reduction of the EEQ that occurred during ozonation can be attributed to the high reactivity of high-potency estrogens with ozone [42]. This conclusion is permitted since estrogenicity is one of the endpoints with a high overlapping of the biological and the chemical BEQ; the latter are calculated by summing up the products of the chemical concentration and the corresponding relative effect potencies [19,43]. Even though estrogenicity decline could not be quantified for GAC, a good EEQ removal potential can be assumed based on a review on toxicity removal by advanced wastewater treatment with ozonation and activated carbon treatment [28]. According to the published data, the median reduction for AC treatment amounted to 75%.

Receptor-mediated estrogenicity is one of the most relevant MOA for endocrine disrupting compounds [19]. Consequently, the significant reduction (median removal > 90%) can be considered a substantial benefit of the multibarrier system.

Studies on the removal of endocrine effects during advanced treatment put more focus on agonistic activity, even though pharmaceuticals like diclofenac belong to the group of hormone receptor antagonists [28]. While the difference was less pronounced for the androgenic receptor (eleven vs. nine studies), 22 studies were done on estrogenicity and seven on anti-estrogenicity. Three of them reported tamoxifen equivalent concentrations < LOQ. Similar results were obtained in the presented study with two samplings covering anti-estrogenicity. Anti-estrogenicity was only measured once in the effluent of the conventional treatment, and it decreased to < LOQ after ozonation. This observed decline contradicts four studies reporting an increase after ozonation, which appeared to correlate with an increasing ozone dose [28]. Contrary to this, an unclear elimination pattern was found on a full-scale ozonation plant, i.e., independent of the ozone dose, formation and elimination were observed during six monitoring campaigns [40].

The anti-androgenicity was in the lower range measured for effluents of other conventional biological WWTPs in Austria [34] and in the Danube River Basin [29]. Despite the calculation with $\frac{1}{2}$ LOQ due to 100% non-detects in the effluent of the advanced treatment stages, a clear removal pattern for anti-androgenicity in ozonation and the multibarrier system was detected during the monitoring campaign (69 and 77%, respectively). The removal was in line with published data for ozonation (81.5%) and activated carbon (62.4%) [28]. In contrast, a current full-scale study with ozonation did not identify a clear removal pattern [40].

The PAH-CALUX belongs to the specific toxicity endpoints which induce xenobiotic metabolism. This endpoint is characterized by a high frequency of occurrence in municipal wastewater [28]. Positive signals for PAH activity were detected in all WWTP effluents investigated in Austria (n = 9) and in the Danube River Basin (n = 12) [29,34]. The BEQ in the present paper were in a similar concentration range, thus, representative for urban WWTPs. The removal efficiency for B[a]P-EQ of approx. 60% during advanced treatment was slightly lower than published values of 79 and 84% for ozonation and activated carbon, respectively, [28]. After all, the multibarrier system could not reduce the activity below the discussed EBT of 6.2 ng B[a]P-EQ/L.

In addition to the PAH CALUX, the PXR CALUX is another bioassay targeting the induction of xenobiotic metabolism. Since the pregnane X receptor is activated by different types of chemically nonrelated compounds, comprising environmental pollutants and pharmaceuticals, this bioassay can be applied for xenobiotic sensing [44]. PAH and PXR activity belonged to the most frequently detected endpoints in the Joint Danube Survey [29], and the Nic-EQ in CAS-OUT were within the same range as the investigated WWTPs. In literature, a median removal of ≥ 78% was reported for combined ozonation–GAC treatment [17]. In the present study, two contrary results were obtained. In a first campaign, 66% were achieved by ozonation, but a negative removal was determined for the multibarrier system after GAC treatment at a low BV of around 1000. During a campaign at a later stage (33,098 BV), the removal efficiency increased from 48% (after ozonation) to 63%. Reasons for these divergent results are not clear; on the one hand, the adsorption capacity of the activated carbon around 1000 BV is still high; on the other hand, the biological activation of the filter can be assumed to be in the start-up phase. Both processes (adsorption and biodegradation) occur in parallel, with a share depending on the treated wastewater, the BV treated, and the substance characteristics. Usually, it is not possible to easily differentiate between these two processes in a GAC filter [33]. However, the data set is too small to conclude the effect of biological activity in the filter on the PXR activity, and further investigations would be needed.

In the monitoring studies in Austria and the Danube river basin [29,34], oxidative stress (Nrf2 CALUX®) was identified in 18 out of 21 conventional WWTP effluent samples. The numbers were in the same order of magnitude as in the presented long-term study, confirming the consistent exceedance of EBTs. For the ozonation process, a median removal of 46% was determined. Additional reduction in the GAC step was observed during three of five sampling campaigns. In contrast, the other campaigns revealed an increase in response and no change to the ozonation. Literature results showed 63% removal for an ozonation plant and > 25 to > 95% (median of 44%) for a combined O3 and GAC treatment [28].

Oxidative stress represents a rather general cellular stress response that can often be detected before cytotoxic effects [37,45]. This is consistent with the present results, with cytotoxic effects more often < LOQ, even before advanced treatment. Like biological treatment, ozonation processes lead to transformation rather than mineralization, which causes the weaker decline observed for oxidative stress response compared to other investigated endpoints, such as estrogenicity or anti-androgenicity.

Figure 4, as a summarizing graph, gives an overview of the removal range for the investigated MOA considering all sampling campaigns irrespective of the specific ozone doses. Genotoxicity and anti-estrogenicity were not integrated due to their lack of occurrence. A median removal of > 80% was achieved only for genotoxicity and cytotoxicity. Estrogenicity was the endpoint with the lowest variations. After ozonation, the 25th percentile removal was > 80%, and after GAC the minimum removal determined was 84%. A removal < 70% can be related to D_{spec} < 0.3 g O_3/g DOC, though. Cytotoxicity seemed to have higher variations, but all results were < LOQ after ozonation and activated carbon treatment, respectively. Thus, the calculated removal based on $\frac{1}{2}$ LOQ can deviate. The same is valid for anti-androgenicity with 100% of the data < LOQ after advanced treatment.

While the evaluation of the BEQ decrease represents the first approach to assess the suitability and performance of the treatment system, the second approach dealt with the comparison of the BEQ with currently discussed MOA-specific EBT (cf. Figures 3b, 5b, 6b and S1b–S3b). Table 2 shows the n-fold exceedance of the median relative to currently discussed EBT values according to the concept suggested by [29].

Table 2. n-fold EBT-exceedance of the median BEQ for all sampling campaigns. The color code refers to the degree of exceedance.

Bioassay	EBT	CAS-OUT	O3-OUT	GAC-OUT	CAS-OUT	O3-OUT	GAC-OUT
					Frequency of Exceedance [n/All]		
ERα	0.1	4	0.4	0.6	14/16	1/14	1/7
anti-AR	14	0.1	0.2	0.2	2/16	0/16	0/7
Nrf2	10	9	7	9	13/13	13/13	5/5
PAH	6.2	23	9	14	8/8	8/8	4/4
PXR	3	13	6	5	3/3	2/3	2/2
BEQ/EBT < 1		1 ≤ BEQ/EBT < 3		3 ≤ BEQ/EBT < 10	10 ≤ BEQ/EBT < 100		BEQ/EBT > 100

A typical pattern for the degree of exceedance could be observed by a decline in response from left to right, following the treatment train. An increase in treatment steps usually resulted in improved water quality even if the BEQ was still exceeded by up to ninefold for selected endpoints other than hormone-mediated endpoints. PAH activity, however, seemed to increase again after GAC treatment, which is most probably due to the smaller data set for GAC than for O3 (four versus eight data points).

Bioassay responses of more than 100 times the EBT are indicative for high risk, but appropriate measures can be taken if chemicals causing the effects are known [24]. Linking in vitro effects and organic micropollutants detected in surface water with mixture-toxicity modeling was only applicable for a limited number of endpoints, among them estrogenic effects [43]. The >100-fold exceedance in WWTP influent for estrogenicity is significantly reduced by conventional wastewater treatment [39], and can be further reduced by advanced treatment.

The amount of CEC currently in use is high, and CEC abatement by conventional treatment is limited. Thus, the additional barrier of advanced treatment technologies should be considered in the future, even if bioassay responses after advanced treatment with a multibarrier system comprising O_3 and GAC were still elevated for endpoints like PAH-like activities and oxidative stress activities.

Bioassays targeting xenobiotic metabolism and repair and defense mechanisms are sensitive tools to detect the occurrence of CEC, since effects can often be identified at concentrations lower than those resulting in cell death or damage [43], as confirmed by literature [31] and the present results. In addition, the activation of metabolism as a toxicokinetic process cannot be considered an adverse effect *per se*, but rather an indication of the presence of bioactive chemicals [10,28]. Thus, wastewater samples can be regarded as subject to the induction of specific endpoints like the aryl hydrocarbon receptor (AhR) targeted by the PAH CALUX and the pregnane X receptor (PXR) as observed in the long-term monitoring. This is not necessarily linked to an adverse outcome, especially when considering the high level of treatment in the multibarrier system and the overall decreasing response trend. Moreover, a slight exceedance of EBT in one or two bioassays is not inherently linked to increased ecological risk [24]. After all, not only WWTP discharges, as point sources, but also diffuse sources, e.g., from agriculture, show effects [47].

5. Conclusions

The following summarizing conclusions can be drawn from the application of toxicological long-term monitoring of an advanced wastewater treatment plant with a battery of in vitro bioassays covering various endpoints along the toxicity pathway:

- Toxicological long-term monitoring delivered a valid basis for assessing the applicability and the performance of a multibarrier system for an advanced treatment.
- The combination of the two approaches applied in the present study, namely the quantification of BEQ decline and comparison of BEQ with currently discussed EBTs, represented a solid means of assessing the final effluent quality of the multibarrier system combining ozonation and granular activated carbon treatment.
- Despite natural variations in wastewater characteristics and other factors influencing CAS treatment efficiency over the 13-month monitoring, the overall removal pattern for various modes of action covering endpoints along the toxicity pathway revealed a decrease in BEQ.
- Even though the positive effect of ozonation resulting in signals below LOQ for some MOA impeded the assessment of the GAC treatment, a combination of O3 and GAC is strongly recommended for advanced treatment in order to follow the multibarrier approach and guarantee a high level of treatment.
- Since the presented toxicological results did not reveal significant differences within the recommended ozone dose range of 0.4–0.7 g O3/g DOC, it can be concluded that potential toxicological requirements should not be limiting for the implementation and operation of multibarrier systems for CEC abatement.
- Even though measures like implementing advanced treatment at WWTPs do not result in a complete removal, advanced treatment represents a relevant step in reducing the toxicological burden for the aquatic environment.
- Effect-based bioassays with their linked EBTs should be used as treatment goals and quality criteria for design, operation and the evaluation of advanced wastewater treatment.

Supplementary Materials: The following are available online at https://www.mdpi.com/article/10.3390/w13223245/s1, Table S1: Frequency of sampling for each sampling point, sorted by specific ozone dose. Bed volumes (BV) are only given for routine monitoring campaigns; Table S2: Frequency of analysis for the applied bioassays at the sampling points including method references; Table S3: BEQs for cytotoxicity, estrogenicity and toxic PAH-like activities along the multibarrier system for each sampling campaign; Table S4: BEQs for anti-androgenicity (Anti-AR), xenobiotic sensing (PXR) and oxidative stress (Nf2) response along the multibarrier system for each sampling campaign; Figure S1: (a) Boxplots showing the range of flutamide equivalents over all campaigns along the multibarrier system; (b) Flu-EQ along the multibarrier system for each sampling campaign; Figure S2: (a) Boxplots showing the range of nicardipine equivalents over all campaigns along the multibarrier system; (b) Nic-EQ along the multibarrier system for each sampling campaign; Figure S3: (a) Boxplots showing the range of curcumin equivalents over all campaigns along the multibarrier system; (b) Cur-EQ along the multibarrier system for each sampling campaign; Figure S4: The range of removal for the investigated MOA after ozonation over the one-year monitoring; Figure S5: The range of removal for the investigated MOA after the multibarrier system (ozonation and GAC) over the one-year monitoring.

Author Contributions: Sample extraction, data curation, writing-original draft preparation: L.T.P.; conceptualization, supervision, writing, review and editing: H.S.; plant setup, plant supervision and sampling: D.R. & S.W.; chemical analysis and supervision of laboratory work: E.S.; project supervision, review: J.K.; bioassay selection, review and editing: P.A.B.; project administration, review and supervision: N.K. All authors have read and agreed to the published version of the manuscript.

Funding: This research was funded by the Austrian Federal Ministry of Agriculture, Regions and Tourism, grant number B601389, and by the Federal Government of Burgenland, grant number A5/WWK.WPO200-10034-2-2016.

Acknowledgments: Acknowledgments are given to the NEREUS COST Action ES1403 for funding the short-term scientific mission for the training at BioDetection Systems bv. Special thanks are also dedicated to the treatment plant operators of the municipality Frauenkirchen for their effort in operating a novel treatment system. The authors also acknowledge Open Access Funding by TU Wien.

Conflicts of Interest: The authors declare no conflict of interest.

References

1. Escher, B.I.; Stapleton, H.M.; Schymanski, E.L. Tracking complex mixtures of chemicals in our changing environment. *Science* **2020**, *367*, 388–392. [CrossRef]
2. Fatta-Kassinos, D.; Bester, K.; Kümmerer, K. *Xenobiotics in the Urban Water Cycle*, 1st ed.; Springer: Dordrecht, The Netherlands; Heidelberg, Germany; London, UK; New York, NY, USA, 2010.
3. Krzeminski, P.; Tomei, M.C.; Karaolia, P.; Langenhoff, A.; Almeida, C.M.R.; Felis, E.; Gritten, F.; Andersen, H.R.; Fernandes, T.; Manaia, C.M.; et al. Performance of secondary wastewater treatment methods for the removal of contaminants of emerging concern implicated in crop uptake and antibiotic resistance spread: A review. *Sci. Total Environ.* **2019**, *648*, 1052–1081. [CrossRef] [PubMed]
4. Directive 2000/60/EC. Directive of the European Parliament and of the Council of 23 October 2000 Establishing a Framework for the Community Action in the Field of Water Policy. *OJ L 327*, 22 December 2000.
5. Directive 91/271/EEC. Directive concerning Urban Waste Water Treatment (UWWTD). *OJ L 135*, 30 May 1991.
6. Commission Implementing Decision (EU) 2015/495. Commission Implementing Decision establishing a Watch List of Substances for Union-Wide Monitoring in the Field of Water Policy Pursuant to Directive 2008/105/EC of the European Parliament and of the Council. *OJ L 78*, 2015.
7. Commission Implementing Decision (EU) 2018/840. Commission Implementing Decision of 5 June 2018 Establishing a Watch List of Substances for Union-Wide Monitoring in the Field of Water Policy Pursuant to Directive 2008/105/EC of the European Parliament and of the Council and Repealing Commission Implementing Decision (EU) 2015/495 (Notified under Document C(29018) 3362). *OJ L 141*, 7 June 2018.
8. Directive 2008/105/EC. Directive of the European Parliament and of the Council of 16 December 2008 on Environmental Quality Standards in the Field of Water Policy, Amending and Subsequently Repealing Council Directives 82/176/EEC, 83/513/EEC, 84/156/EEC, 84/491/EEC, 86/280/EEC and Amending Directive 2000/60/EC of the European Parliament and of the Council. *OJ L 348*, 24 December 2008.
9. Directive 2013/39/EU. Directive of the European Parliament and of the Council of 12 August 2013 Amending Directives 2000/60/EC and 2008/105/EC as Regards Priority Substances in the Field of Water Policy. *OJ L 226*, 24 August 2013.
10. Escher, B.I.; Aït-Aïssa, S.; Behnisch, P.A.; Brack, W.; Brion, F.; Brouwer, A.; Buchinger, S.; Crawford, S.E.; Du Pasquier, D.; Hamers, T.; et al. Effect-based trigger values for in vitro and in vivo bioassays performed on surface water extracts supporting the environmental quality standards (EQS) of the European Water Framework Directive. *Sci. Total Environ.* **2018**, *628–629*, 748–765. [CrossRef] [PubMed]
11. Eggen, R.I.L.; Hollender, J.; Joss, A.; Schärer, M.; Stamm, C. Reducing the Discharge of Micropollutants in the Aquatic Environment: The Benefits of Upgrading Wastewater Treatment Plants. *Environ. Sci. Technol.* **2014**, *48*, 7683–7689. [CrossRef] [PubMed]
12. Connon, R.E.; Geist, J.; Werner, I. Effect-Based Tools for Monitoring and Predicting the Ecotoxicological Effects of Chemicals in the Aquatic Environment. *Sensors* **2012**, *12*, 12741–12771. [CrossRef]
13. Escher, B.I.; Leusch, F.D.L. *Bioanalytical Tools in Water Quality Assessment*; IWA Publishing: London, UK, 2012.
14. Neale, P.A.; Altenburger, R.; Aït-Aïssa, S.; Brion, F.; Busch, W.; de Aragão Umbuzeiro, G.; Denison, M.S.; Du Pasquier, D.; Hilscherová, K.; Hollert, H.; et al. Development of a bioanalytical test battery for water quality monitoring: Fingerprinting identified micropollutants and their contribution to effects in surface water. *Water Res.* **2017**, *123*, 734–750. [CrossRef]
15. Altenburger, R.; Ait-Aissa, S.; Antczak, P.; Backhaus, T.; Barceló, D.; Seiler, T.-B.; Brion, F.; Busch, W.; Chipman, K.; de Alda, M.L.; et al. Future water quality monitoring—Adapting tools to deal with mixtures of pollutants in water resource management. *Sci. Total Environ.* **2015**, *512–513*, 540–551. [CrossRef]

16. Brack, W.; Ait-Aissa, S.; Burgess, R.M.; Busch, W.; Creusot, N.; Di Paolo, C.; Escher, B.I.; Mark Hewitt, L.; Hilscherova, K.; Hollender, J.; et al. Effect-directed analysis supporting monitoring of aquatic environments—An in-depth overview. *Sci. Total Environ.* **2016**, *544*, 1073–1118. [CrossRef]
17. Escher, B.I.; Allinson, M.; Altenburger, R.; Bain, P.A.; Balaguer, P.; Busch, W.; Crago, J.; Denslow, N.D.; Dopp, E.; Hilscherova, K.; et al. Benchmarking Organic Micropollutants in Wastewater, Recycled Water and Drinking Water with In Vitro Bioassays. *Environ. Sci. Technol.* **2014**, *48*, 1940–1956. [CrossRef]
18. Könemann, S.; Kase, R.; Simon, E.; Swart, K.; Buchinger, S.; Schlüsener, M.; Hollert, H.; Escher, B.I.; Werner, I.; Aït-Aïssa, S.; et al. Effect-based and chemical analytical methods to monitor estrogens under the European Water Framework Directive. *TrAC Trends Anal. Chem.* **2018**, *102*, 225–235. [CrossRef]
19. Kase, R.; Javurkova, B.; Simon, E.; Swart, K.; Buchinger, S.; Könemann, S.; Escher, B.I.; Carere, M.; Dulio, V.; Ait-Aissa, S.; et al. Screening and risk management solutions for steroidal estrogens in surface and wastewater. *TrAC Trends Anal. Chem.* **2018**, *102*, 343–358. [CrossRef]
20. Pop, C.-E.; Draga, S.; Măciucă, R.; Niță, R.; Crăciun, N.; Wolff, R. Bisphenol A Effects in Aqueous Environment on Lemna minor. *Processes* **2021**, *9*, 1512. [CrossRef]
21. Wernersson, A.S.; Carere, M.; Maggi, C.; Tusil, P.; Soldan, P.; James, A.; Sanchez, W.; Dulio, V.; Broeg, K.; Reifferscheid, G.; et al. The European technical report on aquatic effect-based monitoring tools under the water framework directive. *Environ. Sci. Eur.* **2015**, *27*, 1–11. [CrossRef]
22. Brack, W.; Dulio, V.; Ågerstrand, M.; Allan, I.; Altenburger, R.; Brinkmann, M.; Bunke, D.; Burgess, R.M.; Cousins, I.; Escher, B.I.; et al. Towards the review of the European Union Water Framework Directive: Recommendations for more efficient assessment and management of chemical contamination in European surface water resources. *Sci. Total Environ.* **2017**, *576*, 720–737. [CrossRef] [PubMed]
23. Escher, B.I.; Neale, P.A.; Leusch, F.D.L. Effect-based trigger values for in vitro bioassays: Reading across from existing water quality guideline values. *Water Res.* **2015**, *81*, 137–148. [CrossRef] [PubMed]
24. Van der Oost, R.; Sileno, G.; Suárez-Muñoz, M.; Nguyen, M.T.; Besselink, H.; Brouwer, A. SIMONI (Smart Integrated Monitoring) as a novel bioanalytical strategy for water quality assessment: Part I—Model design and effect-based trigger values. *Environ. Toxicol. Chem.* **2017**, *36*, 2385–2399. [CrossRef]
25. NORMAN Network; Water Europe. Contaminants of Emerging Concern in Urban Wastewater. *Joint NORMAN and Water Europe Position Paper*. 2019. Available online: https://www.normandata.eu/sites/default/files/files/Publications/Position%20paper_CECs%20UWW_NORMAN_WE_2019_Final_20190910_public.pdf (accessed on 26 September 2021).
26. Kienle, C.; Kase, R.; Werner, I. Evaluation of bioassays and wastewater quality. In *In Vitro and In Vivo Bioassays for the Performance Review in the Project "Strategy MicroPoll"*; Swiss Centre for Applied Ecotoxicology, Eawag-EPFL: Duebendorf, Switzerland, 2011.
27. Bain, P.A.; Williams, M.; Kumar, A. Assessment of multiple hormonal activities in wastewater at different stages of treatment. *Environ. Toxicol. Chem.* **2014**, *33*, 2297–2307. [CrossRef]
28. Völker, J.; Stapf, M.; Miehe, U.; Wagner, M. Systematic Review of Toxicity Removal by Advanced Wastewater Treatment Technologies via Ozonation and Activated Carbon. *Environ. Sci. Technol.* **2019**, *53*, 7215–7233. [CrossRef]
29. Alygizakis, N.A.; Besselink, H.; Paulus, G.K.; Oswald, P.; Hornstra, L.M.; Oswaldova, M.; Medema, G.; Thomaidis, N.S.; Behnisch, P.A.; Slobodnik, J. Characterization of wastewater effluents in the Danube River Basin with chemical screening, in vitro bioassays and antibiotic resistant genes analysis. *Environ. Int.* **2019**, *127*, 420–429. [CrossRef]
30. Ankley, G.T.; Bennett, R.S.; Erickson, R.J.; Hoff, D.J.; Hornung, M.W.; Johnson, R.D.; Mount, D.R.; Nichols, J.W.; Russom, C.L.; Schmieder, P.K.; et al. Adverse outcome pathways: A conceptual framework to support ecotoxicology research and risk assessment. *Environ. Toxicol. Chem.* **2010**, *29*, 730–741. [CrossRef]
31. Escher, B.I.; Dutt, M.; Maylin, E.; Tang, J.Y.M.; Toze, S.; Wolf, C.R.; Lang, M. Water quality assessment using the AREc32 reporter gene assay indicative of the oxidative stress response pathway. *J. Environ. Monit.* **2012**, *14*, 2877–2885. [CrossRef] [PubMed]
32. Rizzo, L.; Malato, S.; Antakyali, D.; Beretsou, V.G.; Đolić, M.B.; Gernjak, W.; Heath, E.; Ivancev-Tumbas, I.; Karaolia, P.; Lado Ribeiro, A.R.; et al. Consolidated vs new advanced treatment methods for the removal of contaminants of emerging concern from urban wastewater. *Sci. Total Environ.* **2019**, *655*, 986–1008. [CrossRef]
33. Bourgin, M.; Beck, B.; Boehler, M.; Borowska, E.; Fleiner, J.; Salhi, E.; Teichler, R.; von Gunten, U.; Siegrist, H.; McArdell, C.S. Evaluation of a full-scale wastewater treatment plant upgraded with ozonation and biological post-treatments: Abatement of micropollutants, formation of transformation products and oxidation by-products. *Water Res.* **2018**, *129*, 486–498. [CrossRef]
34. Braun, R.; Hartmann, C.; Kreuzinger, N.; Lenz, K.; Schaar, H.; Scheffknecht, C. *Untersuchung von Abwässern und Gewässern auf Unterschiedliche Toxikologische Endpunkte*; Bundesministerium für Landwirtschaft, Regionen und Tourismus Vienna: Vienna, Austria, 2021; p. 248.
35. Escher, B.I.; Fenner, K. Recent Advances in Environmental Risk Assessment of Transformation Products. *Environ. Sci. Technol.* **2011**, *45*, 3835–3847. [CrossRef]
36. Dopp, E.; Pannekens, H.; Gottschlich, A.; Schertzinger, G.; Gehrmann, L.; Kasper-Sonnenberg, M.; Richard, J.; Joswig, M.; Grummt, T.; Schmidt, T.C.; et al. Effect-based evaluation of ozone treatment for removal of micropollutants and their transformation products in waste water. *J. Toxicol. Environ. Health Part A* **2021**, *84*, 418–439. [CrossRef] [PubMed]
37. Neale, P.A.; O'Brien, J.W.; Glauch, L.; König, M.; Krauss, M.; Mueller, J.F.; Tscharke, B.; Escher, B.I. Wastewater treatment efficacy evaluated with in vitro bioassays. *Water Res. X* **2020**, *9*, 100072. [CrossRef]

38. Clara, M.; Kreuzinger, N.; Strenn, B.; Gans, O.; Kroiss, H. The solids retention time—A suitable design paramter to evaluate the capacity of wastewater treatment plants to remove micropollutants. *Water Res.* **2005**, *39*, 97–106. [CrossRef] [PubMed]
39. NEREUS. Deliverable 13. White Paper on the existing knowledge with regard to wastewater and biological hazards. In *NEREUS COST Action ES 1403*; Deliverable of WG3; 2018; 20p, Available online: http://www.nereus-cost.eu/wp-content/uploads/2019/12/D13.pdf (accessed on 26 September 2021).
40. Wolf, Y.; Oster, S.; Shuliakevich, A.; Brückner, I.; Dolny, R.; Linnemann, V.; Pinnekamp, J.; Hollert, H.; Schiwy, S. Improvement of wastewater and water quality via a full-scale ozonation plant?—A comprehensive analysis of the endocrine potential using effect-based methods. *Sci. Total Environ.* **2022**, *803*, 149756. [CrossRef]
41. Escher, B.I.; Bramaz, N.; Ort, C. JEM Spotlight: Monitoring the treatment efficiency of a full scale ozonation on a sewage treatment plant with a mode-of-action based test battery. *J. Environ. Monit.* **2009**, *11*, 1836–1846. [CrossRef]
42. Huber, M.M.; Göbel, A.; Joss, A.; Hermann, N.; Löffler, D.; McArdell, C.S.; Ried, A.; Siegrist, H.R.; Ternes, T.A.; von Gunten, U. Oxidation of Pharmaceuticals during Ozonation of Municipal Wastewater Effluents: A Pilot Study. *Environ. Sci. Technol.* **2005**, *39*, 4290–4299. [CrossRef]
43. Neale, P.A.; Ait-Aissa, S.; Brack, W.; Creusot, N.; Denison, M.S.; Deutschmann, B.; Hilscherová, K.; Hollert, H.; Krauss, M.; Novák, J.; et al. Linking in Vitro Effects and Detected Organic Micropollutants in Surface Water Using Mixture-Toxicity Modeling. *Environ. Sci. Technol.* **2015**, *49*, 14614–14624. [CrossRef]
44. Lemaire, G.R.; Mnif, W.; Pascussi, J.-M.; Pillon, A.; Rabenoelina, F.; Fenet, H.l.N.; Gomez, E.; Casellas, C.; Nicolas, J.-C.; Cavaillès, V.; et al. Identification of New Human Pregnane X Receptor Ligands among Pesticides Using a Stable Reporter Cell System. *Toxicol. Sci.* **2006**, *91*, 501–509. [CrossRef] [PubMed]
45. König, M.; Escher, B.I.; Neale, P.A.; Krauss, M.; Hilscherová, K.; Novák, J.; Teodorović, I.; Schulze, T.; Seidensticker, S.; Kamal Hashmi, M.A.; et al. Impact of untreated wastewater on a major European river evaluated with a combination of in vitro bioassays and chemical analysis. *Environ. Pollut.* **2017**, *220*, 1220–1230. [CrossRef]
46. Mišík, M.; Ferk, F.; Schaar, H.; Yamada, M.; Jaeger, W.; Knasmueller, S.; Kreuzinger, N. Genotoxic activities of wastewater after ozonation and activated carbon filtration: Different effects in liver-derived cells and bacterial indicators. *Water Res.* **2020**, *186*, 116328. [CrossRef] [PubMed]
47. Neale, P.A.; Munz, N.A.; Aït-Aïssa, S.; Altenburger, R.; Brion, F.; Busch, W.; Escher, B.I.; Hilscherová, K.; Kienle, C.; Novák, J.; et al. Integrating chemical analysis and bioanalysis to evaluate the contribution of wastewater effluent on the micropollutant burden in small streams. *Sci. Total Environ.* **2017**, *576*, 785–795. [CrossRef] [PubMed]

Communication

Performance of Newly Developed Intermittent Aerator for Flat-Sheet Ceramic Membrane in Industrial MBR System

Hiroshi Noguchi [1,*], Qiang Yin [1], Su Chin Lee [1], Tao Xia [1], Terutake Niwa [1], Winson Lay [2], Seng Chye Chua [2], Lei Yu [2], Yuke Jen Tay [2], Mohd Jamal Nassir [2], Guihe Tao [2], Shu Ting Ooi [3], Adil Dhalla [3] and Chakravarthy Gudipati [3,*]

1. Meiden Singapore Pte Ltd., Singapore 619363, Singapore; roger.yin@meidensg.com.sg (Q.Y.); suchin.lee@meidensg.com.sg (S.C.L.); tony.x@meidensg.com.sg (T.X.); niwa.t@meidensg.com.sg (T.N.)
2. Public Utilities Board, Singapore's National Water Agency, Singapore 228231, Singapore; winson_lay@pub.gov.sg (W.L.); chua_seng_chye@pub.gov.sg (S.C.C.); yu_lei@pub.gov.sg (L.Y.); tay_yuke_jen@pub.gov.sg (Y.J.T.); mohd_jamal_nasir@pub.gov.sg (M.J.N.); tao_guihe@pub.gov.sg (G.T.)
3. Separation Technologies Applied Research and Translation (START) Centre, Nanyang Technological University—NTUitive Pte Ltd., Singapore 639798, Singapore; shu-ting.ooi@outlook.com (S.T.O.); adil.dhalla@ntu.edu.sg (A.D.)
* Correspondence: noguchi.h@meidensg.com.sg (H.N.); chakra@ntu.edu.sg (C.G.)

Abstract: An intermittent aerator was newly developed to reduce energy costs in a flat-sheet ceramic membrane bioreactor (MBR) for industrial wastewater treatment. Large air bubbles were supplied over a short time interval by the improved aerator technology at the bottom of the flat-sheet membrane. Performance tests for the intermittent aerator were carried out in a pilot system with two cassettes immersed in a membrane tank of the 1-MGD demonstration plant at Jurong Water Reclamation Plant (JWRP) in Singapore. Stable operation was achieved at an average flow of 19–22 LMH with every-2-days MC and peak flow of 27 to 33 LMH with daily MC with reduced air flow for membrane aeration. This indicates that energy costs for membrane aeration can be reduced by using the intermittent aerator. Stable MBR operation with a projected 43% reduction in the overall operating costs could be achieved with an improved aerator together with improved MC regime and membrane cassette.

Keywords: ceramic membranes; industrial wastewater; intermittent aerator; water reclamation; MBR

1. Introduction

The membrane bioreactor (MBR) has been widely used for treatment of domestic sewage and industrial wastewater [1,2]. The MBR system has the advantages of smaller footprint, higher quality of product water, stability of bioprocess and shorter retention times compared to conventionally activated sludge processes [3–5]. Membrane aeration is a key operational parameter and has a significant impact on the energy costs of the MBR system [6–8]. Membrane aeration is applied to prevent fouling on the surface of the membrane, and not only provides oxygen to the biomass in the membrane tank, but also scours the membrane surface and maintains solids in suspension to control the fouling layer on the membrane surface. Many researchers have carried out trials to reduce energy for membrane aeration through novel methods such as controlling aeration for bioprocess [5,9–13] and reducing membrane scouring air [14–17]. On–off control of scouring air has been developed to reduce scouring air [14,15] while pulsed air for membrane scouring has been utilized to reduce the required amount of air [16,17]. An intermittent aerator has been developed by MEIDEN and Separation Technologies Applied Research and Translation (START) Centre to reduce the required air amount for membrane scouring, which results in a reduction of energy consumption in the membrane filtration system. PUB, Singapore's National Water Agency, has developed a one million gallons per day (1 MGD) demonstration plant (DEMO) at Jurong Water Reclamation Plant (JWRP), Singapore, to treat high-strength industrial wastewater from an industrial estate. MEIDEN's Ceramic Membrane Bioreactor (CMBR) process has been installed and operational at the plant since 2014 [14]. It has been demonstrated and reported widely in the literature that the CMBR system can produce high-quality water for reuse application [6,7,9,15–18]. A new intermittent aerator with pulsated, lower air flow was developed by MEIDEN and START Centre, for energy reduction in the CMBR system. The performance evaluation of the intermittent aerator was carried out in the DEMO plant at JWRP to assess the feasibility and extent of energy reduction for the CMBR system. The stability of membrane filtration was investigated with the intermittent aerator at reduced air flow compared with that for the conventional continuous aerator.

2. Materials and Methods

2.1. Intermittent Aerator Tank Operation Principles

An intermittent aerator was developed to release large air bubbles at the bottom of the flat-sheet ceramic membrane with a width of 250 mm (Figure 1). The intermittent aerator can release larger bubbles from ϕ24 mm air release holes as compared to the ϕ6 mm air release holes from a conventional continuous aerator. Air is continuously supplied from air pipes at the bottom of the intermittent aerator; it is accumulated in the air chamber of the aerator and is released in 0.75 sec with 3.5 sec of air release interval to produce large air bubbles. Air bubbles from the ϕ24 mm air release holes were split into the gaps of 10 mm to produce mixed air and water flow in the gaps as is shown in Figure 2.

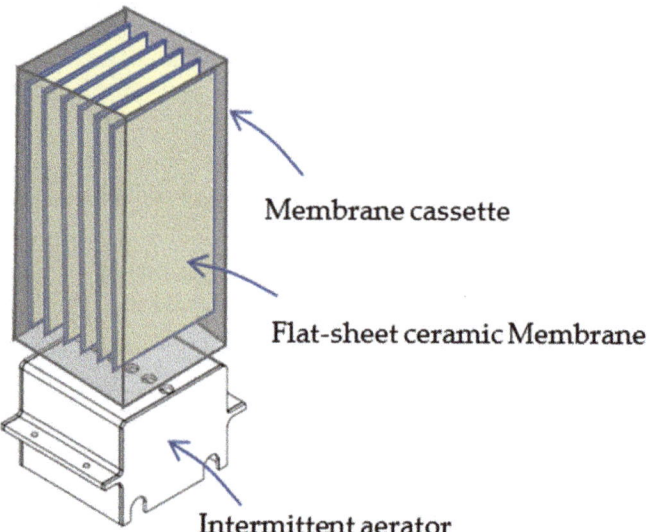

Figure 1. Membrane cassette with the intermittent aerator.

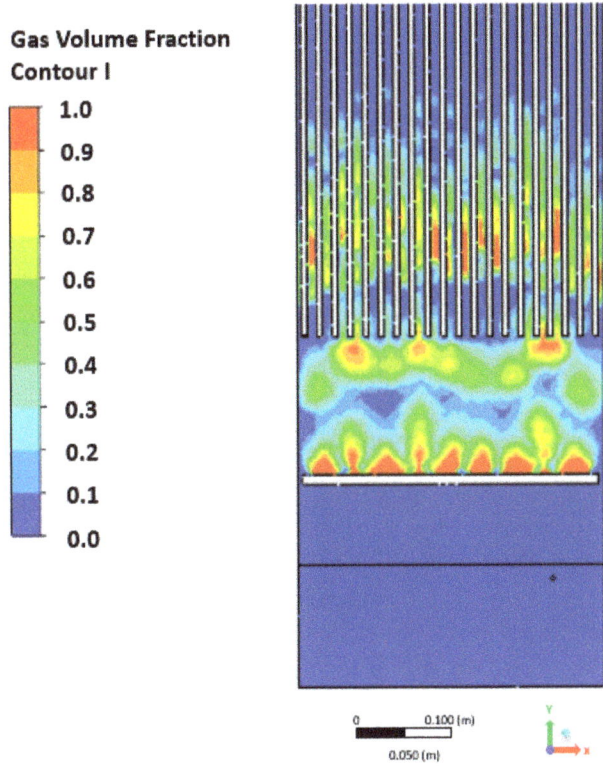

Figure 2. CFD analysis result for air flow with the intermittent aerator cassette.

Multiphase CFD simulations were performed by the Eulerian model in Ansys® Fluent, Release 19.2 (Ansys, Canonsburg, PA, USA) with water as the primary phase and air as the secondary phase. The boundary conditions assumed for the modelling include: (a) symmetry on the size of the membrane cassette, and (b) air entrainment from bottom to top. The material properties used in the simulation are defined in Table 1 below. The air mass flow rate was converted accordingly to correspond with the air flowrate of 2.2 m³/min and the number of holes that exist throughout the air-scouring pipes. To improve the convergence behavior, an initial solution was computed before solving the complete Eulerian multiphase model. In this work, a mass-flow inlet boundary condition was utilized to initialize the flow conditions. It is also more recommended to set the value of the volume fraction close to the value of the volume fraction at the inlet. At the beginning of the solution, a lower time step i.e., 0.001 s was used and recommended in order to reach convergence. In addition, a sufficient number of iterations (i.e., 30) are required to maintain the convergence level at each time step. The simulation was executed until the end time of 9 s to ensure that there has been a flow stability reaching the top part of the two-level stack membrane unit that would be fully submerged inside water.

Table 1. Material properties employed in the CFD simulation.

Material	Density (kg/m³)	Viscosity (kg/m/s)
Water (primary phase)	998.20	0.001003
Air (secondary phase)	1.225	1.7894 e-05

2.2. Test Systems

Performance tests of the new aeration system were carried out at the 1-MGD DEMO MBR plant at JWRP. The DEMO plant consisted of an Up-flow Anaerobic Sludge Blanket (UASB) reactor, aeration and membrane separation tanks. Wastewater from the industrial estate was treated in a UASB reactor followed by an MBR system using flat-sheet ceramic membrane [6]. Two commercial membrane cassettes with flat-sheet ceramic membranes manufactured by MEIDENSHA Corporation were used for the performance tests. Each cassette had 400 membrane sheets with an effective area of 200 m^2. One membrane cassette (Train 1) was equipped with the newly developed intermittent aerator while the other cassette (Train 2) was installed with a conventional continuous aerator, in which air was supplied from 6-mm air release holes on the PVC pipes. The two cassettes were immersed in a membrane tank of the 1-MGD DEMO plant, and they were independently operated by using a control panel. ON–OFF control air supply for the continuous aerator was also carried out in Train 2, for comparison.

2.3. Operating Conditions

Membrane filtration was carried out under similar conditions for Train 1 and Train 2, except for the membrane aeration condition. The filtration/backwash cycle was fixed at 10 min; backwash duration was 30 s with a flow rate of 1.5 Q (Q = filtration flow). Different flux conditions were set to check membrane stability at average and peak flow conditions (Table 2).

Table 2. The operating conditions for the MBR system.

Parameter	Unit	Average Flow	Peak Flow
Membrane		MEIDEN flat-sheet ceramic membrane (pore size: 0.1 μm)	
Membrane air		Train 1: Intermittent aerator (35 m^3/h/train)	
		Train 2: Continuous aerator (66 m^3/h/train)	
Net flux	L.m^{-2}.h^{-1} (LMH)	19–22	27–33
Backwash cycle	min	9.5	9.5
Backwash duration	min	0.5	0.5
Backwash flow	Ratio to filtrate	1.5 Q	1.5 Q
MC Frequency	-	Every 2 days	Daily
NaClO concentration	mg/L	250	250

Maintenance cleaning (MC) was carried out with 250 mg/L sodium hypochlorite (NaClO) every two days for the average flow and daily for the peak flow (Table 2). Average flow setting assumed normal operating conditions with full operation of membrane filtration tanks in the membrane systems. Peak flow setting was for higher flux condition when one or two membrane tanks were out of service during maintenance or recovery cleaning. Long-term operation was carried out with average and continuous flux settings to observe filtration stability and impact of peak flow between average flow operation.

Air flow for the intermittent aerator (Train 1) and the continuous aerator (Train 2) was set to 35 m^3/h/train and that for the continuous aerator (Train 2) was set to 66 m^3/h/train, which corresponded to 5.3 m^3-air/m^3-permeate and 10 m^3-air/m^3-permeate as specific air demand per permeate (SADp) at 33 L/m^2/h (LMH). Air flow setting for the intermittent aerator was 53% of that for conventional continuous aerator.

3. Results and Discussion

3.1. CFD Analysis for the Intermittent Aerator

Figure 2 shows CFD analysis results of air flow by the intermittent aerator. The large air bubbles released from the bottom of the membrane cassette were evenly distributed in between the gaps of 10 mm for the membrane sheets. Multiphase CFD simulations were performed with the Eulerian model in Ansys® Fluent, Release 19.2. As an unstructured solver, Ansys® Fluent used internal data structures to assign an order to the cells, faces, and grid points in a mesh and to maintain contact between adjacent cells. Upon generating the grids, the minimum orthogonal quality of the whole CFD model was up to 0.067 which shows a high quality of the mesh, and that quality plays a significant role in the accuracy and stability of the numerical computation. Ansys Fluent allows the user to evaluate the mesh quality through a quantity called orthogonality. Orthogonal quality was computed for cells using cell skewness and the vector from the cell centroid to each of its faces, the corresponding face area vector, and the vector from the cell centroid to the centroids of each of the adjacent cells. If the orthogonality exceeds the limits the numerical error will be higher, which will lead to solution convergence issues. The minimum orthogonal quality for all types of cells should be more than 0.01, with an average value that is significantly higher.

3.2. Performance with Intermittent Aerator

The trans-membrane pressure (TMP) and the permeate flux data were obtained and collated for five months of continuous operation, to determine the long-term efficiency of the intermittent aerator. Figures 3 and 4 show the profiles of comparison (1) TMP and (2) permeability between the two aerators with permeate flux settings at 22 LMH and 33 LMH, respectively. From Figure 3, base TMP after each MC for the intermittent aerator was lower than a conventional continuous aerator. However, TMP trends became similar for intermittent and continuous aerators after plant shut down on 15th August. Aeration for bioprocess was stopped during the shut down period, which might have caused lower COD or BOD removal in the activated sludge system; this could be attributed to potential membrane fouling when the organics became more dominant after the plant shut down and the reducing impact of the larger bubbles from the intermittent aerator on membrane surface resulted in a similar TMP increase for intermittent and continuous aerators.

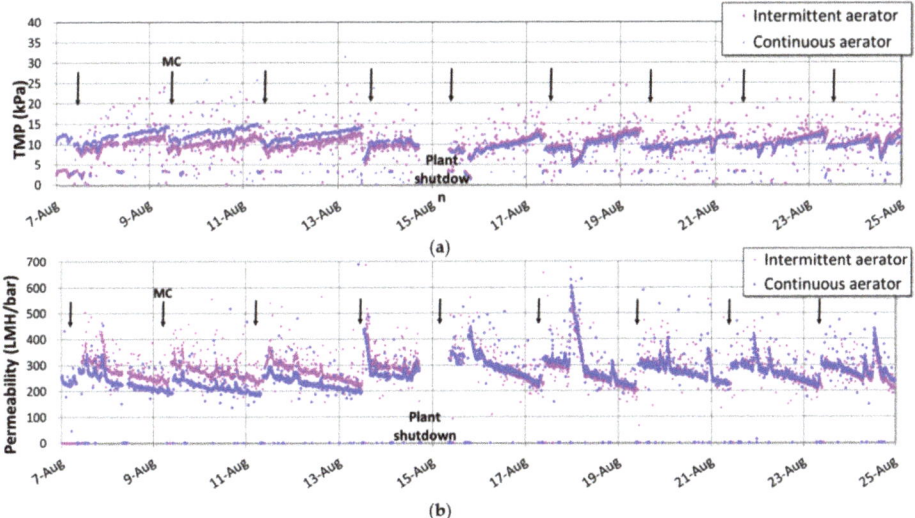

Figure 3. Comparison of aeration methods at 22 LMH flux: (**a**) TMP trend; (**b**) Permeability trend.

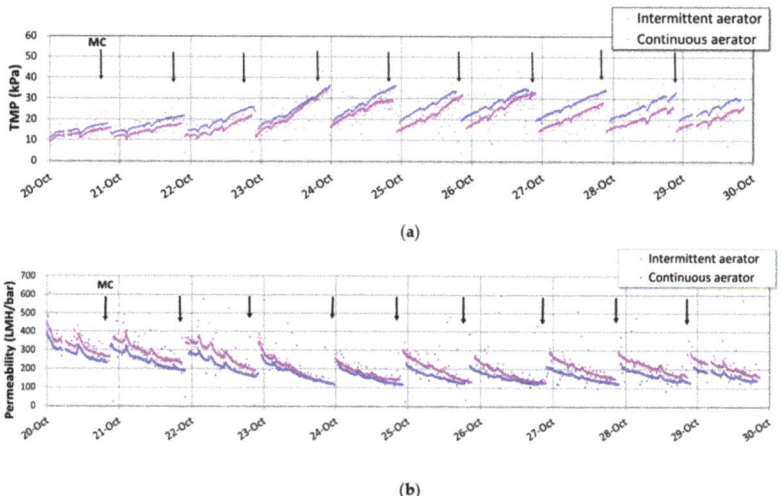

Figure 4. Comparison of aeration methods at a flux of 33 LMH: (**a**) TMP trend; (**b**) Permeability trend.

Figure 4 shows TMP and permeability for the intermittent and continuous aeration methods at flux of 33 LMH. TMP increases between MC for the intermittent aerator, ranging from 5 to 23 kPa. This was much higher than that at 22 LMH which has the TMP ranged from 2 to 5 kPa for the intermittent aerator. Both base TMP and TMP after MC have become stable at around 15 to 17 kPa after 24th October, which indicates that stable operation was achieved with the intermittent aerator at 33 LMH.

From Figure 4, it is shown that the performance of TMP and permeability of the intermittent aeration method surpassed the conventional continuous aeration method at a higher flux of 33 LMH. This can be observed from Figure 4a where TMP increase between MC for the intermittent aeration was lower than the conventional continuous aeration method. The Figure 4b also shows a higher consistency in permeability data using the intermittent aeration method than that of the conventional continuous aeration method.

Membrane fouling occurs on the surface of the membrane by formation of a cake layer and inside of the pores by the accumulation of foulants. As membrane air can remove foulants on the surface of the membrane and foulants inside of the membrane pores can be removed by backwashing and chemical cleaning during MC. As air flow on the surface of the membrane mitigates the cake layer formation [18], the difference in increase in TMP between MC for the intermittent and continuous aerator might be attributed to the difference in formation of cake layer on the surface. Shear force was obtained from the CFD analysis. Average shear force in a membrane cassette for the intermittent aerator was 11.22 µST, while that for the continuous aerator was 6.97 µST. Higher shear force resulting from larger air bubbles from the intermittent aerator can strongly reduce foulants on the surface compared with the continuous aerator. This might result in a lower increase in TMP between MC for the intermittent aerator. Reduced formation of cake layer during filtration can affect the efficiency of backwashing, which might result in lower base TMP for the intermittent aerator. Cake layer formation might become faster at higher flux setting of 33 LMH. This might result in more significant suppression of TMP increase with intermittent aerator than the flux condition of 22 LMH.

3.3. Long-Term Operation

Figure 5a presents the TMP trend graph of the intermittent aerator operation with various flux settings over a period of five months without recovery cleaning (RC). The results indicate that stable operation can be achieved with the intermittent aerator to reduce air consumption while achieving sustainable flux.

SADp was reduced to 5.3 m^3-air/m^3-permeate by using the newly developed intermittent aerator, and longer-term stability was shown in the large-scale MBR plant in JWRP. SADp in full-scale plant ranges from 10 to 50 m^3-air/m^3-permeate [19]. Some researchers reported achievements of lower SADp ranging from 6 to 9 m^3-air/m^3-permeate with polymeric hollow fiber membrane in lab or pilot scale tests for MBR systems [7,10]. Reduction of SADp to 5.3 m^3-air/m^3-permeate for the flat-sheet membrane system was similar to the range achieved for the hollow fiber membrane system, suggesting that similar energy for membrane scouring for the flat-sheet ceramic membranes can be almost the same level as that for hollow fiber polymeric membrane. Achievement of lower SADp at 5.3 m^3-air/m^3-permeate is a significant development for MBR systems, especially for flat-sheet membranes whose footprint is larger than those for hollow fiber membranes. Flat-sheet ceramic membrane can withstand higher shear force by larger air bubbles, which enables lower SADp in MBR system.

Turbidity is an indicator of performance in water and wastewater applications. Thus, it is monitored regularly to ensure treatment systems are operating effectively. Figure 5b shows that the turbidity of the MBR permeate remained constant at ≤0.1 NTU throughout the testing period, regardless of the wide variations in turbidity for raw wastewater that ranged between 30 NTU and 600 NTU. As shown in Figure 5c, due to the industrial nature of influent, the COD of feed water to the MBR fluctuated with values ranging between 150 and 2500 mg/L. Despite the variable characteristics of the influent, the MBR was able to consistently produce a permeate with COD ≤ 50 mg/L. The results showed the robustness of the aerobic biological process. The consistent permeate quality also indicated stable operation of the ceramic membrane system.

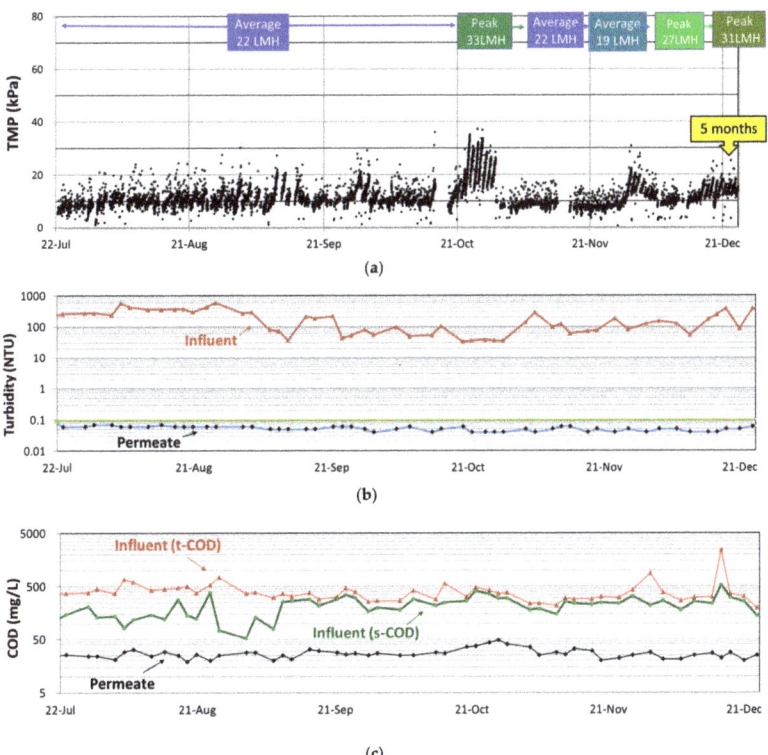

Figure 5. TMP trend, turbidity, and COD of MBR: (**a**) TMP trend with intermittent aerator; (**b**) Turbidity; (**c**) COD.

3.4. Analysis of Operating Costs

Analysis of operation costs was carried out to show the impact of the improved membrane aeration method, and cost analysis was done for large scale MBR with a capacity of 20 MGD as a case study. Energy costs were estimated for the intermittent aerator and compared with the conventional continuous aerator and ON–OFF control of the continuous aerator. The 7 min-ON and 3 min-OFF conditions were used for the analysis of ON–OFF control. Chemical costs included MC with NaClO and recovery cleaning (RC) with NaClO and citric acid. Improved MC regime has been applied in the estimation for developing this improved system with new aeration method. Costs for spare parts were also estimated based on this improved system, which has the improvement in membrane cassette design to eliminate spare parts of tubes and connectors for permeate collection from each membrane sheet. Total life cycle costs were estimated with a forecast of 20 years operation of the MBR system.

Conventional continuous aeration required 66 m^3/h/train, which corresponds to 0.258 kWh/m^3. This results in 5.68 $¢/m^3$ in Singapore Dollars converted with an electrical cost of 0.22 $¢/kWh$. The energy cost was reduced to 3.98 $¢/m^3$ by using 7 min-ON and 3 min-OFF condition, 30% reduction of energy; the energy cost was further reduced to 3.01 $¢/m^3$ with intermittent aerator, which required 35 m^3/h/train. The chemical cost was 2.14 $¢/m^3$, which was obtained from 44 g/m^3 usage of NaClO (29.5 $¢/kg$) and 14 g/m^3 of citric acid (60 $¢/kg$) for MC and RC. This was reduced to 1.67 $¢/m^3$ with improved regime of MC. The cost for spare parts was 0.47 $¢/m^3$ and was reduced to 0.05 $¢/m^3$ with the improved membrane cassette. The total cost was 8.29 $¢/m^3$ and was reduced to 5.55 $¢/m^3$ with ON–OFF control, 4.73 $¢/m^3$ with intermittent aerator together with improvement of MC regime and membrane cassette.

Figure 6 shows analysis of the projected operating costs that cover energy, chemical and parts replacement. Energy costs here relate to membrane air scouring. The current operation with continuous air scouring serves as the reference condition at 100%. It can be noted that about 70% of the overall cost is due to energy for membrane air scouring. Through implementation of the newly developed intermittent aeration technology, higher efficiency could be achieved with reduction of the overall cost by approximately 43% as shown in Figure 6. Chemical costs and costs related to parts replacement could also be reduced, as the improved MBR performance would also reduce the frequency of membrane cleanings. This could result in lower life cycle cost (LCC) for MBR system using flat-sheet ceramic membrane.

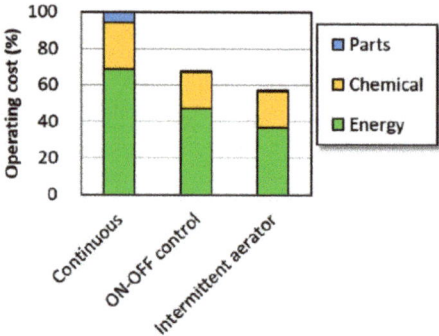

Figure 6. Operating cost comparison between continuous and intermittent aerators.

4. Conclusions

An intermittent aerator for flat-sheet ceramic membrane was developed to release larger bubbles at the bottom of the membrane cassette. CFD analysis was carried out to observe air distribution in the membrane cassette. Membrane filtration stability with the intermittent aerator was investigated in 1-MGD DEMO plant, in which wastewater from an industrial estate was treated by the combined process of UASB and MBR. TMP increase was suppressed with the intermittent aerator compare with that for the conventional continuous aerator. Five months of operation with the intermittent aerator showed long-term stability of the membrane filtration system with the developed aerator. Reduction in operating costs was estimated to be 43% together with an improved chemical cleaning regime and membrane cassette. A performance test with the developed intermittent aerator in the MBR system for domestic sewage treatment will be carried out to investigate energy reduction for the domestic MBR system.

Author Contributions: Conceptualization, H.N. and T.N.; methodology, H.N. and S.T.O.; validation, H.N., Q.Y., T.X. and S.T.O.; formal analysis, H.N.; investigation, H.N., S.C.L., W.L., S.C.C., L.Y., S.C.L., M.J.N. and G.T.; resources, T.N., Y.J.T., A.D. and G.T.; data curation, H.N.; writing—original draft preparation, H.N.; writing—review and editing, C.G. and A.D.; visualization, C.G. and G.T.; supervision, H.N. and T.N.; project administration, H.N. All authors have read and agreed to the published version of the manuscript.

Funding: This research received no external funding.

Institutional Review Board Statement: Not applicable.

Informed Consent Statement: Not applicable.

Data Availability Statement: Not applicable.

Conflicts of Interest: The authors declare no conflict of interest.

References

1. Bonetta, S.; Pignata, C.; Gsaparro, E.; Richiardi, L.; Bonetta, S.; Carraro, E. Impact of wastewater treatment plants on microbiological contamination for evaluating the risks of wastewater reuse. *Environ. Sci. Eur.* **2022**, *34*, 1–13. [CrossRef]
2. Ahmed, M.; Mavukkandy, M.O.; Giwa, A.; Elektorowica, M.; Katsou, E.; Khelifi, O.; Naddeo, V.; Hasan, S. Recent developments in hazardous pollutants removal from wastewater and water reuse within a circular economy. *npj Clean Water* **2022**, *5*, 1–25. [CrossRef]
3. Bera, S.P.; Godhaniya, M.; Kothari, C. Emerging and advanced membrane technology for wastewater treatment: A review. *J. Basic Microbiol.* **2022**, *62*, 245–259. [CrossRef] [PubMed]
4. Asante-Sackey, D.; Rathilal, S.; Tetteh, E.K.; Armah, E.K. Membrane Bioreactors for Produced Water Treatment: A Mini-Review. *Membranes* **2022**, *12*, 275. [CrossRef] [PubMed]
5. Judd, S. *The MBR Book: Principles and Applications of Membrane Bioreactors in Water and Wastewater Treatment*; Elsevier: Oxford, UK, 2006.
6. Niwa, T.; Hatamoto, M.; Yamashita, T.; Noguchi, H.; Takase, O.; Kekre, K.A.; Ang, W.S.; Seah, H.; Yamaguchi, T. Demonstration of a full-scale plant using an UASB followed by a ceramic MBR for the reclamation of industrial wastewater. *Bioresour. Technol.* **2016**, *218*, 1–8. [CrossRef] [PubMed]
7. Noguchi, H.; Niwa, T.; Nakamura, Y.; Agrawal, J.K.; And, W.S.; Kekre, K.; Tao, G. Treatment of industrial used water by UASB + CMBR process to produce industrial grade water for reuse. *Water Pract. Technol.* **2018**, *13*, 424–430. [CrossRef]
8. Kitanou, S.; Ayyoub, H.; El-Ghzizel, S.; Belhamidi, S.; Taky, M.; Elmidaoui, A. Membrane bioreactor for domestic wastewater treatment: Energetic assessment. *Desalin. Water Treat.* **2021**, *240*, 55–62. [CrossRef]
9. Lay, W.; Lim, C.; Lee, Y.; Kwok, B.H.; Tao, G.; Lee, K.S.; Chua, S.C.; Wah, W.L.; Ghani, A.; Seah, H. From R&D to application: Membrane bioreactor technology for water reclamation. *Water Pract. Technol.* **2017**, *12*, 12–24.
10. Yamashita, K.; Itokawa, H.; Hashimoto, T. Demonstration of energy-saving membrane bioreactor (MBR) systems. *Water Sci. Technol.* **2019**, *79*, 448–457. [CrossRef] [PubMed]
11. Mannina, G.; Cosenza, A.; Reboucas, T.F. Aeration control in membrane bioreactor for sustainable environmental footprint. *Bioresour. Technol.* **2020**, *301*, 122734. [CrossRef] [PubMed]
12. Wang, S.; Zou, L.; Li, H.; Zheng, K.; Wang, Y.; Zheng, G.; Li, J. Full-scale membrane bioreactor process WWTPs in East Taihu basin: Wastewater characteristics, energy consumption and sustainability. *Sci. Total Environ.* **2020**, *723*, 137982. [CrossRef] [PubMed]
13. Iorhemen, O.T.; Hamza, R.A.; Tay, J.H. Membrane fouling control in membrane bioreactors (MBRs) using granular materials. *Bioresour. Technol.* **2017**, *240*, 9–24. [CrossRef] [PubMed]
14. Kekre, K.; Ang, W.S.; Niwa, T.; Yamashita, T.; Shiota, H.; Noguchi, H. Production of Industrial Grade Water for Reuse from Industrial Used Water using UASB CMBR Process. *Proc. Water Environ. Fed.* **2015**, *2015*, 150–158. [CrossRef]
15. Asif, M.B.; Zhang, Z. Ceramic membrane technology for water and wastewater treatment: A critical review of performance, full-scale applications, membrane fouling and prospects. *Chem. Eng. J.* **2021**, *418*, 129481. [CrossRef]
16. Wang, C.; Ng, T.C.A.; Ng, H.Y. Comparison between novel vibrating ceramic MBR and conventional air-sparging MBR for domestic wastewater treatment: Performance, fouling control and energy consumption. *Water Res.* **2021**, *203*, 117521. [CrossRef]

17. Asif, M.B.; Li, C.; Ren, B.; Maqbool, T.; Zhang, X.; Zhang, Z. Elucidating the impacts of intermittent *in-situ* ozonation in a ceramic membrane bioreactor: Micropollutant removal, microbial community evolution and fouling mechanisms. *J. Hazard. Mater.* **2021**, *402*, 123730. [CrossRef] [PubMed]
18. Miyoshi, T.; Yamamura, H.; Morita, T.; Watanabe, Y. Effect of intensive membrane aeration and membrane flux on membrane fouling in submerged membrane bioreactors: Reducing specific air demand per permeate (SADp). *Sep. Purif. Technol.* **2015**, *148*, 1–9. [CrossRef]
19. Freeman, B.R.; Peck, S.; Droxsos, N.; Kawashimo, T.; Oda, Y. Advancing the use of integrated membrane bioreactor-reverse osmosis technology to reclaim wastewater through pilot testing. *Ultrapure Water* **2011**, *28*, 16–21.

Article

Reuse of Textile Dyeing Wastewater Treated by Electrooxidation

Cláudia Pinto, Annabel Fernandes, Ana Lopes *, Maria João Nunes, Ana Baía, Lurdes Ciríaco and Maria José Pacheco

Fiber Materials and Environmental Technologies (FibEnTech-UBI), Universidade da Beira Interior, R. Marquês de D'Ávila e Bolama, 6201-001 Covilhã, Portugal; garcia.pinto@ubi.pt (C.P.); annabelf@ubi.pt (A.F.); maria.nunes@ubi.pt (M.J.N.); ana.isabel.mota.baia@ubi.pt (A.B.); lciriaco@ubi.pt (L.C.); mjap@ubi.pt (M.J.P.)
* Correspondence: analopes@ubi.pt

Abstract: Wastewater reuse has been addressed to promote the sustainable water utilization in textile industry. However, conventional technologies are unable to deliver treated wastewater with the quality required for reuse, mainly due to the presence of dyes and high salinity. In this work, the feasibility of electrooxidation, using a boron-doped diamond anode, to provide treated textile dyeing wastewater (TDW) with the quality required for reuse, and with complete recovery of salts, was evaluated. The influence of the applied current density on the quality of treated TDW and on the consecutive reuse in new dyeing baths was studied. The ecotoxicological evaluation of the process towards *Daphnia magna* was performed. After 10 h of electrooxidation at 60 and 100 mA cm^{-2}, discolorized treated TDW, with chemical oxygen demand below 200 (moderate-quality) and 50 mg L^{-1} (high-quality), respectively, was obtained. Salt content was unchanged in both treatment conditions, enabling the consecutive reuse without any salt addition. For the two reuse cycles performed, both treated samples led to dyed fabrics in compliance with the most restrictive controls, showing that an effective consecutive reuse can be achieved with a moderate-quality water. Besides the water reuse and complete salts saving, electrooxidation accomplished an ecotoxicity reduction up to 18.6-fold, allowing TDW reuse without severe ecotoxicity accumulation.

Keywords: wool dyeing wastewater; electrooxidation; boron-doped diamond; ecotoxicity; *Daphnia magna*; salts recovery; wastewater reuse

Citation: Pinto, C.; Fernandes, A.; Lopes, A.; Nunes, M.J.; Baía, A.; Ciríaco, L.; Pacheco, M.J. Reuse of Textile Dyeing Wastewater Treated by Electrooxidation. *Water* 2022, 14, 1084. https://doi.org/10.3390/w14071084

Academic Editors: Martin Wagner and Sonja Bauer

Received: 28 February 2022
Accepted: 28 March 2022
Published: 29 March 2022

Publisher's Note: MDPI stays neutral with regard to jurisdictional claims in published maps and institutional affiliations.

Copyright: © 2022 by the authors. Licensee MDPI, Basel, Switzerland. This article is an open access article distributed under the terms and conditions of the Creative Commons Attribution (CC BY) license (https://creativecommons.org/licenses/by/4.0/).

1. Introduction

The textile industry is one of the most water-intensive industries in the world. In 2015, the worldwide annual consumption of water, in the textile and clothing industry, was estimated to be around 79 billion cubic meters [1]. During the textile manufacturing process, close to 80% of the used water is discharged as wastewater [2]. This process is also responsible for the vast consumption of different chemical products, mainly in the finishing sector (dyeing, bleaching, washing, etc.), and for the generation of large volumes of highly contaminated wastewater, containing residual dyes, dyeing auxiliaries, and high salinity, which must undergo treatment before being discharged into the environment.

The volume and characteristics of the wastewaters generated by the textile industry depend on the type of fabric processed, the industrial processes applied, the type of equipment used and the water consumption [3,4]. Thus, reducing both the water consumption and the discharge of highly contaminated wastewater represents a huge challenge of efficiently managing water with a strategic focus on sustainability. This issue was covered by a large-scale European project AQUAFIT4USE (7th EU Framework Programme), in which one of the objectives was to reduce the use of resources, in particular the use of fresh high-quality water [2]. For the textile finishing sector, the outcomes of this project point to sustainable thinking related to the treatment of textile wastewater for later reuse. One of the proposed approaches was the separation of waste streams, which increases treatability options using individual treatment technologies, and verification of the same quality preservation of the textile material when high-quality process water is replaced by lower quality water in the dyeing processes [2].

Considering that the dyeing process consumes more water than other textile processes and, consequently, generates a larger volume of wastewaters, this specific wastewater is a good candidate to be treated for reuse. However, the water used in textile processing, particularly in the dyeing step, must comply with stricter quality requirements than those applied for discharge into the environment [2,4]. Thus, for its reuse to be economically viable, it is necessary to develop efficient treatment processes, which lead to treated wastewaters with the minimum quality necessary for reuse. On the other hand, the study of new technologies for the treatment of these wastewaters should aim not only to remove the color and the organic matter but also to recover salts and other constituents [5].

Several review papers, found in the literature, focus on the different technologies applied for the remediation of textile wastewaters, from biological treatments to physical and chemical processes, with some integrated processes also [6–9]. Among the described technologies, the electrochemical advanced oxidative processes (EAOPs) are shown to be very effective in the treatment of textile dyeing wastewaters (TDW). EAOPs are based on the electrogeneration of highly reactive hydroxyl radicals, which non-selectively react with organic compounds [10,11]. In particular, the electrochemical oxidation (EO) using a high O_2-overpotential anode, such as a boron-doped diamond (BDD) electrode, has proved to be very effective in dyes removal, and the mineralization of the organic compounds can be achieved [11,12]. It has also the advantage of not producing sludge or concentrates and, in this case, the high salinity of the textile wastewater may be an advantage for the dyeing wastewater reuse [13–15].

Most studies found in the literature are focused on the development of methodologies that allow the treatment of textile wastewaters to comply with discharge limits. However, in recent years, given the evidence on the potential for reuse within the textile industry, with the prospect of moving towards the goal of zero discharge of industrial wastewaters, the number of studies in this area have increased, namely with a focus on TDW reuse in new dyeing process [4,12–20]. However, to obtain an acceptable quality of dyed fabrics, in some of these studies, the treated wastewater is diluted before its reuse [12,13,17]. On the other hand, considering the high amounts of salts necessary to obtain a good dye fixation to the fiber, another important economic and environmental aspect to be considered is the salt reuse, which can be accomplished by using electrochemical oxidation to treat the TDW. In the literature, only a few studies can be found focusing on the reuse of TDW treated by EO. Riera-Torres et al. [21] reused a simulated TDW, prepared with previously hydrolyzed reactive dyes and sodium sulfate, treated by EO, using titanium covered by platinum oxides as anode. The treated wastewater was diluted before its reuse, and for most of the dyes used, the level of dye degradation in the wastewater was considered non-relevant in the direct reuse, considering that the color removal was enough. Orts et al. [14] studied the reuse of TDW containing a trichrome mixture of reactive dyes and sulfate ions, treated by EO, using a DSA anode, and the decolorized treated wastewater was diluted before its reuse. The electrochemical treatment and reuse of industrial reactive dyeing wastewaters were performed by Sala et al. [12], and the discoloration of dyeing wastewater treatment and reuse was effective, allowing practitioners to save 70% of water and 60% of electrolytes. The water volume lost during the dyeing process, by adsorption into the fiber and by evaporation, was restored with fresh water. This strategy was also used in a different study [22], where the residual dyeing and washing wastewaters were electrochemically treated using a Ti anode covered by Pt oxides. Once again, by the addition of 30% of fresh water, to reconstitute the dyebath, the reduction of organic matter of 49% was obtained. The residual oxidant compounds generated during the EO treatment, namely the chloride active species, had to be removed before the treated wastewater reuse.

For the reuse of TDW treated by EO, some reformulations of the dye concentration and auxiliary chemicals may be necessary, especially in light color dyeing operations [23].

Furthermore, multiple reuses of the treated wastewater would require changes in salt and other auxiliary chemicals to achieve the same fabric color as fresh process water [23].

Considering that ecotoxicity is one of the major problems of textile wastewaters, some studies have evaluated this parameter and verified its reduction through electrochemical oxidation using a BDD anode [24]. Due to the TDW complexity, especially when several organic auxiliaries and salts are used, the evaluation of the ecotoxicity of the treated wastewaters cannot be ignored. Still, very few TDW reuse studies have addressed this issue. Although it may not seem a relevant aspect for TDW reuse, for other reuse applications and for final disposal, it may be crucial. Furthermore, it is important to acknowledge the effect of the consecutive reuse cycles on the wastewater ecotoxicity. Another aspect that has not been much explored in previous studies is the influence of the dyeing bath composition on the treatment–reusing cycle, since, besides the dyes and salts, different organic auxiliaries, e.g., equalizer and humectant agents, are very often utilized, at an industrial level, in the dyeing of woolen fibers [4]. These TDW present a very high content of organic matter of different natures and the discoloration of the wastewater may not be enough for the reuse, as the reduction of the organic matter to increase the quality of the treated wastewater is necessary. In this context, it is important to evaluate the influence of using treated TDW with different qualities, and without using any dilution, on the quality of the color of the dyed fabrics.

To fill some of the gaps presented by the previous studies performed, this work aims to evaluate the feasibility of an EO process using a BDD anode to: (i) treat a TDW from a woolen fabric dyeing process using a trichromatic acid dyes combination in the presence of sulfate salt, equalizer and humectant agents, widely used at the industrial level; (ii) provide a treated TDW with complete salt recovery and a quality level that allows its successful consecutive reuse in new dyeing baths; (iii) significantly reduce the ecotoxicity towards *Daphnia magna* of the wastewaters, even after consecutive reuses. The effect of EO operational parameters, namely applied current density and treatment duration, on the quality of the treated TDW, and the influence of this quality level in terms of organic load content on the consecutive reuse in new dyeing baths are also addressed. Furthermore, the ecotoxicity towards *Daphnia magna* of the different dyes utilized and of the dyeing auxiliaries was evaluated, aiming to establish which of the compounds contributed most to the TDW ecotoxicity.

2. Materials and Methods

2.1. Textile Dyeing Wastewater

The dyeing process is the most intensive water and chemical consumer in the textile industry, generating the largest effluent stream with minimally known composition [25]. For these reasons, it has been highlighted as a good candidate for reusing the wastewater after treatment and, considering the high stability under sunlight and resistance to microbial attack and temperature of the synthetic organic dyes, the textile dyeing wastewater is also an optimum matrix to evaluate the performance of the electrochemical oxidation process [3,11,25].

The TDW utilized in this study was obtained from the dyeing process of a 100% wool twill fabric (weight = 351.6 g m^{-2}; finesse of yarn: warp and weft = 100 Tex) with a trichromatic combination of Nylosan acid dyes (Nylosan® Red N-2RBL (C.I. AR 336), Nylosan® Yellow N-3RL (C.I. AO 67), and Nylosan® Blue N-GL (C.I. AB 230)) (Figure 1).

The dyeing process was carried out in a Mathis Labomat type BFA-12 equipment, purchased from Maquicontrolo (Oporto, Portugal), under a fabric:dyeing bath ratio of 1:50 (5 g of wool fabric per 250 mL of dyeing bath). The dyeing bath consisted of an aqueous solution with a composition as described in Table 1 and a pH of 4.5. The Nylosan acid dyes, the Sarabid PAW (equalizer agent), and the SERA WET C-NR (humectant agent) were purchased, respectively, from Clariant (Leça do Balio, Portugal), CHT (Tübingen, Germany) and Dystar (Oporto, Portugal). Sodium sulfate, sodium acetate, and acetic acid were purchased from Sigma-Aldrich (Lisbon, Portugal).

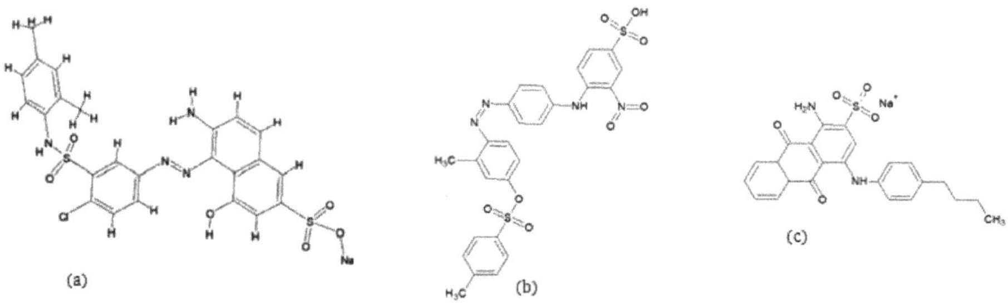

Figure 1. Structure of three Nylosan N acid dyes: (**a**) Red Nylosan N-2RBL; (**b**) Yellow Nylosan N-3RL and (**c**) Acid Blue 230, one of the Blue Nylosan N-GL constituents.

Table 1. Composition of the dyeing bath utilized in the dyeing process.

Constituents	Concentration
Nylosan® Red N-2RBL	80 mg L^{-1} (0.4% of dye/fabric (w/w))
Nylosan® Yellow N-3RL	60 mg L^{-1} (0.3% of dye/fabric (w/w))
Nylosan® Blue N-GL	60 mg L^{-1} (0.3% of dye/fabric (w/w))
Equalizer agent (Sarabid PAW)	200 mg L^{-1} (1% of equalizer/fabric (w/w))
Humectant agent (SERA WET C-NR)	1 g L^{-1}
Sodium sulfate anhydrous, ≥99.0%,	1 g L^{-1} (5% of Na$_2$SO$_4$/fabric (w/w))
Sodium acetate anhydrous, ≥99.0%	Acetate buffer (2 g L^{-1}) pH 4.5
Acetic acid glacial, ≥99.7%	

The dyeing program involved a heating step, from room temperature up to 98 °C, at a heating rate of 2 °C/min, followed by a 30 min period at this final temperature. After this period, the mini-reactors were maintained in the apparatus for cooling to room temperature. After the dyeing process, the wool dyed fabrics were washed at room temperature, to remove the unfixed dye. The TDW obtained from each mini-reactor was collected and combined in a single TDW sample, which was characterized and then utilized in the EO experiments.

2.2. Electrochemical Treatment

The TDW electrochemical treatment was conducted in batch mode, with stirring (300 rpm), using an undivided cylindrical cell, containing 250 mL of TDW. A commercial Si/BDD anode, purchased from Neocoat (La Chaux-de-Fonds, Switzerland), and a stainless-steel cathode, each one with an immersed area of 10 cm^2, were utilized as electrodes. They were placed in parallel, with an inter-electrode gap of 1 cm, and were centered in the electrochemical cell. A GW, Lab DC, model GPS-3030D (0–30 V, 0–3 A), purchased from ILC (Lisbon, Portugal), was used as the power supply.

Before the electrochemical treatment, the TDW was filtered to remove any fiber residues.

In a preliminary set of EO experiments, the influence of the applied current density (j) on the electrochemical treatment performance was evaluated. Assays were run at 30, 60, and 100 mA cm^{-2}, for 10 h.

The EO treatment, for TDW reuse purpose, was performed for 10 h, at 60 and 100 mA cm^{-2}. A set of five EO assays were run at each j studied. The treated TDW obtained from the five assays performed at the same experimental conditions was combined in a single sample, which was characterized and then utilized in a new dyeing process.

2.3. Reuse Experiments

The strategy adopted in this study is summarized in Figure 2. The reuse experiments comprised the dyeing process as described in Section 2.1, but utilizing, in the dyeing bath, the TDW samples treated by EO instead of fresh water. Furthermore, since it was found that the sulfate ion concentration was maintained during the EO treatment, the dyeing bath, in the reuse experiments, was prepared as described in Table 1, but without the addition of sodium sulfate.

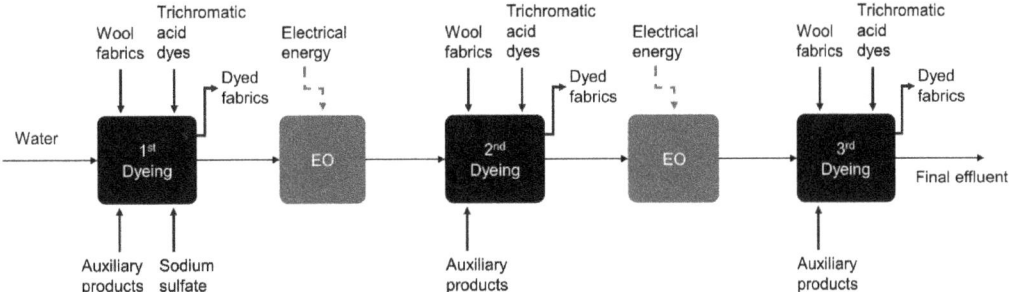

Figure 2. A schematic diagram of the processes.

The TDW generated during the first reuse cycle was collected and submitted to a second EO treatment and dyeing process. The second electrochemical treatment was performed under similar experimental conditions of the first treatment, and the second reuse cycle followed the procedure described above.

2.4. Analytical Methods

The TDW samples, before and after the EO treatments, were characterized in terms of chemical oxygen demand (COD), dissolved organic carbon (DOC), ecotoxicity towards the model organism *Daphnia magna*, sulfate ion concentration, pH, and electrical conductivity. COD determinations followed the closed reflux and titrimetric method, according to the standard procedures [26].

DOC was measured in a Shimadzu TOC-VCPH analyzer, purchased from Izasa Scientific (Carnaxide, Portugal), with samples previously filtered through 0.45 μm membrane filters.

The ecotoxicity towards *Daphnia magna* was evaluated using a commercial Daphtoxkit F microbiotests, DM230921, purchased from Ambifirst (Moita, Portugal), following the OECD/OCDE Guideline 202 [27], by measuring the number of immobilized *Daphnia magna* neonates exposed to different dilutions of the TDW samples. To clarify which constituents most contributed to the ecotoxicity of the TDW, the ecotoxicity of the dyes, the equalizer, and the humectant agents used was also assessed.

Sulfate ion concentration was determined by ion chromatography, using a Shimadzu Prominance LC-20A system with a Shimadzu CDD 10Avp conductivity detector, purchased from Izasa Scientific (Carnaxide, Portugal). An IC I-524A Shodex (4.6 mm ID × 100 mm) anion column was used at 40 °C. The mobile phase was an aqueous solution of 2.5 mM of phthalic acid and 2.3 mM of tris(hydroxymethyl)aminomethane at pH 4, with a flow rate of 1.5 mL min^{-1}.

pH was measured with a HANNA pH meter (HI 931400) and the electrical conductivity (EC) with a Mettler Toledo conductivity meter (SevenEasy S30K), both purchased from MT Brandão (Oporto, Portugal).

The presence of the dyes Nylosan® Yellow N-3RL, Nylosan® Red N-2RBL, and Nylosan® Blue N-G in the TDW samples was also monitored, through UV-vis spectrophotometric measurements at 436, 525, and 620 nm, respectively, utilizing a Shimatzu UV-1800 spectrophotometer, purchased from Izasa Scientific (Carnaxide, Portugal).

The performance of the dyeing process was evaluated in terms of total color difference (ΔE^*) and color fastness to washing of the wool fabric samples dyed with the primary dyeing bath (control fabric), prepared with fresh water, and with the reused dyeing baths. ΔE^* was determined through the CIELab color system, following the procedure described in ISO 105-J03:2009 [28] and utilizing a Spectraflash SF 300X reflectance spectrophotometer, purchased from Datacolor International (Trenton, NJ, USA).

The color fastness to washing was evaluated following the procedure described in ISO 105-C06 A2S:2010 [29]. These washing fastness tests were carried out with a multifiber fabric, which is divided equally by the acetate-cotton-polyamide-polyester-acrylic-wool fibers to which the dyed fabric to be tested was attached.

3. Results and Discussion

3.1. First EO Treatment and Reuse Cycle

The characterization of the TDW obtained from the primary dyeing process (utilizing fresh water) is presented in Table 2. This wastewater presents a high content of organic matter, as can be seen by the values of the COD and DOC concentrations, which can be ascribed to the presence of dyes that were not adsorbed by the fibers and to the dyeing auxiliaries utilized. In fact, the brownish color presented by the TDW and the absorbances at 436, 525, and 620 nm, which are the wavelengths of maximum absorbance of the dyes Nylosan® Yellow N-3RL, Nylosan® Red N-2RBL, and Nylosan® Blue N-GL, respectively, indicate the presence of residual dye content.

Table 2. Characterization of the TDW obtained from the primary dyeing process, before and after treatment by EO at 60 and 100 mA cm^{-2}.

Parameter	Before EO	After EO_60 mA cm^{-2}	After EO_100 mA cm^{-2}
Color (visual)	Light brown	Non-visible	Non-visible
Absorbance at 436 nm	0.157 ± 0.003	0.001 ± 0.001	0.001 ± 0.001
Absorbance at 525 nm	0.110 ± 0.002	0.002 ± 0.001	0.001 ± 0.001
Absorbance at 620 nm	0.086 ± 0.002	0.003 ± 0.002	0.001 ± 0.001
COD (mg L^{-1})	(4.6 ± 0.1) × 10^3	163 ± 8	45 ± 3
DOC (mg L^{-1})	(1.66 ± 0.04) × 10^3	65 ± 2	3.7 ± 0.5
EC$_{50}$-48 h (%)	3.30	19.68	22.30
Toxic units	30.3	5.08	4.48
SO$_4^{2-}$ (mg L^{-1})	672 ± 3	670 ± 5	672 ± 2
pH	4.68 ± 0.03	8.79 ± 0.03	11.4 ± 0.1
EC (mS cm^{-1})	3.12 ± 0.03	4.24 ± 0.04	4.31 ± 0.03

The ecotoxicity towards *Daphnia magna* was assessed and the obtained results confirmed the high toxicity presented by the textile dyeing wastewaters. According to the toxicity classification based on toxic units (TU) reported by Pablos et al. [30], where the toxicity expressed in terms of TU is calculated according to Equation (1), the TDW obtained from the primary dyeing process is classified as very toxic.

$$TU = \frac{100}{EC_{50}(\%)} \quad (1)$$

To clarify which constituents most contributed to the ecotoxicity of the TDW, the ecotoxicity towards *Daphnia magna* of the dyes, and of the equalizer and humectant agents used, was assessed. The obtained results, presented in Table 3, show that the estimated EC$_{50}$–48 h values for all three dyes and equalizer agent are above the values of the initial dyeing bath conditions (Table 1), indicating that these constituents do not contribute to the observed ecotoxicity. However, the humectant agent, with a calculated EC$_{50}$–48 h of 75.18 mg L^{-1}, is used in a concentration ten times higher than the estimated value.

Table 3. Ecotoxicity towards *Daphnia magna* of Nylosan® Yellow N-3RL, Nylosan® Red N-2RBL, Nylosan® Blue N-GL, Sarabid PAW and SERA WET C-NR.

Constituents	EC_{50}–48 h (mg L^{-1})
Nylosan® Red N-2RBL	>200
Nylosan® Yellow N-3RL	114.83
Nylosan® Blue N-GL	140.64
Equalizer agent	>200
Humectant agent	75.18

To establish the best EO operational conditions for TDW reuse purpose, a preliminary set of EO experiments was performed, where the influence of applied current density and treatment duration on the color and organic load removal rates was assessed. Both color and organic load removals increased with j and treatment time, since, at higher currents and treatment times, the production of oxidative species is higher, promoting the enhanced degradation of the organic compounds [11]. At 100 mA cm^{-2}, complete color removal (visual) was accomplished after 8 h assay, but at 60 mA cm^{-2} it took 10 h to attain complete discoloration. At the lowest applied j (30 mA cm^{-2}), after 10 h assay, complete discoloration was not achieved. These observations were in accordance with the removal of the dyes, which was evaluated through absorbance measurements at 436, 525, and 620 nm. As it can be seen in Figure 3, the dye removal rate was higher in the first hours of assay, since more dye molecules were available to be oxidized. As the dye concentration became lower, its oxidation became diffusion controlled and the byproducts formed were preferably oxidized, decreasing the dye removal rate.

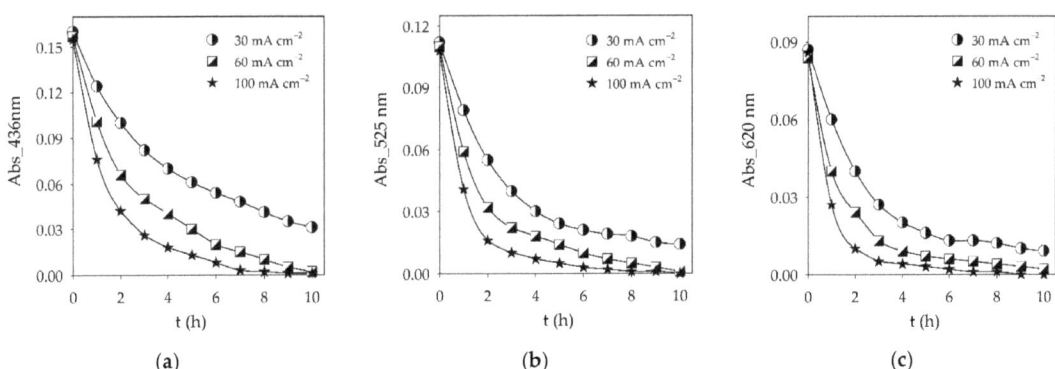

Figure 3. Decay in time of absorbance at (a) 436 nm; (b) 525 nm; and (c) 620 nm, during the preliminary EO experiments performed at different j.

Regarding COD and DOC (Figure 4), they presented similar decays, indicating a high degree of mineralization of the organic compounds for all the j studied. Nevertheless, when comparing COD and DOC removal rates with that of the absorbances at 436, 525, and 620 nm, during the first hours of the assay, COD and DOC removal rates are lower, indicating that the dye degradation occurred more rapidly than the overall organic load removal. This can be explained considering that, for the dye fragmentation (chromophore group breaking), only a primary oxidation stage is required, but, for COD and DOC reductions, more complex multistage oxidative reactions are involved, since the dye oxidation usually results in the formation of different byproducts [31,32].

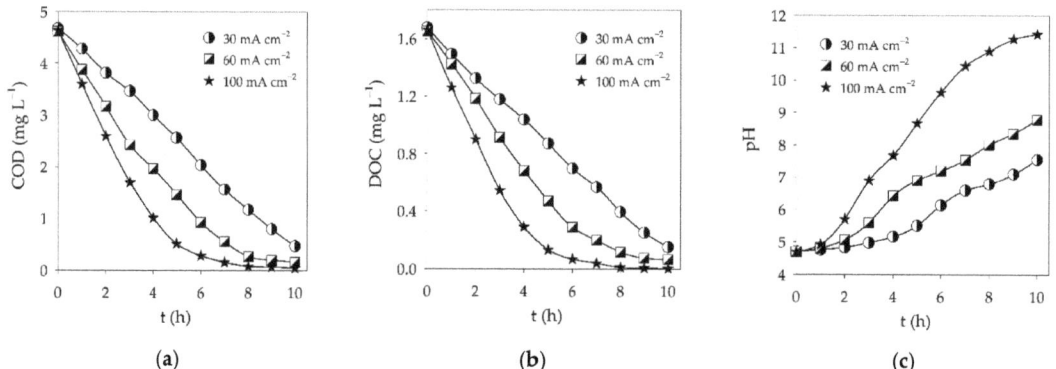

Figure 4. Variation in time of (**a**) COD; (**b**) DOC, and (**c**) pH, during the preliminary EO experiments performed at different j.

Similar results were observed by Solano et al. [32] when studying the electrochemical treatment of a TDW sample, using a BDD anode, and in the presence of sulfate ions. According to these authors, electrolysis under BDD anode, in aqueous media containing sulfate ions, generates peroxodisulfate (Equation (2)), a powerful oxidizing agent that can oxidize organic matter near to the anode surface and in the bulk solution. Even for small sulfate concentrations, as in the case of the TDW under study (672 mg L^{-1}), we found an enhancement in the EO performance, caused, according to the authors, by the production of peroxodisulfate that, together with the reactive oxygen species (ROS) such as hydroxyl radicals, oxidize the organic matter from TDW. In fact, more recent studies on electro-persulfate processes show that, when utilizing anode materials with high oxygen evolution potential, such as BDD, not only can peroxodisulfate be produced from the oxidation of SO$_4^{2-}$ (Equation (2)), but also sulfate radicals can be electrogenerated, either by sulfate ions direct oxidation (Equation (3)) or by oxidation through a hydroxyl radical (Equation (4)) [33]. Moreover, peroxodisulfate can be also obtained through sulfate ions oxidation by hydroxyl radicals (Equation (5)) and sulfate radicals can be generated from the electrochemical activation of peroxodisulfate, according to Equation (6) [33]. According to the literature, the sulfate radical is a strong and highly oxidizing species that presents higher redox potential than that of hydroxyl radicals or peroxodisulfate. It readily reacts at a wide range of pH values, promoting the nonselective oxidation and efficient removal of a wide range of organic compounds [33].

$$2SO_4^{2-} \rightarrow S_2O_8^{2-} + 2e^- \tag{2}$$

$$SO_4^{2-} \rightarrow SO_4^{\bullet -} + e^- \tag{3}$$

$$SO_4^{2-} + OH^\bullet \rightarrow SO_4^{\bullet -} + OH^- \tag{4}$$

$$2SO_4^{2-} + 2OH^\bullet \rightarrow S_2O_8^{2-} + 2OH^- \tag{5}$$

$$S_2O_8^{2-} + e^- \rightarrow SO_4^{\bullet -} + SO_4^{2-} \tag{6}$$

Considering the existence of sulfate ions in the TDW sample under study and that a BDD anode is employed, it can be assumed that the oxidation of the organic compounds occurred in parallel by ROS and reactive sulfate species. It should be noted that, according to Solano et al. [32], the degradation process in the presence of sulfate occurs with significant formation of intermediates, but without formation of organochlorinated or other carcinogenic compounds, as in the case when oxidation is mediated by chloride species.

After 10 h of assay, the COD of the TDW samples was 468, 159, and 46 mg L^{-1}, for the EO experiments run at 30, 60, and 100 mA cm^{-2}, respectively. According to the literature, a high-quality water for reuse purposes in the textile industry should present a maximum COD of 50 mg L^{-1}, with the maximum COD of 200 mg L^{-1} being a moderate-quality water [4]. Considering this criterion and the results obtained for color and organic load removal at the different j studied, the TDW treatment, for reuse purpose in new dyeing processes, was conducted at 60 and 100 mA cm^{-2}, for 10 h, aiming to evaluate the influence of the treated TDW quality (moderate or high) on the performance of the dyeing process.

The characterization of the TDW samples treated by EO at 60 and 100 mA cm^{-2}, for 10 h, which were utilized in the first reuse cycle, is presented in Table 2. Both treated samples were color and dye free. Although the COD of the TDW samples treated at 60 and 100 mA cm^{-2} complied with the moderate-quality and high-quality requirements, respectively, the pH and the electrical conductivity values were higher than those required for wastewater reuse purposes in the textile industry [4]. The pH variation in time for the different j studied, presented in Figure 4c, shows an increase in pH during the EO treatment, which is more pronounced at higher j. This increase in pH during the EO process is well described in the literature and can be attributed to: (i) side reactions that occur at the cathode, such as the reaction described in Equation (7); (ii) sulfate oxidation through hydroxyl radicals (Equations (4) and (5)); and (iii) formation of carbonates and bicarbonates, from the reaction between the hydroxide ions and the CO$_2$ generated during the oxidation of the organic pollutants [33,34]. Such reactions are enhanced by the increase in j, explaining the higher pH values attained by the samples treated at higher j.

$$2H_2O + 2e^- \rightarrow H_2 + 2OH^- \tag{7}$$

According to the requirements for reuse purposes in textile industry [4], pH should not be higher than 8.0. Nevertheless, in the case of the wool fabric dyeing with Nylosan acid dyes, the dyeing bath is required to be at pH 4.5, involving an acidification process through the addition of an acetate buffer. Thus, although the pH of the treated TDW samples, at 60 and 100 mA cm^{-2}, is higher than 8.0, its mandatory correction during the dyeing bath preparation eliminates the possible constraint associated with this high pH. It should be noted that, for most dyes used in the textile industry, pH correction is required.

Regarding the high electrical conductivity presented by the treated TDW samples (>4 mS cm^{-1}), this is mainly due to the sodium and sulfate ions, whose concentration was practically unchanged during the dyeing process and the subsequent EO treatment. In fact, although we assumed there was the generation of reactive sulfate species, from the oxidation of sulfate ions present in the TDW, sulfate ion concentration at the end of the EO treatments was practically unchanged. This is explained by the reaction of reactive sulfate species with dissolved organic matter (DOM) (Equation (8)) [35], and by reactions, such as that described in Equation (9), that occur when there is no DOM available for oxidation [33], which restore the sulfate ions in TDW. Thus, despite not complying with the general requirements for reuse purposes in the textile industry (<1.5 mS cm^{-1}) [4], EO treatment has the advantage of enabling the reuse of the treated TDW without the addition of more sodium sulfate salt.

$$DOM + SO_4^{\bullet -} \rightarrow DOM_{ox}^+ + SO_4^{2-} \tag{8}$$

$$SO_4^{\bullet -} + SO_4^{\bullet -} \rightarrow 2SO_4^{2-} \tag{9}$$

Despite not being a requirement for wastewater reuse purposes in textile industry, ecotoxicity towards *Daphnia magna* was evaluated. It was found that the ecotoxicity was drastically reduced during the EO treatment, indicating that the oxidation products are less toxic than the parent compounds. This fact is important for TDW reuse, since it shows that there is no significant increase in the dyeing bath ecotoxicity when prepared with treated TDW. Furthermore, we found that the reduction in ecotoxicity was slightly more pronounced in the TDW treated at 100 mA cm^{-2}, which is in accordance with the higher oxidation degree of the degradation products at higher j.

The first reuse cycle was performed following the procedure described for the initial dyeing process, but utilizing the treated TDW instead of fresh water, and without adding sodium sulfate. The dyeing process was evaluated in terms of total color difference and color fastness to washing of the wool fabric samples dyed with the primary dyeing bath, prepared with fresh water, and with the reused dyeing baths, with the obtained results presented in Tables 4 and 5.

Table 4. Total color difference (ΔE^*) and differences between L^*, a^*, and b^* obtained from the first reuse cycle.

Parameter	EO_60 mA cm^{-2}	EO_100 mA cm^{-2}
ΔE^*	1.04 ± 0.07	1.00 ± 0.09
ΔL^*	0.96 ± 0.07	0.8 ± 0.1
Δa^*	−0.30 ± 0.02	−0.61 ± 0.05
Δb^*	0.26 ± 0.05	0.15 ± 0.07

Table 5. Color fastness to washing of the wool fabrics dyed using fresh water (primary dyeing) and TDW treated by EO (scale: 1 (poor) to 5 (excellent)).

Dyeing Conditions		Acetate	Cotton	Polyamide	Polyester	Acrylic	Wool
Fresh water (primary dyeing)		5	5	4	5	5	5
Treated TDW	60 mA cm^{-2}	5	5	4	5	5	5
	100 mA cm^{-2}	5	5	4	5	5	5

ΔE^* was calculated through Equation (10), which considers the differences between L^*, a^*, and b^* values of the fabrics dyed with treated TDW and that of the fabrics dyed with fresh water. The lightness/darkness of the color is given by parameter L^*: ΔL^* negative values indicate darker color and ΔL^* positive values indicate lighter color. Parameter a^* gives the color position between red and green: Δa^* negative values indicate greener color and Δa^* positive values indicate redder color. Parameter b^* represents the color position between yellow and blue: Δb^* negative values indicate bluer color and Δb^* positive values indicate yellower color [4]. According to the norm DIN ISO 11664–4:2012–06 [36], only for a ΔE^* above 1.5 are differences in the color between the sample and control fabrics visible. Nevertheless, there are restrictive controls that require a maximum ΔE^* of 1.0 (the smallest value the human eye can detect) [31].

$$\Delta E^* = \sqrt{(\Delta L^*)^2 + (\Delta a^*)^2 + (\Delta b^*)^2} \tag{10}$$

According to the results presented in Table 4, the fabrics obtained from the dyeing process that utilized the treated TDW were slightly lighter, greener, and yellower than the fabrics dyed with fresh water, with the lighter and yellow color more pronounced in the fabrics resultant from the dyeing with the treated TDW at 60 mA cm^{-2} and the greener color more pronounced in the fabrics that utilized the treated TDW at 100 mA cm^{-2}. These variations in the color parameters presented by the fabrics dyed with the treated TDW are, according to the literature, mainly due to the accumulation of byproducts from the degradation of the dyes and auxiliary products, which influence the dyeing process, in either exhaustion or fixation stages [22,31]. In a study performed by López-Grimau et al. [22], positive ΔL^* values were also obtained when reusing electrochemically treated TDW. According to the authors, this is due to a lower dye exhaustion, probably caused by an affinity between the residual hydrolyzed reactive dyes and the fibers or by its reaction with the new dyestuff, decreasing the final depth of shade. The differences found between the fabrics dyed with TDW samples treated at 60 and 100 mA cm^{-2} are probably related to the different byproducts that resulted from the treatments, which are expected to have a higher oxidation degree in the EO treatment performed at 100 mA cm^{-2}. Nevertheless, both dyeing processes, utilizing treated TDW at 60 and 100 mA cm^{-2}, attained ΔE^* values of 1.0, complying with the most restrictive controls that require a maximum ΔE^* of 1.0.

The color fastness to washing was evaluated to verify differences between the primary dyed fabric and the fabrics dyed with the different treated TDW. The color transference was evaluated on a scale between 1 and 5, where 5 corresponds to no color transference. Color fastness to washing results, presented in Table 4, show that the wool fabrics dyed with the treated TDW presented similar behavior to that observed by the fabrics of the primary dyeing, validating the utilization of the treated wastewaters in new dyeing processes, regarding this parameter.

Both ΔE^* and color fastness to washing results showed the suitability of the treated TDW for reuse in new dyeing processes, with no significant differences found between the high-quality (treated by EO at 100 mA cm^{-2}) and moderate-quality (treated by EO at 60 mA cm^{-2}) samples. A similar conclusion was attained in a study performed by Silva et al. [4], where it was found that it is not strictly necessary to meet the reuse requirements to effectively reuse textile wastewater in the textile industry.

3.2. Second EO Treatment and Reuse Cycle

To evaluate the potential consecutive reuse of the treated TDW, the TDW samples obtained from the first reuse cycle were submitted to a second EO treatment, at the same experimental conditions utilized in the first EO treatment, and were then utilized in a second reuse cycle. Table 6 presents the characterization of the TDW samples obtained from the first reuse cycle, before and after the second EO treatment.

Table 6. Characterization of the TDW samples obtained from the first reuse cycle, before and after the second EO treatment.

Parameter	EO_60 mA cm^{-2}		EO_100 mA cm^{-2}	
	After First Reuse	After Second EO	After First Reuse	After Second EO
Color (visual)	Light Brown	Non-visible	Light Brown	Non-visible
Absorbance at 436 nm	0.245 ± 0.003	0.002 ± 0.002	0.287 ± 0.003	0.001 ± 0.001
Absorbance at 525 nm	0.206 ± 0.004	0.002 ± 0.001	0.233 ± 0.004	0.002 ± 0.001
Absorbance at 620 nm	0.187 ± 0.004	0.004 ± 0.002	0.212 ± 0.004	0.002 ± 0.002
COD (mg L^{-1})	(5.8 ± 0.1) × 10^3	168 ± 6	(5.2 ± 0.2) × 10^3	50 ± 2
DOC (mg L^{-1})	(1.88 ± 0.04) × 10^3	66 ± 2	(1.81 ± 0.03) × 10^3	2.23 ± 0.04
EC$_{50}$-48 h (%)	1.46	27.21	1.50	18.62
Toxic units	68.5	3.68	66.7	5.37
SO$_4^{2-}$ (mg L^{-1})	669 ± 5	668 ± 3	670 ± 4	670 ± 2
pH	4.92 ± 0.04	10.70 ± 0.05	4.74 ± 0.02	11.72 ± 0.04
EC (mS cm^{-1})	4.05 ± 0.08	5.75 ± 0.05	3.92 ± 0.08	6.65 ± 0.04

Similar to that described in previous studies [14,18,25], TDW samples obtained from the first reuse cycle presented higher dye contents (given by the absorbances at 436, 525, and 620 nm) in comparison to those of the primary dyeing wastewater. Furthermore, COD and DOC values and the ecotoxicity towards *Daphnia magna* were higher for the TDW samples obtained from the first reuse cycle. According to the literature, this is ascribed to the accumulation of residual organic load and ecotoxicity [18]. Nevertheless, the second EO treatment was effective in the reduction of the organic load and ecotoxicity, with even better results than those obtained with the first EO treatment.

During the second EO treatment, color and dyes were completely removed. COD and DOC absolute removal and ecotoxicity reduction were higher than that attained for the first EO treatment, which can be ascribed to the higher organic load content of the TDW obtained from the first reuse cycle, since, for higher organics concentration, the mass transfer limitations in the electrochemical cell are reduced and thus the EO process is more effective [37]. As observed during the first EO treatment, sulfate ion concentration was practically unchanged during the second EO treatment, enabling the use of the treated

TDW in the second reuse cycle without the addition of sodium sulfate salt being necessary. This unchanged sulfate ion concentration after EO treatment is in agreement with the inability of advanced oxidation processes to remove salinity, which is usually highlighted as a constraint of these processes. Nevertheless, when considering the use of treated TDW in new dyeing baths, it becomes an advantage. According to Bezerra et al. [18], the TDW reuse without requiring salts addition can result in a saving of 10 tons of salt per year, besides reducing the environmental impact, since the disposal and treatment of saline textile wastewaters is one of the major environmental concerns of the textile industry. In the study performed by these authors, where a TDW was decolorized with H_2O_2 catalyzed by UV light, the reuse of the treated TDW without requiring salts addition was not accomplished, with a salt adjustment being necessary.

The second reuse cycle and its performance evaluation followed the procedure applied to the first reuse cycle, utilizing the two TDW samples obtained from the second EO treatment. Similar performance was observed for the first and second reuse cycle. Regarding the total color differences (Table 7), the dyed fabrics obtained from the second reuse cycle presented lower ΔE^* values than those of the dyed fabrics from the first reuse cycle.

Table 7. Total color difference (ΔE^*) and differences between L^*, a^*, and b^* obtained from the second reuse cycle.

Parameter	EO at 60 mA cm^{-2}	EO at 100 mA cm^{-2}
ΔE^*	0.80 ± 0.03	0.80 ± 0.01
ΔL^*	0.17 ± 0.05	0.27 ± 0.01
Δa^*	0.78 ± 0.01	0.71 ± 0.02
Δb^*	−0.01 ± 0.01	0.242 ± 0.003

The color fastness to washing assays (Table 8) showed no significant differences between the fabrics of the primary dyeing and the fabrics dyed with the treated wastewaters from the second EO treatment.

Table 8. Color fastness to washing of the wool fabrics dyed using fresh water (primary dyeing) and TDW after the second EO treatment.

Dyeing Conditions		Acetate	Cotton	Polyamide	Polyester	Acrylic	Wool
Fresh water (primary dyeing)		5	5	4	5	5	5
Treated TDW	60 mA cm^{-2}	5	5	4–5	5	5	5
	100 mA cm^{-2}	5	5	4–5	5	5	5

4. Practical Implications of the Study

The United Nations 2030 Sustainable Development Agenda is "a universal call to action to end poverty, protect the planet and improve the lives and prospects of everyone, everywhere" [38]. The work developed in this study is fully aligned with this agenda, namely with Goal 6—Ensure availability and sustainable management of water and sanitation for all, and Goal 12—Ensure sustainable consumption and production patterns, since it presents a viable treatment solution for textile dyeing recalcitrant wastewater and allows its reuse in new dyeing processes. Many other studies have also presented viable treatment solutions to obtain wastewaters with good quality to be reutilized. However, this study went further, since it proved that salts can be completely recovered and reutilized in new textile dyeing processes, without losing dye uptake by the wool fibers. In addition, a good quality wool dyeing was attained, regardless of the use of a high- or moderate-quality water for preparing the dyeing baths. In fact, this was one of the main objectives in this study, together with the ecotoxicological evaluation towards *Daphnia magna* and the potential consecutive reuse of the treated wastewater. Of course, this is "a drop in the ocean", and many more studies should be performed to validate the reuse of treated wastewaters with different quality levels and without salts addition/readjustment to the dyeing bath, namely from electrochemical oxidation with other anode materials. This is mainly a challenge for the dyeing process of other type of fibers, such as cotton or synthetics. Furthermore, in textile factories where the dyeing wastewater is not separated from the other textile processes, the reuse of the wastewaters with salts recovery, without membrane processes, is an even bigger challenge. If even today the engine that moves the world is based on purely economic reasons, it is necessary to establish a new paradigm that, although always considering economic constraints, promotes the sustainable reuse of all natural resources.

5. Conclusions

Electrochemical oxidation treatment of textile dyeing wastewaters, utilizing a BDD anode, is an effective strategy to obtain moderate-quality (COD $\leq$ 200 mg L^{-1}) or a high-quality (COD $\leq$ 50 mg L^{-1}) water for reuse purposes in the textile industry. This strategy not only provides a reduction in water consumption, but also saves salts, since it allows the complete recovery of the salts utilized in the dyeing process, eliminating the need of salts addition in the subsequential dyeing baths. Furthermore, it promotes a drastic reduction of the wastewater ecotoxicity towards *Daphnia magna*, allowing wastewater reuse without severe ecotoxicity accumulation.

By varying the applied current density and treatment time, it is possible to obtain treated wastewaters with different qualities. High applied current densities result in higher color and organic load removal rates and in more oxidized byproducts. However, when comparing the performance of the dyeing processes that utilized treated wastewaters at 60 and 100 mA cm^{-2}, which complied with the moderate- and high-quality requirements, respectively, no significant differences were found. Both treated samples lead to dyed fabrics with ΔE^* values of 1.0 and with color fastness to washing results similar to that obtained by the fabrics of the primary dyeing (utilizing fresh water), complying with the most restrictive controls of the textile industry. Thus, it is concluded that, to efficiently reuse textile dyeing wastewater, a high-quality water, with a COD lower than 50 mg L^{-1}, is not required.

Electrochemical oxidation process showed also to be feasible for the consecutive reuse of the treated wastewater. The increase in the organic load of the dyeing wastewater, caused by the accumulation of residual organic load from the previous treatment, is not a constraint for the subsequential treatment and reuse. In fact, higher organic load removals were attained during the second treatment and lower ΔE^* values were observed in the fabrics from the second reuse cycle.

Author Contributions: Conceptualization, A.L. and M.J.P.; Data curation, A.F. and M.J.P.; Formal analysis, A.F.; Investigation, C.P., M.J.N. and A.B.; Methodology, A.F. and M.J.P.; Project administration, A.L.; Resources, M.J.P.; Supervision, A.F. and M.J.P.; Validation, A.L. and M.J.P.; Visualization, A.L. and L.C.; Writing—original draft, C.P. and M.J.N.; Writing—review & editing, A.F., A.L., L.C. and M.J.P. All authors have read and agreed to the published version of the manuscript.

Funding: This research was funded by FUNDAÇÃO PARA A CIÊNCIA E A TECNOLOGIA, FCT, project UIDB/00195/2020, PhD grants SFRH/BD/132436/2017 and COVID/BD/151965/2021 awarded to M.J. Nunes, and contract awarded to A. Fernandes, and by INSTITUTO NACIONAL DE GESTÃO DE BOLSAS DE ESTUDO, INAGBE, Ph.D. grant awarded to C. Pinto.

Institutional Review Board Statement: Not applicable.

Informed Consent Statement: Not applicable.

Data Availability Statement: Data sharing is not applicable to this article.

Acknowledgments: The authors are very grateful for the support given by research unit Fiber Materials and Environmental Technologies (FibEnTech-UBI), on the extent of the project reference UIDB/00195/2020, funded by the Fundação para a Ciência e a Tecnologia, IP/MCTES through national funds (PIDDAC).

Conflicts of Interest: The authors declare no conflict of interest. The funders had no role in the design of the study; in the collection, analyses, or interpretation of data; in the writing of the manuscript, or in the decision to publish the results.

References

1. Šajn, N. *Environmental Impact of the Textile and Clothing Industry: What Consumers Need to Know*; European Parliamentary Research Service (EPRS) European Parliament: Brussels, Belgium, 2019.
2. Vajnhandl, S.; Valh, J.V. The status of water reuse in European textile sector. *J. Environ. Manag.* **2014**, *141*, 29–35. [CrossRef] [PubMed]
3. Yaseen, D.A.; Scholz, M. Textile dye wastewater characteristics and constituents of synthetic effluents: A critical review. *Int. J. Environ. Sci. Technol.* **2019**, *16*, 1193–1226. [CrossRef]
4. Silva, L.G.; Moreira, F.C.; Souza, A.A.; Souza, S.M.; Boaventura, R.A.; Vilar, V.J. Chemical and electrochemical advanced oxidation processes as a polishing step for textile wastewater treatment: A study regarding the discharge into the environment and the reuse in the textile industry. *J. Clean. Prod.* **2018**, *198*, 430–442. [CrossRef]
5. Holkar, C.R.; Jadhav, A.J.; Pinjari, D.V.; Mahamuni, N.M.; Pandit, A.B. A critical review on textile wastewater treatments: Possible approaches. *J. Environ. Manag.* **2016**, *182*, 351–366. [CrossRef]
6. Samsami, S.; Mohamadizaniani, M.; Sarrafzadeh, M.-H.; Rene, E.R.; Firoozbahr, M. Recent advances in the treatment of dye-containing wastewater from textile industries: Overview and perspectives. *Process Saf. Environ. Prot.* **2020**, *143*, 138–163. [CrossRef]
7. Ramos, M.; Santana, C.; Velloso, C.; da Silva, A.; Magalhães, F.; Aguiar, A. A review on the treatment of textile industry effluents through Fenton processes. *Process Saf. Environ. Prot.* **2021**, *155*, 366–386. [CrossRef]
8. Zhang, Y.; Shaad, K.; Vollmer, D.; Ma, C. Treatment of Textile Wastewater Using Advanced Oxidation Processes—A Critical Review. *Water* **2021**, *13*, 3515. [CrossRef]
9. Hama Aziz, K.H.; Mahyar, A.; Miessner, H.; Mueller, S.; Kalass, D.; Moeller, D.; Khorshid, I.; Rashid, M.A.M. Application of a planar falling film reactor for decomposition and mineralization of methylene blue in the aqueous media via ozonation, Fenton, photocatalysis and non-thermal plasma: A comparative study. *Process Saf. Environ. Prot.* **2018**, *113*, 319–329. [CrossRef]
10. Moreira, F.C.; Boaventura, R.A.R.; Brillas, E.; Vilar, V.J.P. Electrochemical advanced oxidation processes: A review on their application to synthetic and real wastewaters. *Appl. Catal. B Environ.* **2017**, *202*, 217–261. [CrossRef]
11. Brillas, E.; Martínez-Huitle, C.A. Decontamination of wastewaters containing synthetic organic dyes by electrochemical methods. An updated review. *Appl. Catal. B Environ.* **2015**, *166–167*, 603–643. [CrossRef]
12. Sala, M.; Gutiérrez-Bouzán, M.C. Electrochemical treatment of industrial wastewater and effluent reuse at laboratory and semi-industrial scale. *J. Clean. Prod.* **2014**, *65*, 458–464. [CrossRef]
13. Sala, M.; López-Grimau, V.; Gutiérrez-Bouzán, C. Photo-Electrochemical Treatment of Reactive Dyes in Wastewater and Reuse of the Effluent: Method Optimization. *Materials* **2014**, *7*, 7349–7365. [CrossRef]
14. Orts, F.; del Río, A.I.; Molina, J.; Bonastre, J.; Cases, F. Study of the Reuse of Industrial Wastewater After Electrochemical Treatment of Textile Effluents without External Addition of Chloride. *Int. J. Electrochem. Sci.* **2019**, *14*, 1733–1750. [CrossRef]
15. Rosa, J.; Fileti, A.M.F.; Tambourgi, E.B.; Santana, J. Dyeing of cotton with reactive dyestuffs: The continuous reuse of textile wastewater effluent treated by Ultraviolet/Hydrogen peroxide homogeneous photocatalysis. *J. Clean. Prod.* **2015**, *90*, 60–65. [CrossRef]
16. Chiarello, L.M.; Mittersteiner, M.; de Jesus, P.C.; Andreaus, J.; Barcellos, I.O. Reuse of enzymatically treated reactive dyeing baths: Evaluation of the number of reuse cycles. *J. Clean. Prod.* **2020**, *267*, 122033. [CrossRef]
17. Criado, S.P.; Gonçalves, M.J.; Tavares, L.B.B.; Bertoli, S.L. Optimization of electrocoagulation process for disperse and reactive dyes using the response surface method with reuse application. *J. Clean. Prod.* **2020**, *275*, 122690. [CrossRef]
18. Bezerra, K.C.H.; Fiaschitello, T.R.; Labuto, G.; Freeman, H.S.; Fragoso, W.D.; da Costa, S.M.; da Costa, S.A. Reuse of water from real reactive monochromic and trichromic wastewater for new cotton dyes after efficient treatment using H_2O_2 catalyzed by UV light. *J. Environ. Chem. Eng.* **2021**, *9*, 105731. [CrossRef]
19. Buscio, V.; García-Jiménez, M.; Vilaseca, M.; López-Grimau, V.; Crespi, M.; Gutiérrez-Bouzán, C. Reuse of Textile Dyeing Effluents Treated with Coupled Nanofiltration and Electrochemical Processes. *Materials* **2016**, *9*, 490. [CrossRef]
20. Núñez, J.; Yeber, M.; Cisternas, N.; Thibaut, R.; Medina, P.; Carrasco, C. Application of electrocoagulation for the efficient pollutants removal to reuse the treated wastewater in the dyeing process of the textile industry. *J. Hazard. Mater.* **2019**, *371*, 705–711. [CrossRef]
21. Riera-Torres, M.; Gutierrez-Bouzan, M.C.; Morell, J.V.; Lis, M.J.; Crespi, M. Influence of electrochemical pre-treatment in dyeing wastewater reuse for five reactive dyes. *Text. Res. J.* **2011**, *81*, 1926–1939. [CrossRef]
22. López-Grimau, V.; Gutiérrez-Bouzán, M.D.C.; Valldeperas, J.; Crespi, M. Reuse of the water and salt of reactive dyeing effluent after electrochemical decolorisation. *Color. Technol.* **2011**, *128*, 36–43. [CrossRef]
23. Leshem, E.N.; Pines, D.S.; Ergas, S.J.; Reckhow, D.A. Electrochemical Oxidation and Ozonation for Textile Wastewater Reuse. *J. Environ. Eng.* **2006**, *132*, 324–330. [CrossRef]

24. Nunes, M.J.; Sousa, A.C.A.; Fernandes, A.; Pastorinho, M.R.; Pacheco, M.J.; Ciríaco, L.; Lopes, A. Understanding the efficiency of electrochemical oxidation in toxicity removal. In *Advances in Chemistry Research*; Taylor, J.C., Ed.; Nova Science Publishers: New York, NY, USA, 2019; Volume 58, pp. 1–66.
25. Mohan, N.; Balasubramanian, N.; Basha, C.A. Electrochemical oxidation of textile wastewater and its reuse. *J. Hazard. Mater.* **2007**, *147*, 644–651. [CrossRef]
26. Eaton, A.; Clesceri, L.; Rice, E.; Greenberg, A.; Franson, M.A. *Standard Methods for Examination of Water and Wastewater*, 21st ed.; American Public Health Association: Washington, DC, USA, 2005.
27. OECD. *Guideline for Testing of Chemicals–Daphnia sp. Acute Immobilization Test*; OECD: Paris, France, 2004.
28. *ISO 105-J03:2009*; Textiles: Tests for Colour Fastness. Part J03: Calculation of Colour Differences. ISO: Basel, Switzerland, 2009.
29. *ISO 105-C06 A2S:2010*; Textiles: Tests for Colour Fastness. Part C06: Colour Fastness to Domestic and Commercial Laundering. ISO: Basel, Switzerland, 2010.
30. Pablos, M.; Martini, F.; Fernández, C.; Babín, M.; Herraez, I.; Miranda, J.; Martínez, J.; Carbonell, G.; San-Segundo, L.; García-Hortigüela, P.; et al. Correlation between physicochemical and ecotoxicological approaches to estimate landfill leachates toxicity. *Waste Manag.* **2011**, *31*, 1841–1847. [CrossRef]
31. Hu, E.; Shang, S.; Tao, X.; Jiang, S.; Chiu, K.L. Regeneration and reuse of highly polluting textile dyeing effluents through catalytic ozonation with carbon aerogel catalysts. *J. Clean. Prod.* **2016**, *137*, 1055–1065. [CrossRef]
32. Solano, A.; de Araújo, C.K.C.; de Melo, J.V.; Peralta-Hernandez, J.M.; da Silva, D.R.; Martínez-Huitle, C.A. Decontamination of real textile industrial effluent by strong oxidant species electrogenerated on diamond electrode: Viability and disadvantages of this electrochemical technology. *Appl. Catal. B Environ.* **2013**, *130–131*, 112–120. [CrossRef]
33. Fernandes, A.; Nunes, M.; Rodrigues, A.; Pacheco, M.; Ciríaco, L.; Lopes, A. Electro-Persulfate Processes for the Treatment of Complex Wastewater Matrices: Present and Future. *Molecules* **2021**, *26*, 4821. [CrossRef]
34. Fernandes, A.; Gomes, A.C.; Pereira, C.; Magdziak, A.; Pacheco, M.J.; Ciríaco, L.; Simões, R.; Lopes, A. Influence of Molecular Size on the Electrochemical Oxidation of Fractioned Cork Boiling Wastewater. *ChemElectroChem* **2019**, *6*, 1722–1731. [CrossRef]
35. Giannakis, S.; Lin, K.-Y.A.; Ghanbari, F. A review of the recent advances on the treatment of industrial wastewaters by Sulfate Radical-based Advanced Oxidation Processes (SR-AOPs). *Chem. Eng. J.* **2021**, *406*, 127083. [CrossRef]
36. *DIN ISO 11664-4:2012-06*; German Version of International Organization for Standardization (ISO) 11664-4:2008, Colorim-Etry-Part 4: CIE 1976 L*a*b* Colour Space. German Institute for Standardization (DIN): Berlin, Germany, 2012.
37. Fernandes, A.; Pacheco, M.J.; Ciríaco, L.; Lopes, A. Anodic oxidation of a biologically treated leachate on a boron-doped diamond anode. *J. Hazard. Mater.* **2012**, *199–200*, 82–87. [CrossRef]
38. UN General Assembly. Resolution Adopted by the General Assembly on 25 September 2015. *Transforming Our World: The 2030 Agenda for Sustainable Development*. 2015. Available online: http://www.un.org/ga/search/view_doc.asp?symbol=A/RES/70/1 &Lang=E (accessed on 16 March 2022).

Article

Photo-Catalytic Remediation of Pesticides in Wastewater Using UV/TiO₂

Mohamed H. EL-Saeid [1,*], Modhi O. Alotaibi [2,*], Mashael Alshabanat [3,*], Khadiga Alharbi [2], Abeer S. Altowyan [4] and Murefah Al-Anazy [3]

[1] Chromatographic Analysis Unit, Soil Science Department, College of Food & Agricultural Sciences, King Saud University, P.O. Box 2460, Riyadh 11451, Saudi Arabia
[2] Department of Biology, College of Science, Princess Nourah Bint Abdulrahman University, P.O. Box 88828, Riyadh 11671, Saudi Arabia; kralharbi@pnu.edu.sa
[3] Department of Chemistry, College of Science, Princess Nourah Bint Abdulrahman University, P.O. Box 88828, Riyadh 11671, Saudi Arabia; mmalanazy@pnu.edu.sa
[4] Department of Physics, College of Science, Princess Nourah Bint Abdulrahman University, P.O. Box 88828, Riyadh 11671, Saudi Arabia; asaltowyan@pnu.edu.sa
* Correspondence: elsaeidm@ksu.edu.sa (M.H.E.-S.); mouotaebe@pnu.edu.sa (M.O.A.); mnalshbanat@pnu.edu.sa (M.A.)

Citation: EL-Saeid, M.H.; Alotaibi, M.O.; Alshabanat, M.; Alharbi, K.; Altowyan, A.S.; Al-Anazy, M. Photo-Catalytic Remediation of Pesticides in Wastewater Using UV/TiO₂. *Water* **2021**, *13*, 3080. https://doi.org/10.3390/w13213080

Academic Editor: Chengyun Zhou

Received: 28 July 2021
Accepted: 21 October 2021
Published: 2 November 2021

Publisher's Note: MDPI stays neutral with regard to jurisdictional claims in published maps and institutional affiliations.

Copyright: © 2021 by the authors. Licensee MDPI, Basel, Switzerland. This article is an open access article distributed under the terms and conditions of the Creative Commons Attribution (CC BY) license (https://creativecommons.org/licenses/by/4.0/).

Abstract: One of the most serious environmental concerns worldwide is the consequences of industrial wastes and agricultural usage leading to pesticide residues in water. At present, a wide range of pesticides are used directly to control pests and diseases. However, environmental damage is expected even at their low concentration because they are sustained a long time in nature, which has a negative impact on human health. In this study, photolysis and photocatalysis of the pesticides dieldrin and deltamethrin were tested at two UV wavelengths (254 and 306 nm) and in different test media (distilled water, wastewater, and agricultural wastewater) to examine their ability to eliminate pesticides. TiO₂ (0.001 g/10 mL) was used as a catalyst for each treatment. The purpose was to determine the influence of UV wavelength, exposure time, and catalyst addition on the pesticide decomposition processes in different water types. Water was loaded with the tested pesticides (2000 µg) for 12 h under UV irradiation, and the pesticide concentrations were measured at 2 h intervals after UV irradiation. The results showed a clear effect of UV light on the pesticides photodegradations that was both a wavelength- and time-dependent effect. Photolysis was more effective at λ = 306 nm than at λ = 254 nm. Furthermore, TiO₂ addition (0.001 g/10 mL) increased the degradation at both tested wavelengths and hence could be considered a potential catalyst for both pesticide degradations. Deltamethrin was more sensitive to UV light than dieldrin under all conditions.

Keywords: photolysis; catalysis; degradation; pesticides; UV; wastewater; agricultural wastewater

1. Introduction

Agrochemicals are substances that are commonly used in agriculture to protect crops and ensure their productivity [1]. Such chemicals are commonly applied to eliminate pests (such as rodents), and include pesticides—namely, insecticides, fungicides, and herbicides—and un-wanted plants [2]. In public health, agrochemicals are used to combat human disease vectors such as mosquitoes; they are also used against crop-damaging epidemics in the agricultural sector [3,4] that also offer producers an efficient means to manage crop pests that decrease yield and threaten food security [5]. While some are used at a crop's initial production stage, others are generally used on edible plant parts before harvest or even during storage. Therefore, crop-based agrochemicals such as pesticides are dissolved in water, and the crop are sprayed in the fields.

Many new pesticides have been introduced over the last few decades, which have toxic effect in the short and/or long term [3,4]. Commercial pesticide formulations often include additional compounds (such as solvents and surfactants) to increase their activity; solvents and other co-formulants often increase the environmental impact of the formulation as well.

The potential harmful effects of used pesticide on both humans and the environment have received growing attention from the community and expert authorities. Numerous studies have focused on health or environmental concerns from accidental or intentional pesticide exposure, specifically those highly toxic to mammals or found in the environment. The pesticide's risks should be reduced to their minimum via careful regulation and appropriate user guidance. However, the positive effects of pesticide use should not be overlooked. When rational, careful use of pesticides in combination with other technologies is considered in integrated pest management systems, their usage is likely to be justified [6].

On the other hand, fresh fruit and vegetable growers use various water sources, including surface water sources (such as rivers or lakes) that are potentially contaminated by chemical pollutants [7]. Thus, the water used in agricultural production increasingly might have the potential to introduce pathogenic viruses into fresh produce supply chains. The Codex Committee on Pesticide Residues' Code of Hygienic Practice for Fresh Fruits and Vegetables (adopted in 2003; revision in 2010 (new Annex III for Fresh Leafy 334 Vegetables), from the Netherlands, March 2010 [7], pointed out the importance of using 'clean water' for fresh produce cultivation, particularly if water is applied before harvest and in close contact with the edible plant part. Environmental protection programs seek to encourage a reduction in the use of pesticides as a precaution product in growing crops [8].

One of the most abundant water contaminants is pesticide residue, and despite pesticides' economic advantages, such as high crop yields, their potential health hazards are still unknown. Water bodies are polluted by toxic chemicals resulting from human activities in industry, agriculture, and housing [9]. The residues of industrial and agricultural areas contaminated with pesticides are dumped unattended into the nearby water bodies, although most of them are not degradable in water, and thus the aquatic environment becomes threatened [10]. The effects of pollutants are generally characterized by alteration in the animal physiological behavior, and therefore affect survival, reproduction, and growth.

Most Europeans are concerned primarily with the long-term or chronic consequences of low exposure levels through various pathways, especially residues in food crops, as well as through pesticide fraction losses from the target areas [11].

Dieldrin and deltamethrin are environmental pollutants with long-term adverse effects. The use of dieldrin has been banned in most countries around the world primarily for environmental reasons. The widespread use of dieldrin and its ecological persistence have resulted in survival in the environment [12], with bioaccumulation in the food chain due to their low volatile, chemically stable, and lipophilic properties [13]. The half-life of dieldrin is about 5 years [14]. In addition, it will take 25 years for 90% to disappear [15] and remains for 60 years when not exposed to sunlight. In addition, it was used for termite control until about 1985, which means it is still in the basement of the most houses right now and in the soils of agricultural fields with these pesticides [16]. In a previous study published in 2005 [17], Saqib et al. identified residues of DDT, DDE, aldrin, dieldrin, and deltamethrin in fish tissues in Haleji Lake, and more different pesticide compounds were identified in Kalri Lake, possibly because of runoff from surrounding agricultural farms [17]. On the other hand, in Saudi Arabia at Al-Qassim, high pesticide concentrations might be linked to intensive agricultural activity [18]. Researchers [19] stated that dieldrin and deltamethrin are pesticides found in fruits and palm in Riyadh market, Saudi Arabia, which might suggest contamination of irrigation water. The search for a potential tool such as photolysis for the above-mentioned pesticide residue degradation is important.

Recent studies are starting to shed light on disposal strategies for pesticides in contaminated water that would lead to better effluent water quality and have focused on the possibility of analyzing pollutants and removing pollution by available and less expensive methods [20–25]. Currently, several processes have been developed to reduce harmful pollutants in wastewater, including advanced oxidation processes (AOP) [26], activated sludge treatments [27], electro-removal [28], ozonation [29], sunlight [30], UV radiation [31], and combined/integrated methods [32]. Catalyst methodologies have also been used to improve the pesticide disposal mechanism. Photodegradation is induced by the action of light and is attributed to chemical reactions arising by photoionization. Researchers [33] explained that one of the most important abiotic transformations of pesticides in the aquatic environment is photolysis, where the high energy of solar rays causes characteristic reactions such as bond separation, rotation, and rearrangement. Photolysis with the aid of catalytic compounds could be beneficial when UV radiation is applied. UV radiation has adequate energy for chemical bond breakdown; the high-energy photons cause ionization.

Dieldrin and deltamethrin are environmental pollutants that are prohibited in most countries around the world, but they are still used, leading to contamination of many environments, such as soil, sediment, and groundwater [16,34]. Therefore, looking for solution to reduce their negative impact is an urgent issue. UV has been well known as water disinfectant for microbial removal [35,36], and as an efficient technique for the treatment of wastewater [37]. However, new studies to develop the photo-remediation by UV radiation as an effective method for water treatment systems are needed.

The purpose of this work was to obtain information that is currently lacking regarding the photo-remediation of the pesticides dieldrin and deltamethrin, since their high levels in contaminated water are expected and attributed to pollution as consequences to industrial wastes and agricultural usage. Therefore, in this study, different types of water contaminated with dieldrin and deltamethrin were irradiated at two UV wavelengths, with and without a catalyst, to observe the effect of the potential catalyst and identify the most effective UV wavelength and time span for maximum pesticide decomposition.

The importance of this study is in the remediation of dieldrin- and deltamethrin-polluted water taken from the local environment, including treated wastewater and agricultural wastewater. It will provide new data and potential breakthroughs to scientists, especially those working in environmental pollution and water remediation.

2. Materials and Methods

2.1. Study Samples

Three water samples were used as targets in the study: distilled water, wastewater, and agriculture wastewater. The distilled water used was obtained from a Millipore distilled water system (College of Food and Agricultural Sciences, King Saud University, Riyadh, Saudi Arabia).

First, 5 L of the wastewater sample was taken in dark glass container from treatment plants at Al Mansuriyah, Riyadh (pH = 7.33, EC = 1.77 µS/cm, TDS = 1133 mg/L, turbidity = 1.89 NTU) where the treatment process was performed by activated sludge method using tertiary treatment, and 5 L of the agriculture wastewater was taken from the Al-Kharj agricultural region (pH = 8.42, EC = 2.48 µS/cm, TDS = 1579 mg/L, turbidity = 3.24 NTU). The crops produced by the farms were corn, grains, dates, and some vegetables and leafy crops.

Both samples were transferred under cooling within 2 h to the analysis and experimental lab. Wastewater samples were analyzed for pesticide residues within 24 h and then the treatment of remediation began.

Pesticide residues in the collected wastewater samples were analyzed before spiking with 2000 ppb concentration of dieldrin and deltamethrin.

2.2. Standards and Reagents

Dieldrin, deltamethrin (Table 1 and Figure 1), calibration, and injection standards (99.9% purity) were purchased from AccuStandard, Inc., New Haven, CT, USA as individual or mixture standards at concentrations of 10 µg/mL. All internal standards were 13C 12-labelled (13C-labelled compound use allowed for the analysis to be quantified without clean up). All solvents used for the extraction and analysis of pesticides were analysis-grade residues (99.9% purity) and were obtained from Fisher Scientific (Fair Lawn, NJ, USA). QuEChERS kits were purchased from Phenomenex (Torrance, CA, USA). Titanium dioxide (TiO_2) as a photocatalyst was from Sigma-Aldrich Chemie GmbH, Germany (molecular weight: 79.87, CAS Number: 13463-67-7, 718467nanopowder, 21 nm primary particle size (TEM), ≥99.5% trace metals basis).

Table 1. Chemical properties of pesticides.

Pesticides	Molecular Formula	Molecular Weight	Solubility	Type	Effective against
Dieldrin	$C_{12}H_8Cl_6O$	380.895 g/mol	soluble in water	organochlorine insecticide	controlling locusts, vectors, tropical disease, and termites
Deltamethrin	$C_{22}H_{19}Br_2NO_3$	505.206 g/mol	soluble in water	pyrethroid ester pesticide	controlling malaria vectors

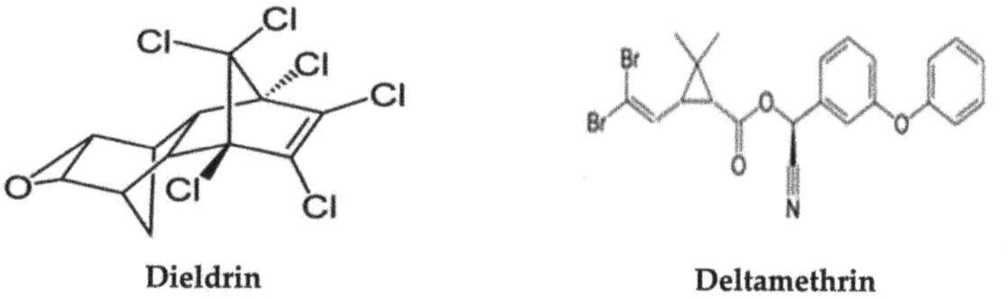

Dieldrin   **Deltamethrin**

Figure 1. A structure of Dieldrin and deltamethrin.

2.3. Sample Remediation by UV Photolysis (UV)

The water was photo-treated using ultraviolet radiation at 254 and 306 nm wavelengths for the two pesticides' decomposition. Boekel UV Crosslinker (BUV) model 234100-2: 230 VAC, 175 W, 0.8 A was applied with four 254 nm lamps and Boekel Scientific, 855 Pennsylvania Blvd. Feasterville, PA, USA with four 306 nm lamps. The lamps and water samples were at a 15 cm distance at 1071 µWcm^{-2} intensity of UV irradiation. Each pesticide (approximately 2000 µg/L) was loaded into the water and incubated for 12 h under UV lighting. Samples were taken for pesticide quantity residue measurement at 2 h intervals to identify the correct UV wavelength and the photolysis process exposure time. Furthermore, the same procedure was repeated for each pesticide with the addition of 0.001 g TiO_2 to each 10 mL water sample to study the effect of the catalyst.

2.4. Samples Extraction and Cleanup by QuEChERS

First, 10 mL of the water sample was transferred into a 50 mL centrifuge tube and vortexed briefly. After that, 10 mL acetonitrile was added to each sample and shaken using a vortex for 5 min to extract the pesticides, using a Spex Sample Prep Geno/Grinder 2010 operated at 1500 rpm. Next, the contents of an ECQUEU750CT-MP (citrate salts) Mylar pouch were added to each centrifuge tube. The samples were then shaken for at least 2 min and centrifuged for 5 min at ≥3500 rcf. A 1 mL aliquot of supernatant was transferred to a 2 mL CUMPSC18CT ($MgSO_4$, PSA, C18) dSPE tube. The samples were shaken in a vortex for about 1 min, then centrifuged for 2 min at high rcf (e.g., ≥5000). The purified supernatant was filtered through a 0.2 µm syringe filter directly into a GC sample vial, and thereafter the sample was kept for further analysis.

2.5. Analysis by Triple-Quadrupole Gas Chromatography Mass Spectrometry (GCMSMSTSQ 8000/SRM)

The analysis was carried out using the latest Thermo Scientific™ TSQ 8000™ triple-quadrupole GC-MS/MS system equipped with the Thermo Scientific™ TRACE™ 1310 GC with SSL Instant Connect™ SSL module and Thermo Scientific™ TriPlus™ RSH auto sampler (Waltham, MA, USA). The transition conditions are presented in Table 2.

Table 2. GCMSMSTQD 8000 SRM instrumental conditions.

GC Trace Ultra Conditions		TSQ Quantum MS/MS Conditions	
Column	TR-Pesticide 30 m × 0.25 mm × 0.25 μm	Ionization mode	EI
Injector	Splitless	Electron energy	70 eV
Injected volume	1 μL	Emission current	50 μA
Injector temperature	225 °C	Q1/Q3 resolution	0.7 u (FWHM)
Carrier gas	Helium, 1.2 mL/min	Collision gas	Argon
Oven program	80 °C hold 1 min 15 °C/min to 160 °C hold 1 min 2.2 °C/min to 230 °C hold 1 min 5 °C/min to 290 °C hold 5 min, Run time: 57.15 min	Operating mode	Selected reaction monitoring (SRM)
Transfer line temperature	280 °C	Collision gas pressure	1 mTorr
		Polarity	Positive

2.6. QAQC Strategies and Method Performance

For quality analysis and quality control, samples were prepared in triplicate, blanked, and spiked. Certified reference material (CRM) was prepared and processed with each batch (5–10 samples) analyzed. QuEChERS and GCMSMSTSQ 8000/SRM method limit detection (LOD) and limit quantification (LQD), repeatability, reproducibility, accuracy, and precision were also determined for each pesticide (Table 3).

Table 3. Parameters of retention time, LOD, LOQ, recovery%, and GCMSTQD target mass of SRM scanning mode.

Name	RT min	Mass	Product Mass	Collision Energy m/z	LOQ ng/ml	LOD ng/mL	r^2	Recovery %	SD
Deltamethrin	21.4	176	124	9	3.6	1.2	0.8034	102.4	8.3
Dieldrin	30.5	279	243	10	7.9	5.3	0.9486	105.5	7.1

3. Results and Discussion

3.1. Photolysis Process

The photolysis process of dieldrin and deltamethrin was examined in the current study for three different types of water: distilled water (DW), wastewater (WW). and agricultural wastewater (Ag.WW) using UV radiation at varied wavelengths with and without a catalytic agent (TiO_2). The results indicated that the amount of both dieldrin and deltamethrin decreased gradually with increasing time after photolysis. The pesticides' degradation rate and reduction (%) was calculated as the variation between the concentration after treatment in relation to that before treatments. The reduction percentage in the concentration of pesticide residues after degradation process for dieldrin and deltamethrin in DW reached 39.35% and 73.6% at 254 nm and 43.95% and 76.55% at 306 nm, respectively, after 12 h of treatment. As for WW, reduction percentages of 43.3% and 83.8% at 254 nm and 49.3% and 84.35% were recorded after 12 h for dieldrin and deltamethrin, respectively, at 306 nm. Furthermore, in Ag.WW, the percentages of reduction of dieldrin and deltamethrin after 12 h were 58.8% and 46.8% at 254 nm versus 52.9% and 37.3% at 306 nm, respectively. It was observed that the longer UV wavelength (306 nm) had a higher capacity for pesticide degradation compared to 254 nm. The dieldrin and deltamethrin amounts in DW, WW, and Ag.WW samples were decreased with increased UV exposure time as indicated in Figures 2–4, and therefore, a time-dependent reduction was noted.

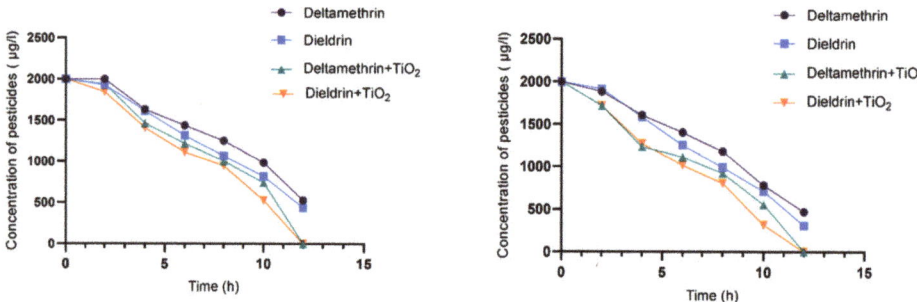

Figure 2. Concentration of pesticides (µg/L) versus exposure time of UV radiation for photolysis process at 254 nm (**left**) and 306 nm (**right**) in distilled water with and without the catalytic agent.

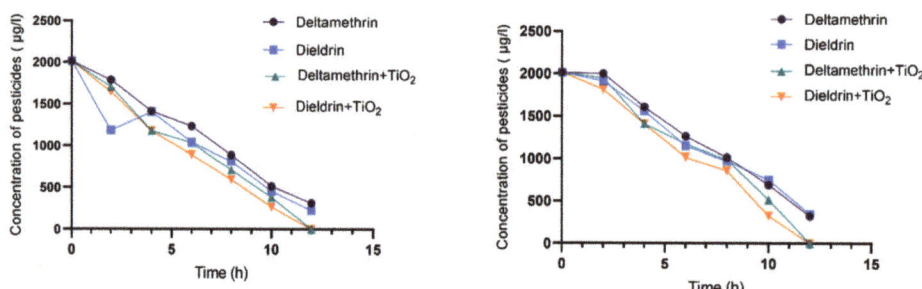

Figure 3. Concentration of pesticides (µg/L) versus exposure time of UV radiation for photolysis process at 254 nm (**left**) and 306 nm (**right**) in wastewater with and without the catalytic agent.

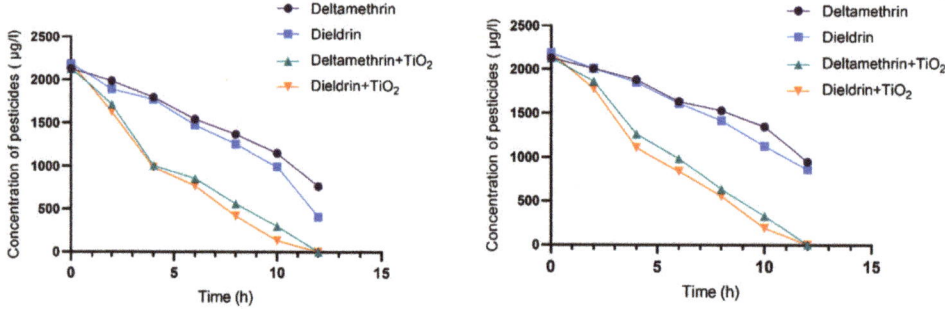

Figure 4. Concentration of pesticides (µg/L) versus exposure time of UV radiation for photolysis process at 254 nm (**left**) and 306 nm (**right**) in agricultural water with and without catalytic agent.

Experimental results indicated the abilities of the two tested wavelengths to promote pesticide photolysis. At both wavelengths investigated, deltamethrin was more degradable than dieldrin. In DW and WW, both pesticide degradations were observed after 4 h treatment at 254 nm; after 4 h at 306 nm, deltamethrin degradation was faster than that of dieldrin. However, in Ag.WW, both pesticide quantities were reduced from 2000 µg/L after the first 2 h at both wavelengths, suggesting higher UV efficiency in Ag.WW. After 2 h, the dieldrin amount detected was the same at both wavelengths, and a higher degradation was observed for deltamethrin, especially at 306 nm.

The longer UV wavelength (306 nm) showed a higher capacity for pesticide degradation compared with 254 nm, which is consistent with previous findings [38,39]; other pesticides are known to be degraded under UV exposure, suggesting that UV reactor usage might be a suitable approach for pesticide photolysis [40,41]. For example, in a previous study, the photolysis rate of deltamethrin and bifenthrin, another pyrethroid, under UV irradiation at 237, 240, and 246 nm was investigated by Tariq et al. [42]. Their findings revealed that deltamethrin was highly degradable in a time-dependent manner when subjected to UV irradiation in organic solvents. In the absence of UV light, the organophosphorus pesticide degradation rate was insignificant, indicating the significant role of UV in pesticide degradation [43].

This trend indicates that longer wavelengths lead to faster degradation as compared to that at shorter wavelengths, especially for deltamethrin. The destructive effect of UV on molecular bonds is well known; therefore, UV exposure should lead to increased pesticide degradation as time increases. Increased UV irradiation time increases the formation of free radicals in water, potentially leading to decomposition pesticide poisoning [44,45]. Furthermore, the difference in the degradation levels between deltamethrin and dieldrin may be due to differences in their structures.

3.2. Photocatalysis Process

The photo-remediation was performed with photocatalyst to study the effect of adding a catalytic amount of TiO_2 on the pesticide photodegradations [41,44]. Photocatalytic pesticide residue breakdown by oxidation processes (AOP) is a modern approach that uses photons to degrade pesticides to H_2O, CO_2, and inorganic compounds with no side effects [46]. However, catalyst type is an important factor in pesticide photodegradation [47]. Titanium dioxide (TiO_2) was used as a photocatalyst because it is effective in the decomposition of organic compounds and is more photochemically stable in water [48]. It is considered as a beneficial material for wastewater treatment because of its safe character; it is used in different applications, mainly in environmental remediation [49].

The catalytic effect of TiO_2 (0.001 g/10 mL) added to aqueous media was tested when dieldrin and deltamethrin were exposed to ultraviolet irradiation. The degradation of pesticides with different wavelengths and exposure times was observed. The results are displayed in Figures 2–4, indicating degradation in a time-dependent manner, as was noticed for degradation without catalysis for both tested pesticides.

When TiO_2 was applied in DW, UV treatment led to complete disappearance of pesticides at 306 nm at the end of treatment time; however, deltamethrin disappeared at 254 nm and only 49% of dieldrin was identified after 12 h.

In addition, after adding the catalyst, it was noticed that the deltamethrin pesticide was not detected in all samples of water media of DW, WW, and Ag.WW after 12 h at any of the tested wavelengths, but only 49.9%, 49.1%, and 40.5% of dieldrin were detected in water media of DW, WW, and Ag.WW, respectively, at 254 nm, as well as only 32.6% and 24.3% of dieldrin were detected in WW and Ag.WW, respectively, when 306 nm was tested; however, after 12 h, no pesticides residues were detected in DW.

Furthermore, it was observed that the percentage of pesticide degradation was higher when the catalyst was present compared to the previous experiments without the catalyst.

Hydrolysis levels in the presence of the catalyst were higher because the final concentrations of the pesticides were low compared to those after remediation without the catalyst. Hence, this catalyst has a role in improving the photolysis process. Thus, pesticides can be effectively destroyed by photocatalysis in the presence of TiO_2 suspensions. The photo-remediation at both wavelengths of UV rays, with and without the catalyst, of the pesticides in different aqueous media as a function of time are displayed in Figures 2–4.

Effect of the addition of the photocatalyst on the degradation process has been reported in previous studies. Degradation of the compounds azinphos methyl, azinphos ethyl, disulfoton, dimethoate, and fenthion was detected in TiO_2 suspensions under UV irradiation [50]. Deltamethrin degradation increased in the presence of catalytic Cu [42].

The same trend was also observed by Burrows et al. [51], who evaluated the degradation ratio of the pesticide malathion by applying natural solar illumination. The 2% WO_3/TiO_2 photocatalyst displayed the best photocatalytic efficiency [52]. Phosalone photodegradation effectiveness was influenced by irradiation time and the amount of TiO_2 present [43]. According to Liu et al. [53], the TiO_2/HZSM-11 (30%) catalyst was effective in solution; it maintained its photocatalytic ability after many cycles, and it could be removed easily from the treated solution and reused immediately, giving it a great advantage for photocatalytic wastewater treatment. A recent study noted that the breakdown of the pesticides profenofos and triazophos was enhanced by TiO_2/Ce application on the leaves of *Brassica chinensis* [54]. Nguyen and Juang [55] noted that TiO_2 use increased UV efficiency in p-chlorophenol degradation. Additionally, such a catalyst might be efficient under solar radiation, conserving electrical energy and consequently becoming an option for environmental remediation. Degradation rates were different in all studied conditions since wavelengths, exposure time, and solvent systems might affect the photodegradation [56]. It is worth noting the mechanism of photodegradation with and without a catalyst, since variations were noted between both conditions. In the remediation process without a photocatalyst, the pesticide molecules become excited by absorption of light energy of the UV radiation, causing homolysis, heterolysis, or photoionization. Whereas, in the process with a photocatalyst, the UV light energy will be absorbed by a semiconductor catalyst (titanium dioxide) to be photoexcited. However, a photoexcitation of the semiconductor catalyst occurs when the adsorbed light energy is greater than or at least equal to that of the gap between conduction and valence bands in the catalyst, leading to electron excitation to the conduction band (e−) and a positive hole (h+) in the valance band. Thus, oxidation–reduction reactions of the pesticide can be started by the radiation on the surface of semiconductor photocatalyst [52]. On the other hand, hydrogen peroxide (H_2O_2) as a powerful oxidant can be added to TiO_2 catalyst to enhance the effectiveness of the treatment by generating electrons, which leads to avoid the recombination of (e−)–(h+) pairs formed in the photocatalytic remediation [57]. This addition could reduce the effectiveness of the degradation process by modifying the photocatalyst surface by H_2O_2 adsorption [58] and the inhibition of generated (h+) and reaction with hydroxyl radicals [59].

Additionally, studies on photoelectrochemical and catalysts applied for advanced treatment of wastewater are still at early stages despite the growing scientific and practical interest in this technology. Several recent studies have demonstrated that advanced oxidation processes (AOP) are more efficient for wastewater treatment, such as electrocatalysis, electro-fenton or photocatalysis, because hydroxyl radicals (OH•) are strong oxidizing agents that are generated from AOP under mild conditions.

Thus, the AOP have recently attracted the attention of researchers because they allow for the continuous electrocatalytic generation of strong oxidizing species under mild conditions. Moreover, energy can be saved by using sunlight in photovoltaic electrolysis systems and using a catalyst to speed up reactions [60]. Generally, UV is a well-known disinfection process normally used for drinking water treatment via their breakdown of water H-O bond. Consequently, water breakdown provides the strong oxidant HO• that has high potential as a redox and organic pollutant oxidizer [61,62]; therefore, UV are efficient in pesticides' removal or reduction from water. Interestingly, although the mode of action for UV in photolysis could be the same in relation to both tested pesticides in the current study, pesticides in varied media with different organic components responded differently. The type of dissolved organic materials in water may affect the UV absorption and it is expected that higher organic compounds in water lead to a high ability in UV absorbance and, therefore, high degradation ability is expected. Ag.WW was approved as a good medium for pesticide removal when the catalyst was added, and such a finding could be explained by the fact that Ag.WW had high organic constituents that could be good substrates for pesticide residues' conjugation. Since organic molecules have a high tendency towards UV absorbance, high degradation ability is therefore expected for the organic-pesticides' conjugate [63].

4. Conclusions

As consequences of industrial wastes and agricultural usage, pesticide residues in water are considered as one of the most serious environmental problems worldwide and, therefore, an efficient method for their elimination is needed. This study demonstrated the efficiency of photocatalytic agents for analyzing pesticide residues, and this is the first study on pesticide degradation (dieldrin and deltamethrin) using UV in three different water media collected from Saudi Arabia. UV radiation was used at 254 and 306 nm to induce photodegradation with and without photocatalytic TiO_2. The results showed that UV use led to successful pesticide photolysis. For both tested pesticides, UV at 306 nm increased photolysis in a time-dependent manner. The catalyst increased the efficiency of UV irradiation at both wavelengths. The photolysis conditions were effective for both insecticides. Deltamethrin showed a higher degradation than dieldrin under all studied conditions. The obtained results in this study are very encouraging, so further kinetics studies of photo-remediation are recommended.

Author Contributions: Conceptualization, M.H.E.-S. and M.A.; Data curation, K.A., A.S.A. and M.A.; Formal analysis, M.O.A. and M.A.; Methodology, M.H.E.-S. and M.A.; Project administration, M.H.E.-S.; Resources, M.O.A., K.A. and A.S.A.; Supervision, M.H.E.-S., M.O.A. and M.A.; Writing—original draft, M.O.A., K.A., A.S.A. and M.A.-A.; Writing—review and editing, M.H.E.-S., M.O.A. and M.A. All authors have read and agreed to the published version of the manuscript.

Funding: This research was supported by the Chair of Environmental Pollution Research at Princess Nourah bint Abdulrahman University (Grant no. EPR023).

Data Availability Statement: All data supporting our findings are contained within the manuscript. Further details can be provided upon written request to the corresponding author.

Conflicts of Interest: The authors declare no conflict of interest.

References

1. Cesco, M.; Lucini, L.; Miras-Moreno, B.; Borruso, L.; Mimmo, T.; Pii, Y.; Puglisi, E.; Spini, G.; Taskin, E.; Tiziani, R.; et al. The hidden effects of agrochemicals on plant metabolism and root-associated microorganisms. *Plant Sci.* **2021**, *311*, 111012. [CrossRef]
2. WHO. Health Topics: Pesticides. 2020. Available online: https://www.who.int/topics/pesticides/en/ (accessed on 1 February 2020).
3. Eddleston, M.; Bateman, D.N. Pesticides. *Medicine* **2012**, *40*, 147–150. [CrossRef]
4. Eddleston, M. Pesticides. *Medicine* **2016**, *44*, 193–196. [CrossRef]
5. Verhaelen, K.; Bouwknegt, M.; Rutjes, S.; Husman, A. Persistence of human norovirus in reconstituted pesticides- Pesticide application as a possible source of viruses in fresh produce chains. *Int. J. Food Microbiol.* **2013**, *160*, 323–328. [CrossRef]
6. Cooper, J.; Dobson, H. The benefits of pesticides to mankind and the environment. *Crop Prod.* **2007**, *26*, 1337–1348. [CrossRef]
7. Codex. Code of Hygienic Practice for Fresh Fruits and Vegetables; (Revised 2010 (New Annex III for Fresh Leafy Vegetables)). 2003. Available online: https://www.ifsh.iit.edu/sites/ifsh/files/departments/ssa/pdfs/codex2003_053e.pdf (accessed on 21 October 2021).
8. Oller, I.; Malato, S.; Sánchez-Peérez, J.A.; Maldonado, M.I.; Gasso, R. Detoxification of wastewater containing five common pesticides by solar AOPs–biological coupled system. *Catal. Today* **2007**, *129*, 69–78. [CrossRef]
9. Begum, G. Carbofuran insecticide induced biochemical lalterations in liver and muscle tissues of the fish Clarias batrachus (Linnaeus) and recovery response. *Aquat. Toxicol.* **2004**, *66*, 83–92. [CrossRef] [PubMed]
10. Mahmood, A.; Malik, R.; Li, J.; Zhan, G. Levels, distribution profile, and risk assessment of polychlorinated biphenyls (PCBs) in water and sediment from two tributaries of the River Chenab, Pakistan. *Environ. Sci. Pollut. Res.* **2014**, *21*, 7847–7855. [CrossRef]
11. Fantke, P.; Friedrich, R.; Jolliet, O. Health impact and damage cost assessment of pesticides in Europe. *Environ. Int.* **2012**, *49*, 9–17. [CrossRef]
12. Costa, L. The neurotoxicity of organochlorine and pyrethroid pesticides (Chapter 9). *Handb. Clin. Neurol.* **2015**, *131*, 135–148. [CrossRef]
13. Hatcher, J.; Richardson, J.; Guillot, T.; McCormack, A.; Di Monte, A.; Jones, D.; Pennell, K.; Miller, G. Dieldrin exposure induces oxidative damage in the mouse nigrostriatal dopamine system. *Exp. Neurol.* **2007**, *204*, 619–630. [CrossRef]
14. Sava, V.; Velasquez, A.; Song, S.; Sanchez-Ramos, J. Dieldrin Elicits a Widespread DNA Repair and Antioxidative Response in Mouse Brain. *J. Biochem. Mol. Toxicol.* **2007**, *21*, 3. [CrossRef]
15. Hashimoto, Y. Dieldrin Residue in the Soil and Cucumber from Agricultural Field in Tokyo. *J. Pestic. Sci.* **2005**, *30*, 397–402. [CrossRef]
16. Maldonado-Reyes, A.; Montero-Ocampo, C.; Solorza-Feria, O. Remediation of drinking water contaminated with arsenic by electro-removal process using different metal electrodes. *Environ. Monit.* **2007**, *9*, 1241–1247. [CrossRef] [PubMed]

17. Saqib, T.; Naqvi, S.; Siddiqui, P.; Azmi, M. Detection of pesticide residues in muscles, liver and fat of 3 species of Labeo found in Kalri and Haleji lakes. *J. Environ. Biol.* **2005**, *26*, 433–438.
18. Al-Wabel, M.; El-Saeid, M.H.; Usman, A.R.; Al-Turki, A.M.; Ahmad, M.; Hassanin, A.S.; El-Naggar, A.H.; Alenazi, K.K. Identification, Quantification, and Toxicity of PCDDs and PCDFs in Soils from Industrial Areas in the Central and Eastern Regions of Saudi Arabia. *Bull. Environ. Contam. Toxicol.* **2016**, *96*, 622–629. [CrossRef]
19. El-Saeid, M.; Al-Dosari, S. Monitoring of pesticide residues in Riyadh dates by SFE, MSE, SFC, and GC techniques. *Arab. J. Chem.* **2010**, *3*, 179–186. [CrossRef]
20. Ahmed, T.; Rafatullah, M.; Ghazali, A.; Sulaiman, O.; Hashim, R.; Ahmad, A. Removal of pesticides from water and wastewater by different adsorbents: A review. *J. Environ. Sci. Health Part C Environ. Carcinog. Ecotoxicol. Rev.* **2010**, *28*, 231–271. [CrossRef]
21. Ali, I.; Gupta, V.K. Advances in water treatment by adsorption technology. *Nat. Protocol.* **2006**, *1*, 2661–2667. [CrossRef] [PubMed]
22. Ali, I.; Khan, T.A.; Asim, M. Removal of arsenic from water by electrocoagulation and electrodialysis techniques. *Sepn. Purif. Rev.* **2011**, *40*, 25–42. [CrossRef]
23. Ali, I.; Khan, T.; Asim, M. Removal of arsenate from groundwater by electrocoagulation method. *Environ. Sci. Pollut. Res.* **2012**, *19*, 1668–1676. [CrossRef]
24. Ali, I.; Basheer, A.; Mbianda, X.; Burakov, A.; Galunin, E.; Burakova, I.; Mkrtchyan, E.; Tkachev, A.; Grachev, V. Graphene based adsorbents for remediation of noxious pollutants from wastewater. *Environ. Int.* **2019**, *127*, 160–180. [CrossRef]
25. Saleh, I.; Zouari, N.; Al-Ghouti, M. Removal of pesticides from water and wastewater: Chemical, physical and biological treatment approaches. *Environ. Technol. Innov.* **2020**, *19*, 101029. [CrossRef]
26. Oturan, M.; Aaron, J. Advanced Oxidation Processes in Water/Wastewater Treatment: Principles and Applications. A Review. *Crit. Rev. Environ. Sci. Tech.* **2014**, *44*, 2577–2641. [CrossRef]
27. Zhou, J.; Liu, Z.; She, P.; Ding, F. Water removal from sludge in a horizontal electric field. *Dry. Technol.* **2001**, *19*, 627–638. [CrossRef]
28. Baghirzade, B.; Yetis, U.; Dilek, F. Imidacloprid elimination by O_3 and O_3/UV: Kinetics study, matrix effect, and mechanism insight. *Environ. Sci. Pollut. Res.* **2021**, *28*, 24535–24551. [CrossRef] [PubMed]
29. Shifu, C.; Gengyu, C. Photocatalytic degradation of organophosphorus pesticides using floating photocatalyst $TiO_2 \cdot SiO_2$/beads by sunlight. *Sol. Energy* **2005**, *79*, 1–9. [CrossRef]
30. Zhang, R.; Yang, Y.; Huang, C.; Zhao, L.; Sum, P. Kinetics and modeling of sulfonamide antibiotic degradation in wastewater and human urine by UV/H_2O_2 and UV/PDS. *Water Res.* **2016**, *103*, 283–292. [CrossRef] [PubMed]
31. Cassano, D.; Zapata, A.; Brunetti, G.; Del Moro, G.; Di Iaconi, C.; Oller, I.; Malato, S.; Mascolo, G. Comparison of several combined/integrated biological-AOPs setups for the treatment of municipal landfill leachate: Minimization of operating costs and effluent toxicity. *Chem. Eng. J.* **2011**, *172*, 250–257. [CrossRef]
32. Katagi, T. Direct photolysis mechanism of pesticides in water. *J. Pestic. Sci.* **2018**, *43*, 57–72. [CrossRef]
33. Matsumoto, K.; Kawanaka, Y.; Yun, S.J.; Oyaizu, H. Bioremediation of the organochlorine pesticides, dieldrin and endrin, and their occurrence in the environment. *Appl. Microbiol. Biotechnol.* **2009**, *84*, 205–216. [CrossRef] [PubMed]
34. Ismail, B.S.; Mazlinda, M.; Tayeb, M.A. The Persistence of Deltamethrin in Malaysian Agricultural Soils. *Sains Malays.* **2015**, *44*, 83–89. [CrossRef]
35. Lehtola, M.J.; Miettinen, I.T.; Vartiainen, T.; Rantakokko, P.; Hirvonen, A.; Martikainen, P.J. Impact of UV disinfection on microbiallyavailable phosphorus, organic carbon, and microbial growth in drinking water. *Water Res.* **2003**, *37*, 1064–1070. [CrossRef]
36. Li, X.; Cai, M.; Wang, L.; Niu, F.; Yang, D.; Zhang, G. Evaluation survey of microbial disinfection methods in UV-LED water treatment systems. *Sci. Total. Environ.* **2019**, *659*, 1415–1427. [CrossRef] [PubMed]
37. Sanz, J.; Lombraña, J.I.; Ma De Luis, A.; Varona, F. UV/H_2O_2 chemical oxidation for high loaded effluents: A degradation kinetic study of las surfactant wastewaters. *Environ. Technol.* **2008**, *247*, 903–911. [CrossRef]
38. El-Saeid, M.; Al-Turki, A.; Nadeem, M.; Hassanin, A.; Al-Wabel, M. Photolysis degradation of polyaromatic hydrocarbons (PAHs) on surface sandy soil. *Environ. Sci. Pollut. Res.* **2015**, *22*, 9603–9616. [CrossRef]
39. EL-Saeid, M.H.; Alotaibi, M.O.; Alshabanat, M.; AL-Anazy, M.M.; Alharbi, K.R.; Altowyan, A.S. Impact of Photolysis and TiO_2 on Pesticides Degradation in Wastewater. *Water* **2021**, *13*, 655. [CrossRef]
40. Aaron, J.; Oturan, M. New photochemical and electrochemical methods for the degradation of pesticides in aqueous media. *Turk. J. Chem.* **2001**, *25*, 509–520.
41. Shayeghi, M.; Dehghani, M.H.; Alimohammadi, M.; Goodini, K. Using Ultraviolet Irradiation for Removal of Malathion Pesticide in Water. *J. Arthropod-Borne Dis.* **2012**, *6*, 45–53. [PubMed]
42. Tariq, S.R.; Ahmed, D.; Farooq, A.; Rasheed, S.; Mansoor, M. Photodegradation of bifenthrin and deltamethrin—effect of copper amendment and solvent system. *Environ. Monit. Assess.* **2017**, *189*, 71. [CrossRef]
43. Daneshvar, N.; Hejazi, M.; Rangarangy, B.; Khataee, A. Photocatalytic Degradation of an Organophosphorus Pesticide Phosalone in Aqueous Suspensions of Titanium Dioxide. *J. Environ. Sci. Health Part B* **2004**, *39*, 285–296. [CrossRef] [PubMed]
44. Xie, Q.; Chen, J.; Shao, J.; Chen, C.; Zhao, H.; Hao, C. Important role of reaction field in photodegradation of deca-bromodiphenyl ether: Theoretical and experimental investigations of solvent effects. *Chemosphere* **2009**, *76*, 1486–1490. [CrossRef] [PubMed]
45. Saqib, N.; Adnan, R.; Shah, I. A mini-review on rare earth metal-doped TiO_2 for photocatalytic remediation of wastewater. *Environ. Sci. Pollut. Res.* **2016**, *23*, 15941–15951. [CrossRef]

46. Thiruvenkatachari, R.; Vigneswaran, S.; Shik, M. A review on UV/TiO$_2$ photocatalytic oxidation process. *Korean J. Chem. Eng.* **2008**, *25*, 64–72. [CrossRef]
47. Mecha, A.C.; Chollom, M.N. Photocatalytic ozonation of wastewater: A review. *Environ. Chem. Lett.* **2020**, *18*, 1491–1507. [CrossRef]
48. Ramos-Delgado, N.A.; Gracia-Pinilla, M.A.; Maya-Trevino, L.; Hinojosa-Reyes, L.; Guzman-Mar, J.L.; Hernández-Ramírez, A. Solar photocatalytic activity of TiO$_2$ modified with WO$_3$ on the degradation of an organophosphorus pesticide. *J. Hazard. Mater.* **2013**, *263*, 36–44. [CrossRef]
49. Yavg, V. Photocatalytic Degradation of Selected Organophosphorus Pesticides Using Titanium Dioxide and UV Light. In *Titanium Dioxide: Material for a Sustainable Environment*; Additional information is available at the end of the chapter; Petsas, A.S., Vagi, M.C., Eds.; IntechOpen: London, UK, 2018; p. 241. [CrossRef]
50. Montañez, J.; Gómez, S.; Santiago, A.; Pierella, L. TiO$_2$ Supported on HZSM-11 Zeolite as Efficient Catalyst for the Photodegradation of Chlorobenzoic Acids. *J. Braz. Chem. Soc.* **2015**, *26*, 1191–1200. [CrossRef]
51. Burrows, H.; Canle, M.; Santaballa, J.; Steenken, S. Reaction pathways and mechanisms of photodegradation of pesticides. *J. Photochem. Photobiol. B Biol.* **2002**, *67*, 71–108. [CrossRef]
52. Konstantinou, I.; Albanis, T. Photocatalytic transformation of pesticides in aqueous titanium dioxide suspensions using artificial and solar light: Intermediates and degradation pathways. *Appl. Catal. B Environ.* **2003**, *42*, 319–335. [CrossRef]
53. Liu, S.; Liu, G.; Feng, Q. Al-doped TiO$_2$ mesoporous materials: Synthesis and photodegradation properties. *J. Porous Mater.* **2010**, *17*, 197–206. [CrossRef]
54. Liu, X.; Zhan, Y.; Zhang, Z.; Pan, L.; Hu, L.; Liu, K.; Zhou, X.; Bai, L. Photocatalytic Degradation of Profenofos and Triazophos Residues in the Chinese Cabbage, Brassica chinensis, Using Ce-Doped TiO$_2$. *Catalysts* **2019**, *9*, 294. [CrossRef]
55. Nguyen, A.T.; Juang, R. Photocatalytic degradation of p-chlorophenol by hybrid H$_2$O$_2$ and TiO$_2$ in aqueous suspensions under UV irradiation. *J. Environ. Manag.* **2015**, *147*, 271–277. [CrossRef] [PubMed]
56. Sarkouhi, M.; Shamsipur, M.; Hassan, J. Metal ion promoted degradation mechanism of chlorpyrifos and phoxim. *Arab. J. Chem.* **2016**, *9*, 43–47. [CrossRef]
57. Miguel, N.; Ormad, M.P.; Mosteo, R.; Overlleiro, J. Photocatalytic Degradation of Pesticides in Natural Water: Effect of Hydrogen Peroxide. *Int. J. Photoenergy* **2012**, *2012*, 371714. [CrossRef]
58. Pelizzetti, E. Concluding remarks on heterogeneous solar photocatalysis. *Sol. Energy Mater. Sol. Cells* **1995**, *38*, 453–457. [CrossRef]
59. Malato, S.; Fernández-Ibáñez, P.; Maldonado, M.I.; Blanco, J.; Gernjak, W. Decontamination and disinfection of water by solar photocatalysis: Recent overview and trends. *Catal. Today* **2009**, *147*, 1–59. [CrossRef]
60. Mousset, E.; Dionysiou, D.D. Photoelectrochemical reactors for treatment of water and wastewater: A review. *Environ. Chem. Lett.* **2020**, *18*, 1301–1318. [CrossRef]
61. Yang, L.; Zhang, Z. Degradation of six typical pesticides in water by VUV/UV/chlorine process: Evaluation of the synergistic effect. *Water Res.* **2019**, *161*, 439–447. [CrossRef]
62. Lopez-Alvarez, B.; Villegas-Guzman, P.; Peñuela, G.A.; Torres-Palma, R.A. Degradation of a Toxic Mixture of the Pesticides Carbofuran and Iprodione by UV/H$_2$O$_2$: Evaluation of Parameters and Implications of the Degradation Pathways on the Synergistic Effects. *Water Air Soil Pollut.* **2016**, *227*, 215. [CrossRef]
63. Konstantinou, I.K.; Zarkadis, A.K.; Albanis, T.A. Photodegradation of Selected Herbicides in Various Natural Waters and Soils under Environmental Conditions. *J. Environ. Qual.* **2001**, *30*, 121–130. [CrossRef]

Article

Carbamazepine Removal by Clay-Based Materials Using Adsorption and Photodegradation

Ilil Levakov [1], Yuval Shahar [1,2] and Giora Rytwo [1,2,*]

1. Environmental Physical Chemistry Laboratory, MIGAL-Galilee Research Institute, Kiryat Shmona 1101602, Israel; ilill@migal.org.il (I.L.); yuvalsha1991@gmail.com (Y.S.)
2. Environmental Sciences & Water Sciences Departments, Tel Hai College, Upper Galilee 1220800, Israel
* Correspondence: rytwo@telhai.ac.il or giorarytwo@gmail.com; Tel.: +972-4-7700-516

Abstract: Carbamazepine (CBZ) is one of the most common emerging contaminants released to the aquatic environment through domestic and pharmaceutical wastewater. Due to its high persistence through conventional degradation treatments, CBZ is considered a typical indicator for anthropogenic activities. This study tested the removal of CBZ through two different clay-based purification techniques: adsorption of relatively large concentrations (20–500 µmol L^{-1}) and photocatalysis of lower concentrations (<20 µmol L^{-1}). The sorption mechanism was examined by FTIR measurements, exchangeable cations released, and colloidal charge of the adsorbing clay materials. Photocatalysis was performed in batch experiments under various conditions. Despite the neutral charge of carbamazepine, the highest adsorption was observed on negatively charged montmorillonite-based clays. Desorption tests indicate that adsorbed CBZ is not released by washing. The adsorption/desorption processes were confirmed by ATR-FTIR analysis of the clay-CBZ particles. A combination of synthetic montmorillonite or hectorite with low H$_2$O$_2$ concentrations under UVC irradiation exhibits efficient homo-heterogeneous photodegradation at µM CBZ levels. The two techniques presented in this study suggest solutions for both industrial and municipal wastewater, possibly enabling water reuse.

Keywords: carbamazepine; adsorption; clay minerals; organoclays; advanced oxidation processes; photocatalysis; water reuse

1. Introduction

Carbamazepine (CBZ) is one of the most common emerging contaminants released into the aquatic environment through domestic and pharmaceutical wastewater [1]. It is mainly used for epilepsy and bipolar disorder treatments and is considered a typical indicator for anthropogenic activities due to its high persistence through degradation processes in regular wastewater treatments. The removal of carbamazepine during conventional wastewater treatment processes was found to be neglectable and didn't exceed 10% [1–3]. Therefore, carbamazepine is found worldwide in surface water, groundwater, soil, and even drinking water with various concentrations of up to 10 µg L^{-1} [4–7], and is expected to be found at higher levels in industrial wastewaters related to its manufacture and use (pharmaceutical and hospitals wastewater). Although no significant health risks were found associated with the exposure to carbamazepine residues in drinking water, several studies examined the negative side effect of consuming carbamazepine medicinally during pregnancy [8–10]. In addition, the ecotoxicity of carbamazepine for different aquatic species was demonstrated in many studies, revealing potential risks such as an increase in mortality rate, inhibition of growth, reproduction, and mobility [11–15]. The official regulations regarding carbamazepine in drinking water are limited and not available in most countries [7,16,17].

Over the years, various treatment approaches were tested aiming for efficient removal of carbamazepine from natural water bodies and wastewater. As mentioned above, removal by conventional technologies was found to be neglectable, but few advanced approaches were able to remove CBZ with relatively high efficiency. For example, integrating biological modification with activated sludge increased the removal rate [18–20], specific microorganisms were found more efficient for its degradation [21,22], and enzymatic degradation including immobilization of the enzymes for increasing their operational stability [23]. In addition, advanced physicochemical treatment technologies such as nanofiltration (NF) and reverse osmosis (RO) were found to be effective for CBZ removal [24,25] even though the treatment of the concentrated brine afterward should be considered, and studies on that were not reported. Advanced oxidation processes [26,27] were an efficient option. Despite the high efficiency of those approaches, several problems limit their applicability, such as high costs, the toxicity of by-products in the oxidation process, biofouling, and the negative influence of natural organic matter during the removal by RO and NF [28–31].

Adsorption is one of the widest-used approaches for CBZ removal. Activated carbon is one of the common methods for adsorbing organic pollutants in drinking water, including pharmaceutically active compounds. The removal of carbamazepine by carbon-based sorbents was tested at various conditions with relatively high sorption capacities of up to 2 mmol g^{-1} [32–34]. Different clays and organoclay are also used as potential adsorbents for pharmaceutical pollution as a low-cost and effective technique. Adsorption of carbamazepine on various clays was examined in several studies. Adsorption studies on montmorillonite are inconclusive, while some studies present poor-to-low adsorption capacities (0–0.02 mmol g^{-1}) [35–38] and others report considerably higher values (up to 0.15 mmol g^{-1}) [39,40]. Most studies have ascribed adsorption of CBZ on montmorillonite to Van der Waals interactions between the aromatic rings and the clay surface, and hydrogen bonds coordinating between oxygen atoms and exchangeable cations [36,38,40]. Khazri et al. (2017) have demonstrated S-type isotherms in pharmaceuticals adsorption [40], meaning low affinity at low concentration but following initially adsorbed molecules, promotes subsequently increased adsorption by Van der Waals forces between the pollutant aromatic moieties themselves. Some studies indicate that the adsorption occurred only on the clay's external surface and CBZ did not enter the clays' interlayer space [36].

Photocatalysis, an advanced oxidation process (AOP), is an additional approach for removing carbamazepine from contaminated water bodies. While photolysis techniques without a catalyst provided relatively poor removal performances, the addition of a catalyst was found to improve the degradation rate significantly [7,31,41]. The most common heterogeneous photocatalysts for CBZ photodegradation are catalytic grade oxides such as TiO_2, ZnO, and MoS_2 [27,42,43]. The main disadvantages of using such conventional catalysts are the difficulties with separation of the particles and their reuse, weak stability, and low quantum efficiency [44–46]. Therefore, modification and improvement of the bare catalysts are of high interest in water treatment research. In recent years, several studies have demonstrated the applicability of using various clay-based materials as improved photocatalysts that provide efficient and stable reactions [44,47–50]. Furthermore, irradiation of UVC light with a combination of a homogeneous catalyst as hydrogen-peroxide and clay-based heterogeneous catalysts may deliver an effective advanced oxidation process, as was recently shown for BPS [51]. Thus, low price, high adsorption capability, stable structure for regeneration, and distinctive spatial structure for possible modification may turn clay-based materials into optimal potential efficient photocatalysts.

This study reports the removal of carbamazepine through two different clay-based purification techniques, aiming for different purposes: adsorption of relatively large concentrations (up to 500 µmol L^{-1}), focusing on industrial effluents and brine from filtration devices and photocatalysis of lower concentrations (<20 µmol L^{-1}), aiming for complete removal in domestic wastewater. The first part of the article includes CBZ adsorption isotherms on natural or modified clay minerals (several smectites, sepiolite, and hydrophobically modified montmorillonite). Fit to Langmuir, dual-mode and Sips models [52,53] was evaluated, and interpretation of the sorption mechanism is presented based on measuring the cations released during the processes, FTIR, and colloidal particle charge of CBZ-clay. The second part includes photocatalysis of carbamazepine using UVC irradiation and combinations of homo- and heterogeneous synthetic clay catalysts. Comparison with high quality catalytic grade TiO_2, which is considered a "gold standard" of heterogeneous catalysts [54–56] was performed. The main objective of the research was to suggest a solution for carbamazepine removal, a persistent emerging contaminant, in an effective, low-cost, and reliable manner, at both mM and µM concentrations, that might be applied to reuse both industrial and municipal wastewater.

2. Materials and Methods

2.1. Adsorption Experiments

2.1.1. Materials

Clay minerals used for the adsorption study were S9 "Pangel" sepiolite purchased from Tolsa SA (Madrid, Spain), bentonite (commercial montmorillonite, CAS: 1302-78-9) from Sigma-Aldrich (Rehovot, Israel), and Ca-montmorillonite prepared from SWy-1 clay (purchased from the Source Clays Repository of The Clay Minerals Society, Chantilly, VA, USA) using a batch procedure [57]. Thiamine hydrochloride (B1, CAS: 67-03-8), benzalkonium (bzk) solution (50% in H_2O, CAS: 63449-41-2), and carbamazepine (CAS: 298-46-4) were supplied by Sigma-Aldrich (Israel).

2.1.2. Clay Minerals Modification and Organoclay Preparation

All clay and organoclay matrices were prepared according to the same procedure with a concentration of 1% (10 g clay L^{-1}). For the clay suspensions, 1 g of the relevant clay was gradually added to 100 mL of double-distilled water (DDW) while stirring with magnetic stirring until homogenous suspensions were obtained. For the organoclay suspensions, different organic cations were added to the homogeneous clay suspensions and the complex was agitated for 24 h in order to reach equilibrium. The bentonite-B1 organoclay (bent-B1) was prepared by the addition of 175 g of thiamine powder to the bentonite suspension for a final load of 0.66 mmol thiamine g^{-1} of clay. The bentonite-bzk organoclay (bent-bzk) was prepared by adding 5.358 mL of benzalkonium solution (50%) for a total load of 0.6 mmol g^{-1}. Previous studies showed that at those loads there is no release of the adsorbed organocations. Adsorption of the organic modifier was estimated by mass balance, after measuring concentrations in the liquid phase. To ensure complete removal of non-strongly bound modifiers from the complexes, the suspensions were centrifuged (3000 rpm for 30 min) and 90% of the supernatant was replaced with DDW. This procedure was repeated three times. Concentrations of the modifiers in supernatants were below detection limits.

2.1.3. Adsorption Isotherms

Several batch experiments were conducted to study the adsorption of carbamazepine on various clay-based matrices. Each experiment included a set of different carbamazepine concentrations added to the relevant clay or organoclay with three replicates for each concentration. The experiments were conducted in 50-mL plastic tubes with a constant amount of adsorbent (0.25–1 g L^{-1}, Table 1), added concentrations of carbamazepine ranging from 0 to 0.45 mM (0–106 mg L^{-1}), and DDW added to a final volume of 50 mL. Details on concentrations of carbamazepine and the relevant matrices in each experiment are described in Table 1. After the addition of carbamazepine, the samples were kept at room temperature (23 ± 1 °C) on an orbital shaker (100 rpm) for 24 h to reach equilibrium. The equilibrium was confirmed by additional sampling and analysis after 48 h. To separate the solids from the supernatant, 2 mL from each tube were sampled to Eppendorf vials and centrifuged at 15,000 rpm for 25 min in a SciLogex D2012 Eppendorfs (Rocky Hill, CT, USA) centrifuge. The concentration of carbamazepine and thiamine (in bent-B1 clay) were measured in the liquid phase using a diode-array HP 8452A UV–Vis spectrophotometer (Hewlett-Packard Company, Palo Alto, CA, USA) and determined by the absorbance

(OD = optical density) at 286 nm and 237 nm, respectively. As a preliminary experiment, CBZ spectrum was measured under different pH values and found stable through the range of 1.3–13 (results not shown). The adsorbed amount of CBZ was estimated by mass balance, thus subtracting the remaining concentration in the supernatant from the initial addition of carbamazepine. Average and standard deviation values were calculated from the triplicates, for the measured concentration in equilibrium (X-axis in the isotherm) and the adsorbed CBZ (Y-axis in the isotherm).

Table 1. Detailed adsorption experiments' conditions.

Adsorbent Type	Adsorbent Concentration (g L^{-1})	Carbamazepine Addition	
		(mmol g^{-1})	(mM)
Bentonite	0.25–0.4	0–2.0	0–0.43
Sepiolite	1.0	0–0.44	0–0.44
Ca-montmorillonite	0.3–0.4	0–1.5	0–0.45
Bentonite-B1 organoclay	0.5–0.6	0–1.0	0–0.47
Bentonite-benzalkonium organoclay	1	0–0.44	0–0.44

2.1.4. FTIR Analysis

All sample suspensions were lyophilized (Christ Alpha 1-2 LD Plus, Germany) and the solids were analyzed in an attenuated total reflectance Fourier transforms infrared (ATR-FTIR) spectrophotometer. Analysis was performed on a Nicolet iS10 FTIR (Nicolet Analytical Instruments, Madison, WI, USA), using a SMART ATR device with a diamond crystal plate (Thermo Fisher Scientific, Madison, WI, USA) within a range of wavenumbers of 4000 to 500 cm^{-1}. Spectra were recorded at 4 cm^{-1} nominal resolution with mathematical corrections yielding a 1.0 cm^{-1} actual resolution and averaged value from 50 measurements. The absorption intensity at different wavenumbers along the spectrum was quantified using TQ analyst EZ 8.0.2.97 software (Thermo Fisher Scientific). Quantification of adsorbed carbamazepine was based on the ratio between the absorbance intensities of specific absorption bands describing the sorbent in the case and CBZ [58]. In order to assess the stability of the CBZ adsorption, a release test was conducted subsequently to the adsorption experiments. The test was performed on three concentrations along the adsorption isotherms of CBZ to bentonite, Ca-SWy1, and bent-B1. For desorption experiments, CBZ loaded samples were washed three times, followed by centrifugation (3000 rpm for 30 min) and replacement of 90% of the supernatant with DDW. Washed samples were lyophilized and analyzed again in the ATR-FTIR spectrophotometer.

2.1.5. Particle Charge Density Measurements

The colloidal charge of the particles was measured using a particle charge detector (PCD) (BTG Mütek, PCD-05, Eclépens, Switzerland). The PCD was connected to an automatic titration unit (PCD-05 Travel Titrator) with polyelectrolytes. The particle charge measurement is based on electrodes measuring the colloidal charge of a suspension agitated mechanically by a piston, in combination with titration of a charge-compensating polyelectrolyte [59,60]. Poly-DADMAC (poly-diallyl-dimethyl-ammonium chloride) or PES-NA (sodium-polyethylensulfonate) were used as cationic and anionic polyelectrolytes according to the charge of the particles. Each measurement required 10 mL of the homogeneous suspension and colloidal charge results were normalized according to the mass of clay or organoclay in each case.

2.1.6. Exchangeable Cations Measurements

Examination of possible cations' exchange on the clay interlayers was evaluated by measuring the changes in cations concentrations in each of the supernatants of the adsorption points along the isotherm. Several major cations (calcium, sodium, potassium, and magnesium) were measured by inductively coupled plasma-optical emission spectrometry (ICP-OES) analysis. The analysis was performed with a Thermo Scientific IRIS Intrepid II XDL ICP-OES (Thermo Electron Corporation, Waltham, MA, USA). All samples were filtered (0.2 μm) and HNO$_3$ was added to a final concentration of 2%. Multielement standard solution (multi-3 for ICP, 49596 Sigma-Aldrich) was used for calcium, magnesium, potassium, and sodium calibration, which calibration curves of 1–10, 0.2–2, 0.1–1, and 0.5–5 mg L^{-1} respectively.

2.1.7. Adsorption Models

As a first approximation, the fit of all adsorption isotherms to the Langmuir adsorption equation (Equation (1))

$$Cs = \frac{S_{max} K_L C_L}{1 + K_L C_L} \quad (1)$$

was tested. Cs is the sorbed concentration (mmol g^{-1}), C_L is the solution concentration (mM), S_{max} defines the number of adsorption sites per mass of sorbent (mmol g^{-1}), and K_L is the Langmuir adsorption coefficient (mM^{-1}). We also tested fit to the Dual-mode model (DMM) (Equation (2)), which combines the Langmuir equation and a linear equation to simulate the combination of a site-specific adsorption mechanism and partitioning mechanism occurring simultaneously [53,61]. Nevertheless, the linear equation component was found to be neglectable, therefore only results of the Langmuir model are reported.

Since some of the sorbents exhibit type V sigmoidal (S) adsorption isotherms, we tested also suitable models for such behavior as BET, Klotz, and Sips equations [62]. From those models we chose Sips equations since it was the only model that showed improved fit.

Sips model is a hybrid combination of Langmuir and Freundlich equations that can describe Type V behavior [63]. Sips model can be described by Equation (2).

$$Cs = \frac{S_{max}(K_L C_L)^{B_s}}{1 + (K_L C_L)^{B_s}} \quad (2)$$

in which B_s is known as "Sips model exponent" [64]. The equation appears in the literature with different notations, and in some cases K_L is not included in the exponent [63]. We adopted the notation in Equation (2) [62] since it keeps the units of K_L identical to Langmuir equation.

The non-linear curve fitting was performed using scipy.optimize.curve_fit functions from the SciPy package (version 1.4.1), Python (Python 3.7.13), https://scipy.org/citing-scipy/ (accessed on 20 June 2022). The function calculating the specific model parameters (S_{max}, K_L and B_S) for each isotherm.

2.2. Photocatalysis Experiments

2.2.1. Materials

Catalyst-grade industrial high-quality TiO$_2$ (Hombikat®, American Elements, Los Angeles, CA, USA) and a 30% (9.79 M) concentrated H$_2$O$_2$ solution were purchased from Merck\Sigma-Aldrich (Merck KGaA, Darmstadt, Germany). SYn-1 Barasym SSM-100 synthetic mica-montmorillonite was obtained from the Source Clays Repository of The Clay Minerals Society (Chantilly, VA, USA), whereas Laponite-RD was provided by BYK-Chemie GmbH (Wesel, Germany).

2.2.2. Methods

Degradation of CBZ from a 21.2 µM (5 mg L^{-1}) solution was measured in batch experiments in 100-mL UV-C-transparent quartz glass (refractive Index n = 1.5048), 5.3-cm diameter beaker placed in a Rayonet RMR-600 mini photochemical chamber reactor (Southern New England Ultraviolet Company, Branford, CT, USA), as described in previous studies [65,66]. The photoreactor was equipped with eight RMR 2537A lamps (254 nm wavelength), each lamp emitting an average irradiance flux of 19 W m^{-2} at 254 nm, equivalent to an overall intensity of 152 W m^{-2}, as measured in the center of the chamber using a Black Comet SR spectrometer with an F400 UV–VIS–SR-calibrated fiber optic probe equipped with a CR2 cosine light receptor (StellarNet Inc., Tampa, FL, USA). The Black Comet SR spectrometer was also used to measure the spectrum of the solutions during experiments using a 20 mm pathlength DP400 dip probe cuvette (StellarNet Inc., Tampa, FL, USA) placed inside the beaker. The solutions were constantly mixed with an external stirrer (VELP Scientifica, Usmate Velate, Italy) rotating at 100 rpm. Spectra were measured using the SpectraWiz software (StellarNet Inc., Tampa, FL, USA) every 10–20 s for approximately 20–60 min (depending on the experiment). A short MP4 clip showing the experimental setup is available as Supplementary Material. The measurement procedure led to >150–400 data points for each experiment. Data were transformed to comma-separated values (CSV) files, and absorption of the net absorption of CBZ at 286 nm (ε_{286} = 12,826 M^{-1} cm^{-1}) was evaluated and downloaded after correcting the baseline. To allow comparison between parameters in different reaction mechanisms, the "relative dimensionless concentration at time t" [A$_{(t)}$] was evaluated [67] as $C_t/C_0 = OD_t/OD_0$ (the ratio of actual to initial concentration, or actual to initial light absorbance); thus A_0 = 1. Analysis of the data was performed as described in Section 2.2.3.

HPLC Chromatography measurements were kindly performed as outsourcing by Dr. Sara Azarred at Shamir Research Institute, to confirm that quantification using UV-Visible measurements yields indeed reliable results. HPLC measurements indeed confirmed (results not shown) that direct UV-Visible spectroscopy measurements are accurate and effective at the required concentrations range (0.1–5 mg L^{-1}, ~0.5–25 µM) and presence of by-products does not influence the measurements.

The experiments performed included homogenous photocatalysis of CBZ with different concentrations (0.5–2.5 mg L^{-1}, 14.7–73.5 µM) of H_2O_2, heterogeneous photocatalysis of CBZ with various concentrations (0.2–1 mg L^{-1}) of TiO_2, barasym or laponite, and combined hetero-homogeneous photocatalysis of CBZ with 0.5 or 2.0 mg L^{-1} H_2O_2 and heterogeneous catalysts (TiO_2, barasym, laponite) at 0.2 mg L^{-1}.

2.2.3. Analysis of the Data

In order to compare the efficacy of the different photocatalysis processes, an evaluation of the pseudo order, the kinetic rate, and the half-life of each process was performed. Calculations were done using the procedure extensively described in previous studies [51,68]. Considering the rate of change in concentration follows Equation (3):

$$v = \frac{d[A]}{dt} = -k_a[A]^{n_a} \tag{3}$$

where v is the reaction rate, k_a is the apparent rate coefficient, A is the concentration of the pollutant in case, and n_a is the apparent or "pseudo" reaction order [69], the concentration at time t can be calculated if the kinetic rate coefficient k_a and the pseudo-order n_a are known(as long as $n_a \neq 1$), using:

$$[A]_{(t)} = \left(\frac{1}{\frac{1}{[A_0]^{n_a-1}} + (n_a-1)k_a t} \right)^{\frac{1}{n_a-1}} \tag{4}$$

"Half-life time" ($t_{1/2}$), defined by the time it takes for the concentration of a reactant to reach half of its initial value [69], are easy-to-compare parameters, even in cases where pseudo orders are completely different. Half-life times can be evaluated by solving mathematically Equations (3) and (4) to the case were $[A]_{(t)} = 0.5$, yielding for $n_a \neq 1$

$$t_{\frac{1}{2}} = \frac{2^{n_a-1} - 1}{(n_a - 1)k_a[A_0]^{n_a-1}} \tag{5}$$

It should be emphasized that except for "first-order" processes, half-life times strongly depends on the initial concentration, as seen in Equation (5). This should be considered when comparing the efficiency of processes, and the use of a constant initial concentration of pollutants is important.

Pseudo-orders and the kinetic coefficient that exhibits the best fit to each of the treatments were found as described in previous studies by a "bootstrap" [70,71] procedure based on choosing five random sets of 20 values from the several hundreds of data points in each experiment, and fitting the optimal parameters using the Solver tool in Excel® software [68].

3. Results and Discussion

3.1. Adsorption Isotherms

Adsorption isotherms of carbamazepine on the different sorbents are presented in Figure 1A. The adsorption was tested as described in Section 2.1.3 on five clay-based adsorbents, including three raw clays: (bentonite, Ca-SWy1, and sepiolite), and two organoclays prepared as described in Section 2.1.2 (bent-bzk and bent-B1). The adsorption isotherms were also described by the Langmuir adsorption model, and the parameters S_{max} and K_L were evaluated (Table 2). R^2 and RMSE values between the observed and the calculated adsorption results indicate a good fit with the Langmuir model (Table 2, $R^2 > 0.99$). It is important to mention that the following Langmuir models are relevant mainly for the high CBZ adsorption results (Figure 1A). While the adsorption isotherm on bentonite and Ca-SWy1 at low CBZ concentration (Figure 1B) presented a Type V isotherm behavior, that did not fully fit the standard adsorption models and required a more advanced model to describe the slight S-shape measured. For that purpose we tested several models suitable for Type V isotherms as BET, Sips, and Klotz [62]). The only relative improvement in the fit was obtained by the Sips model that is described in Section 2.1.7. It should be emphasized that as shown in Table 2, the improvement in the fit when compared with Langmuir model is minimal, and in the case of the organoclays (BZK- and B1 bentonites) there is no improvement at all, and the exponent in the Sips model (B_s) for those sorbents is close to 1.

Adsorption of CBZ on sepiolite is negligible (Figure 1A). This is not obvious, since several non-charged molecules and oils are adsorbed in large amounts on this clay [72,73]. As for the smectites, organoclay based on BZK exhibited the lowest adsorption, whereas raw bentonite and Ca-SWy-1 adsorb more than 0.5 mmol g^{-1}. Such effect is unusual since smectites usually have a low affinity to non-charged organic molecules. However, it should be emphasized that at low CBZ concentration adsorption to those clays is almost zero (see Figure 1B, which shows adsorption isotherms at low CBZ concentrations), yielding an "S-type" (or "Type V") isotherm [74]. This indicates that the direct interaction between the clay surface and the pollutant is low, and only after obtaining some coverage of the clay surface, adsorption increase. A similar observation was made for several pharmaceuticals in previous studies [40]. On the other hand, bent-B1 maximum adsorption is lower (app. 0.25 mmol g^{-1}, see Figure 1A) but it is also very efficient at low concentrations (see Figure 1B). The affinity of carbamazepine to bent-B1 increased considerably as demonstrated also in a higher calculated Langmuir K_L value, 16.9 compared to 2.89 mM^{-1} on raw bentonite (Table 2). Due to the neutral charge of carbamazepine, it is assumed that more effective adsorption will be observed on the neutral charge particles such as bent-B1 and bent-bzk organoclays. However, higher adsorption capacities were found for montmorillonite clays. Studies reporting adsorption of CBZ on clay minerals mention lower adsorption values as 0.02–0.2 mmol g^{-1} [35,36,39]. Few studies have evaluated the adsorption on modified and organo-clays with even lower adsorbed amounts, from 0–0.05 mmol g^{-1} [75–78].

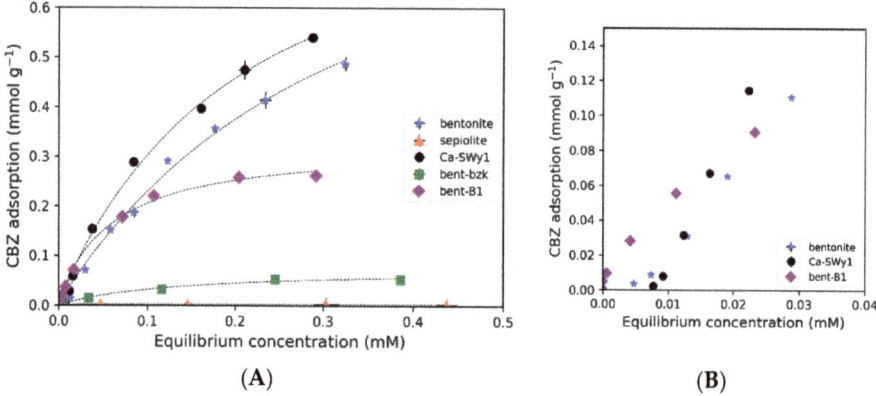

Figure 1. Adsorption isotherms of (**A**) carbamazepine on bentonite (blue asterisk), sepiolite (red triangle), Ca-SWy1 (black circle), bent-bzk (green square), and bent-B1 (purple diamond). Error bars represent triplicate standard deviations. The dashed lines represent Langmuir model predictions according to the estimated parameters detailed in Table 2. (**B**) Isotherms at low concentration of carbamazepine on bentonite, Ca-SWy1, and bent-B1.

Table 2. Langmuir and Sips equations' parameters for carbamazepine adsorption on clays and organoclays.

Adsorbent Type	Model	S_{max} (mmol g^{-1})	K_L (mM^{-1})	B_s	R^2	RMSE
Bentonite	Langmuir	1.023	2.89	-	0.994	0.014
	Sips	0.663	6.41	1.35	0.997	0.009
Ca-SWy1	Langmuir	0.936	4.79	-	0.995	0.015
	Sips	0.721	8.10	1.22	0.996	0.013
Bentonite-Benzalkonium organoclay	Langmuir	0.076	7.51	-	0.987	0.002
	Sips	0.079	6.67	0.95	0.986	0.003
Bentonite-B1 organoclay	Langmuir	0.327	16.91	-	0.997	0.005
	Sips	0.318	18.20	1.05	0.998	0.004

Despite bent-B1 advantage in adsorption at low concentration, some thiamine (B1) was released from the bent-B1 complex during the adsorption process. The concentration of the released B1 was very low (averagely 0.02 mM) with relatively constant values through different CBZ adsorption concentrations. Moreover, thiamine is considered a non-hazardous component that can provide a safe use under high standard regulations [79]. However, such instability in the behavior of the sorbent should be taken into consideration. B1 release interfered with UV Visible measurements, and in order to overcome this problem a simple mathematical spectra separation was performed, based on two well-known spectrums of the components (B1 and carbamazepine) at known concentrations, and calculating the mix spectrum by superposition [80] using the "Solver"® optimization tool of the Microsoft Excel computer program.

3.2. Colloidal Charge

The influence of CBZ adsorption on the electrokinetic colloidal charge of bentonite, Ca-SWy1, and bent-B1 organoclay was evaluated (Figure 2) using a particle charge detector (PCD) as described in Section 2.1.5. The initial colloidal charge of bentonite was significantly more negative than Ca-SWy1 due to the influence of the divalent calcium ions increasing neutralization on the Stern layer. Thus, in raw bentonite, the presence of monovalent sodium ions resulted in a more negative electrokinetic charge of the clay particles. The colloidal charge of the organoclay (bent-B1) was around zero due to the exchange of B1 with the inorganic ions, forming a non-charged surface. Similar values were observed in organoclays with organocations added at amounts near the cation exchange capacity of the clay as in B1- [52], berberine- [81], crystal violet, and tetraphenyl-phosphonium [82] smectites. For the raw bentonite and Ca-SWy-1, the increase in the electrokinetic colloidal charge, making it closer to neutralization, is attributed to the adsorption of the hydrophobic carbamazepine molecule. In raw bentonite, the increase in charge is accompanied by a cationic exchange of sodium with additional calcium arriving from the carbamazepine solution. Those results will be further discussed in the next chapters (Sections 3.3 and 3.4).

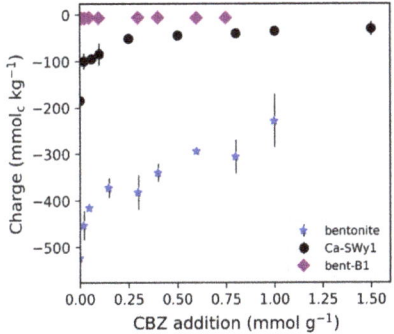

Figure 2. Colloidal charge of bentonite (blue asterisk), Ca-SWy1 (black circle), and bent-B1 organoclay (purple diamond) at 0–1.5 mmol g^{-1} CBZ addition.

3.3. Cation Exchange in Raw Bentonite

The large adsorbed CBZ amounts on raw bentonite, which is, as most natural clay minerals, negatively charged, led to the assumption that CBZ may probably behave as a cation, exchanging other exchangeable cations from the raw clay. To test this assumption, ion exchange processes were examined as part of the adsorption mechanisms evaluation. Raw bentonite is reported to have a cation exchange capacity (CEC) of 0.8 mmole g^{-1}, whereas the cations composition is similar to that measured in SWy-1 and SWy-2 clays, and thus about 30% of the CEC is Na$^+$, while almost all the rest are divalent cations [57]. The release of inorganic cations due to the adsorption of CBZ was measured as described in Section 2.1.6. It is interesting to mention that apparently according to our measurements, carbamazepine as purchased from Sigma may contain traces of calcium in it. This assumption is based on Ca concentrations measured in "pure" CBZ solutions using ICP-OES analysis and reinforced by FTIR measurements reported in Section 3.4 that exhibit a strong peak ascribed to CaCO$_3$. An additional explanation to the presence of such Ca traces in the CBZ stock could be related to a possible contamination during the laboratory work. In any case, Figure 3A represents the release of sodium and calcium ions into the liquid phase as a function of the adsorption of carbamazepine. The release was calculated by subtracting the ions in the initial clay solution from the various solution concentrations after carbamazepine adsorption. The release of exchangeable sodium cations from the bentonite is linearly correlated to the amount of adsorbed carbamazepine (Figure 3A). Despite this linear correlation, CBZ adsorption is not explained by the sodium release and the ratio between carbamazepine adsorption and Na release was higher than 2. Thus, this is not a mere exchange CBZ/Na$^+$. Moreover, since calcium was added to the solution through the carbamazepine stock solution (as mentioned above), a comparison between the calcium addition and the release of sodium was conducted (Figure 3B), resulted in a linear correlation with a 1:1 ratio as for mmole$_c$. Thus, the hypothesis that CBZ exchanges Na$^+$ appears wrong, and a more reasonable explanation for the release of sodium cations is a Na$^+$/Ca^{2+} exchange process (Ca^{2+} coming from the CBZ solution) considering there is a strong preference in clays for divalent cations [83], and the adsorption of carbamazepine did not include ions exchange on the clay's negative sites. To confirm this assumption, we performed the adsorption isotherms on Ca homoionic SWy-1 montmorillonite (Ca-SWy1), where all the CEC was *a-priori* saturated with Ca^{2+}. As shown in Figure 1A CBZ adsorption to Ca-SWy1 reaches similar and even slightly larger values than for raw bentonite, with similar Langmuir and Sips S_{max} values.

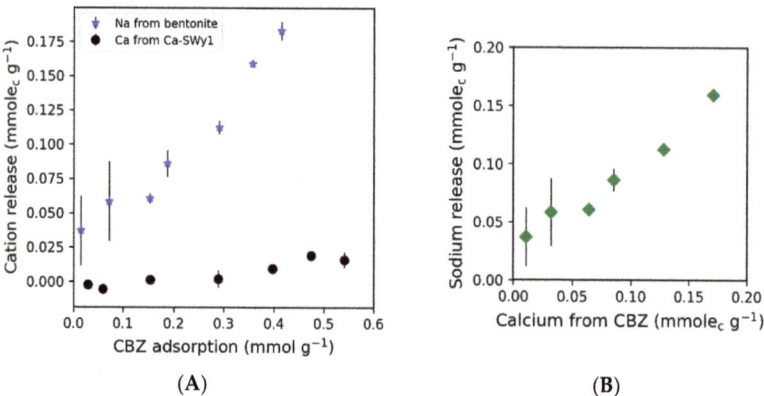

Figure 3. (**A**) Cations released to supernatant from bentonite (blue asterisk) or Ca-SWy1 (black circle) as a function of the adsorbed CBZ. (**B**) Na$^+$ measured due to the addition of Ca^{+2} from CBZ solution (green diamonds).

3.4. FTIR Analysis

One of the hypotheses to explain the relatively large adsorbed amounts of neutral CBZ on negatively charged clays was that as the matter of fact it undergoes degradation on the surface of the mineral, as was observed for example in the case of tri-methyl aryl dyes "adsorbed" on Texas vermiculite (VTx-1) [84]. In order to test that, FTIR spectra of dried CBZ-clays were measured using an ATR device as described in Section 2.1.4. The rationale aimed to confirm and evaluate the adsorption of carbamazepine on the absorbent particles by identifying the relevant structural group in the measured samples, whereas CBZ degradation will lead to different functional group vibrations. Figure 4A shows the spectrum of CBZ, raw bentonite, and CBZ-bentonite at several carbamazepine amounts. CBZ-bentonite samples exhibit five absorption bands that were not observed in raw bentonite, at approximately 1640, 1570, 1490, and 1460–1435 cm^{-1}. Those absorption bands were correlated to typical peaks of functional groups of CBZ carbamazepine, as observed in the carbamazepine spectrum sample, and are known from the literature [85]. The only absorption band in raw bentonite in this region is at ~1630 cm^{-1} and ascribed to O-H deformation of hygroscopic water [86]. While the three CBZ absorption bands in the range 1400–1500 cm^{-1} (ascribed to C=C vibrations) are in almost the same place for adsorbed and raw CBZ, bands ascribed to C-N bond (at app. 1600 cm^{-1}) and to the amide group (NH$_2$ scissoring/C=O stretching, at ~1670 cm^{-1}) appear shifted to lower energies. This may indicate that the interaction between CBZ molecules and the clay is via the amide group.

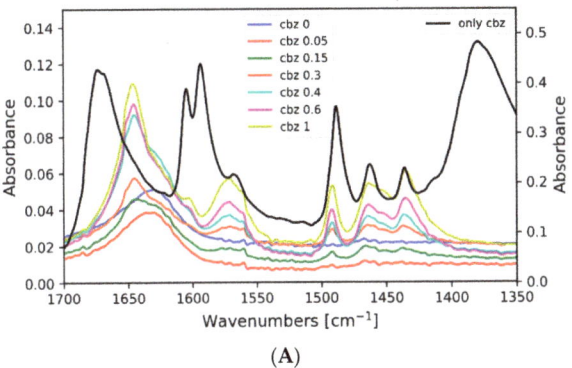 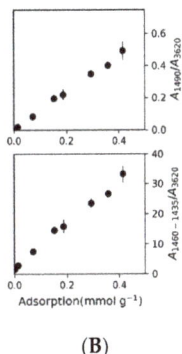

Figure 4. (**A**) ATR-FTIR spectra of CBZ (black, OD values on the right axis), bentonite raw clay (blue), and bentonite with added CBZ (in mmol g^{-1}) as denoted in the legend (OD values on the left axis). (**B**) Normalized absorption bands' height at 1490 and area at 1460–1435 cm^{-1} compared to the adsorption results as measured by the mass balance during the adsorption experiments.

An increase in all absorption bands in the range 1400–1700 cm^{-1} is observed accordingly to the adsorption process as measured in the experiment (Figure 4A). Relative quantification of the CBZ absorbed can be performed according to Section 2.1.4, by calculating the ratio between the intensity or the area of CBZ absorption bands, to that of the structural O-H band related to the clay at 3620 cm^{-1}, where CBZ has no absorption at all. Subsequently, those ratios were compared to the amount of carbamazepine adsorbed to the clay as measured in the previous adsorption experiment (Figure 1). Figure 4B represents the comparison of two normalized absorption bands to the adsorption results, including the absorption bands' height at 1490 and their area at 1460–1435 cm^{-1}. The linear correlation with very good fit ($R^2 > 0.99$) between the two data sets confirms the adsorption process as the reason for CBZ decrease in the equilibrium solution.

An additional absorption band was observed in the raw carbamazepine spectrum at approximately 1370 cm^{-1} and was not observed in any of the CBZ-bentonite samples (Figure 4A). According to the literature, these absorption bands may represent the presence of calcite ($CaCO_3$) [86] in the carbamazepine or as a consequence of Ca impurities in our stock solution as described above. Presence of calcite is confirmed by a small and sharp absorption band at 870 cm^{-1} (results not shown) in the CBZ spectrum. The absence of this peak in all CBZ-bentonite spectra indicates that calcite was released to the liquid phase, as correlated to the increase in calcium ions that was observed after adsorption (Figure 3). The stability of the CBZ adsorption was examined by a release test (as described in Section 2.1.4). The results of the washed and original samples were compared, and no significant differences were observed in the absorption bands and the calculated ratios (results not shown). Hence, the adsorptions of CBZ to bentonite, Ca-SWy1, and bent-B1 were confirmed as stable processes without unexpected CBZ release.

3.5. Photocatalytic Degradation of Carbamazepine

3.5.1. Homogenous Photocatalysis with Increasing Concentrations of H_2O_2

Figure 5 shows the photodegradation of a 21.2 µM (5 mg L^{-1}) CBZ solution, under UVC irradiation as described in Section 2.2.2, with 0.5–2.5 mg L^{-1} (14.7–73.5 µM) H_2O_2 as homogenous catalysts. Such H_2O_2 concentration is in the range that was used recently for photodegradation of caffeine [65], but considerably lower than was usually used for photo- [87] or radio-catalysis [88] of CBZ. It can be seen that CBZ is stable under UVC radiation and does not undergo any photolysis. At the initial CBZ concentration used in this study (21.2 µM) very low H_2O_2 concentration (0.5 mg L^{-1}) exhibits almost no degradation, as in photolysis. Increasing the H_2O_2 concentrations to 1.0 mg L^{-1} changes the pseudo-order as evaluated using Equations (3) and (4) from n = 0 to n = 0.75, while $t_{1/2}$ as evaluated using Equation (5) changes from 219 to 15.6 min. Further increase of H_2O_2 to 2 or 2.5 mg L^{-1} reduces $t_{1/2}$ further to 7.6 and 6.4 min, respectively. Pseudo orders and half lifetimes of all experiments are shown also in Table 3.

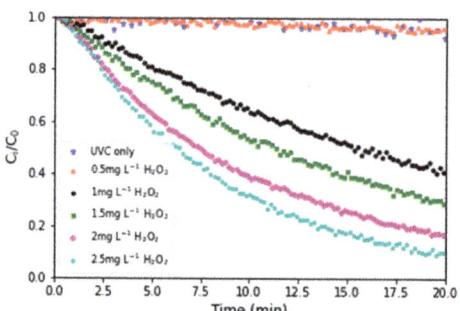

Figure 5. Photodegradation of 21.2 μM (5 mg L^{-1}) carbamazepine solution under UVC irradiation at several concentrations of hydrogen peroxide.

Table 3. Pseudo-orders and half-life times for photodegradation experiments.

Heterogeneous Catalyst		H_2O_2 (mg L^{-1})	Pseudo-Order n_a	Half-Life $t_{1/2}$ (min)
Type	(mg L^{-1})			
None	0	0	0	212.1 ± 1.56%
		0.5	0	219.3 ± 2.77%
		1.0	0.76 ± 4.31%	15.6 ± 0.62%
		1.5	0.93 ± 3.39%	11.4 ± 0.89%
		2.0	1.01 ± 2.65%	7.60 ± 1.00%
		2.5	0.82 ± 2.13%	6.39 ± 0.85%
TiO$_2$,	0.2	0	0	189.31 ± 2.53%
	0.4	0	0	122.2 ± 1.97%
	1	0	0	120.7 ± 2.66%
	0.2	0.5	0	68.0 ± 0.86%
	0.2	2.0	0.90 ± 3.10%	5.90 ± 1.39%
Barasym	0.2	0	0	296.7 ± 1.68%
	1.0	0	0	215.9 ± 1.74%
	0.2	0.5	1.24 ± 5.24%	33.2 ± 0.41%
	0.2	2.0	0.84 ± 1.99%	6.90 ± 0.92%
Laponite	0.2	0	No degradation	-
	1.0	0	No degradation	-
	0.2	0.5	1.04 ± 4.27%	37.0 ± 0.54%
	0.2	2.0	0.82 ± 3.27%	6.80 ± 1.03%

3.5.2. Heterogenous Photocatalysis with TiO$_2$, Barasym and Laponite

Table 3 shows pseudo orders and half-lifetime for the photodegradation of a 21.2 μM (5 mg/L) CBZ solution, under UVC irradiation, with 0–1 mg L^{-1} of commercial catalytic grade TiO$_2$, Barasym SSM-100 (synthetic montmorillonite) or Laponite® (synthetic hectorite) as heterogeneous catalysts. Synthetic clay minerals were chosen to avoid impurities present in natural minerals. High quality catalytic grade TiO$_2$ was chosen as a "gold standard" since it is widely used for the photodegradation of organic refractory pollutants [89]. In most heterogeneous photocatalysis studies, the catalyst concentration is from tens to thousands mg L^{-1} [90]. We chose to test relatively low concentrations of 0.2, 0.4 and 1 mg L^{-1}, based on our previous studies with BPS and ofloxacin [51,68]. It can be seen that the clay minerals when added alone have almost no effect (Table 3). TiO$_2$ indeed has some photocatalytic effect, but even at a 1 mg L^{-1} is not very effective ($t_{1/2}$ = 121 min, n = 0). Previous studies dealing with CBZ photodegradation used three orders of magnitude higher concentrations of TiO$_2$ as a heterogeneous catalyst and obtained half-life times of 10–20 min — considerably shorter than the present study [91,92].

3.5.3. Hetero-Homogeneous Photocatalysis

In previous studies [51,68], we have shown that a combination of low concentrations of both heterogeneous and homogeneous catalysts may yield synergistic effects and speed up the photodegradation of priority pollutants such as BPS or ofloxacin. To test this effect on CBZ, we performed photodegradation experiments of a 21.2 µM (5 mg L^{-1}) CBZ solution (Figure 6), under UVC irradiation, with a low concentration (0.2 mg L^{-1}) of the heterogeneous catalysts used in Section 3.5.2, at two hydrogen peroxide levels: (a) high (2.0 mg L^{-1}) and (b) low (0.5 mg L^{-1}).

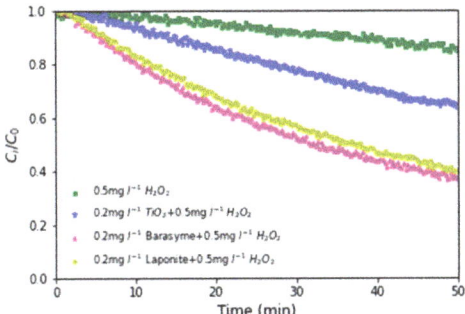

Figure 6. Photodegradation of a 21.2 µM (5 mg L^{-1}) carbamazepine under UVC irradiation, with 0.5 mg L^{-1} H$_2$O$_2$ alone or combined with 0.2 mg L^{-1} TiO$_2$, barasym or laponite.

At 2 mg L^{-1} H$_2$O$_2$ concentration (results summarized in Table 3) the homogeneous catalyst already yields low $t_{1/2}$ values of less than 8 min. The addition of clays as heterogeneous catalysts makes almost no difference. As for TiO$_2$, it should be emphasized that when added alone at a low concentration (0.2 mg L^{-1}) almost no degradation is observed (Table 3), but its addition to 2.0 mg L^{-1} H$_2$O$_2$ slightly speeds up the process ($t_{1/2}$ changes from 6.4 to 5.9), and a small change in the pseudo-order is observed.

At a low homogeneous catalyst concentration of 0.5 mg L^{-1}, the influence of all heterogeneous catalysts is significant (Table 3): While with no heterogeneous catalyst $t_{1/2}$ at that hydrogen peroxide amount is about 219 min, the addition of 0.2 mg L^{-1} of TiO$_2$, barasym or laponite lowers $t_{1/2}$ to 68.0, 33.2 and 37.0 min, respectively. The pseudo-order also changes, especially for the clay minerals.

4. Conclusions

CBZ removal at relatively large (>1 mM) concentrations in industrial effluents and nanofiltration brines, or at very low (<10 µM) concentrations, should be removed in order to enable water reuse. Montmorillonite clays and organoclays may provide an efficient solution for adsorption for industrial effluents containing relatively high carbamazepine concentrations. Batch experiments have shown different adsorption capabilities of carbamazepine on various montmorillonite-based adsorbents, and in raw bentonite it might reach 0.5 mmole g^{-1}. The adsorption process was confirmed by ATR-FTIR analysis on the clay particles, and CBZ desorption was not observed, validating the stability of the sorbent-CBZ complexes. While B1-bentonite organoclay exhibits high affinity at low concentrations, an S-shape isotherm was observed for the raw clays, indicating low affinity at low adsorbed CBZ. However, the maximum capacity of raw montmorillonites is higher than for organoclay. In raw montmorillonites apparently, the initial coverage of the surface with CBZ molecules promotes enhanced adsorption of additional molecules probably by π-π interactions. According to the results, CBZ adsorption occurred only on clay surfaces, and the pollutant does not enter the internal pores of acicular clays as sepiolite. The high adsorption capability to the smectite raw clays on one hand and the low affinity at low concentrations, on the other hand, may provide a good solution for high concentrated contaminated solutions such as pharmaceutical wastewater or nano/microfiltration brine. Moreover, a combined implementation of raw montmorillonite clay and bent-B1 organoclay together can offer the advantages of raw clay for the high concentration and organoclay for the low carbamazepine concentrations.

As for AOP processes, it should be emphasized that since such processes are very specific, the description hereby focuses on the conditions of this study as for radiation intensity and CBZ concentrations. This study focuses on a relatively high initial concentration for municipal wastewater, even though there are several reports reaching such levels [7,93]. Results indicate that direct photolysis with no additional catalyst does not yield CBZ degradation. The addition of hydrogen peroxide as homogeneous photocatalysts is very effective, but only above certain levels. ($H_2O_2 > 1$ mg L^{-1}). The heterogeneous catalysts tested (including the catalytic grade TiO_2) at concentrations ranging 0–1 mg L^{-1} do not yield almost any CBZ degradation. Combined with high concentrations of H_2O_2 (2.0 mg L^{-1}) very effective degradation is observed, but this is mostly due to the homogeneous process. However, when combining a low amount of heterogeneous catalysts with very low H_2O_2 amounts (0.5 mg L^{-1}) a synergistic effect is observed, and two treatments that each of them by itself are completely ineffective lead to a relatively effective process. The advantage of low amounts of catalysts is obvious, considering that the reuse of water will require the removal of the remaining catalysts—both H_2O_2, and clays or TiO_2. It is worthwhile to emphasize that the by-products in the photodegradation process were not measured and identified at this current research stage, although previous studies have shown their presence during CBZ degradation [94,95]. Additional study is required in order to evaluate the by-products and will be conducted in the near future using LCMS-MS.

The search for more effective water techniques is driving researchers toward AOPs. However, we should recall the limitations of AOPs in general and photocatalytic water treatment devices in particular: Since processes are very specific, the challenge of dealing efficiently with multiple low-concentration priority pollutants is far from being achieved. E.L. Cates [96] strongly criticizes the pursue for "new applications, improved catalysts, and reaction mechanisms", and summarizes that "it is time to stop patting ourselves on the back for laboratory 'successes' that clearly turn a blind eye on fatal implementation hurdles". In this sense, probably a combination of processes such as adsorption on specifically tailored sorbents followed by advanced oxidation devices (or vice versa) may yield more efficient and feasible water treatment technologies for both industrial and municipal wastewater.

Supplementary Materials: The following supporting information can be downloaded at: https://www.mdpi.com/article/10.3390/w14132047/s1.

Author Contributions: Conceptualization, G.R.; methodology, G.R., I.L. and Y.S.; software, I.L. and Y.S.; validation, G.R., I.L. and Y.S.; formal analysis, G.R., I.L. and Y.S.; investigation, G.R., I.L. and Y.S.; resources, G.R.; data curation, I.L. and Y.S.; writing—original draft preparation, G.R., I.L. and Y.S.; writing—review and editing, I.L. and G.R.; visualization, G.R., I.L. and Y.S.; supervision, G.R.; project administration, G.R.; funding acquisition, G.R. All authors have read and agreed to the published version of the manuscript.

Funding: This research was partially funded by CSO-MOH (Israeli Ministry of Health), in the frame of the collaborative international consortium (REWA) financed under the 2020 AquaticPollutants Joint call of the AquaticPollutants ERA-NET Cofund (GA N° 869178).

Institutional Review Board Statement: Not applicable.

Informed Consent Statement: Not applicable.

Data Availability Statement: Additional details on the raw data can be obtained by contacting the authors.

Acknowledgments: The authors would like to thank the European Commission and AKA (Finland), CSO-MOH (Israel), IFD (Denmark) and WRC (South Africa) for funding in the frame REWA international consortium (additional details in "funding" paragraph). REWA is an integral part of the activities developed by the Water, Oceans and AMR JPIs. The authors are also thankful to Chen Barak for all the technical support, Sara Azerrad (from the Shamir Research Institute) for the HPLC confirmation measurements, and the whole team of the Hydrogeology and Examination of Soil Fertility Lab at MIGAL Research Institute.

Conflicts of Interest: The authors declare no conflict of interest.

References

1. Clara, M.; Strenn, B.; Kreuzinger, N. Carbamazepine as a possible anthropogenic marker in the aquatic environment: Investigations on the behaviour of Carbamazepine in wastewater treatment and during groundwater infiltration. *Water Res.* **2004**, *38*, 947–954. [CrossRef]
2. Bahlmann, A.; Brack, W.; Schneider, R.J.; Krauss, M. Carbamazepine and its metabolites in wastewater: Analytical pitfalls and occurrence in Germany and Portugal. *Water Res.* **2014**, *57*, 104–114. [CrossRef]
3. Ekpeghere, K.I.; Sim, W.J.; Lee, H.J.; Oh, J.E. Occurrence and distribution of carbamazepine, nicotine, estrogenic compounds, and their transformation products in wastewater from various treatment plants and the aquatic environment. *Sci. Total Environ.* **2018**, *640–641*, 1015–1023. [CrossRef]
4. Bahlmann, A.; Carvalho, J.J.; Weller, M.G.; Panne, U.; Schneider, R.J. Immunoassays as high-throughput tools: Monitoring spatial and temporal variations of carbamazepine, caffeine and cetirizine in surface and wastewaters. *Chemosphere* **2012**, *89*, 1278–1286. [CrossRef]
5. Dvory, N.Z.; Livshitz, Y.; Kuznetsov, M.; Adar, E.; Gasser, G.; Pankratov, I.; Lev, O.; Yakirevich, A. Caffeine vs. carbamazepine as indicators of wastewater pollution in a karst aquifer. *Hydrol. Earth Syst. Sci.* **2018**, *22*, 6371–6381. [CrossRef]
6. Li, W.C. Occurrence, sources, and fate of pharmaceuticals in aquatic environment and soil. *Environ. Pollut.* **2014**, *187*, 193–201. [CrossRef]
7. Hai, F.I.; Yang, S.; Asif, M.B.; Sencadas, V.; Shawkat, S.; Sanderson-Smith, M.; Gorman, J.; Xu, Z.Q.; Yamamoto, K. Carbamazepine as a Possible Anthropogenic Marker in Water: Occurrences, Toxicological Effects, Regulations and Removal by Wastewater Treatment Technologies. *Water* **2018**, *10*, 107. [CrossRef]
8. Cummings, C.; Stewart, M.; Stevenson, M.; Morrow, J.; Nelson, J. Neurodevelopment of children exposed in utero to lamotrigine, sodium valproate and carbamazepine. *Arch. Dis. Child.* **2011**, *96*, 643–647. [CrossRef]
9. Jentink, J.; Dolk, H.; Loane, M.A.; Morris, J.K.; Wellesley, D.; Garne, E.; De Jong-van Den Berg, L. Intrauterine exposure to carbamazepine and specific congenital malformations: Systematic review and case-control study. *BMJ* **2010**, *341*, 1261. [CrossRef]
10. Atkinson, D.E.; Brice-Bennett, S.; D'Souza, S.W. Antiepileptic medication during pregnancy: Does fetal genotype affect outcome? *Pediatr. Res.* **2007**, *62*, 120–127. [CrossRef]
11. Hillis, D.G.; Antunes, P.; Sibley, P.K.; Klironomos, J.N.; Solomon, K.R. Structural responses of Daucus carota root-organ cultures and the arbuscular mycorrhizal fungus, Glomus intraradices, to 12 pharmaceuticals. *Chemosphere* **2008**, *73*, 344–352. [CrossRef]
12. van den Brandhof, E.J.; Montforts, M. Fish embryo toxicity of carbamazepine, diclofenac and metoprolol. *Ecotoxicol. Environ. Saf.* **2010**, *73*, 1862–1866. [CrossRef]
13. Han, G.H.; Hur, H.G.; Kim, S.D. Ecotoxicological risk of pharmaceuticals from wastewater treatment plants in Korea: Occurrence and toxicity to Daphnia magna. *Environ. Toxicol. Chem.* **2006**, *25*, 265–271. [CrossRef]
14. Jos, A.; Repetto, G.; Rios, J.C.; Hazen, M.J.; Molero, M.L.; Del Peso, A.; Salguero, M.; Fernández-Freire, P.; Pérez-Martín, J.M.; Cameán, A. Ecotoxicological evaluation of carbamazepine using six different model systems with eighteen endpoints. *Toxicol. Vitr.* **2003**, *17*, 525–532. [CrossRef]
15. Ferrari, B.; Paxéus, N.; Giudice, R.L.; Pollio, A.; Garric, J. Ecotoxicological impact of pharmaceuticals found in treated wastewaters: Study of carbamazepine, clofibric acid, and diclofenac. *Ecotoxicol. Environ. Saf.* **2003**, *55*, 359–370. [CrossRef]
16. Lofgren, H.; De Boer, R. Pharmaceuticals in Australia: Developments in regulation and governance. *Soc. Sci. Med.* **2004**, *58*, 2397–2407. [CrossRef]
17. Vergili, I. Application of nanofiltration for the removal of carbamazepine, diclofenac and ibuprofen from drinking water sources. *J. Environ. Manag.* **2013**, *127*, 177–187. [CrossRef]
18. Serrano, D.; Suárez, S.; Lema, J.M.; Omil, F. Removal of persistent pharmaceutical micropollutants from sewage by addition of PAC in a sequential membrane bioreactor. *Water Res.* **2011**, *45*, 5323–5333. [CrossRef]
19. Li, X.; Hai, F.I.; Nghiem, L.D. Simultaneous activated carbon adsorption within a membrane bioreactor for an enhanced micropollutant removal. *Bioresour. Technol.* **2011**, *102*, 5319–5324. [CrossRef]
20. Tiwari, B.; Sellamuthu, B.; Ouarda, Y.; Drogui, P.; Tyagi, R.D.; Buelna, G. Review on fate and mechanism of removal of pharmaceutical pollutants from wastewater using biological approach. *Bioresour. Technol.* **2017**, *224*, 1–12. [CrossRef]

21. Marco-Urrea, E.; Pérez-Trujillo, M.; Vicent, T.; Caminal, G. Ability of white-rot fungi to remove selected pharmaceuticals and identification of degradation products of ibuprofen by Trametes versicolor. *Chemosphere* **2009**, *74*, 765–772. [CrossRef]
22. Zhang, Y.; Geißen, S.U. Elimination of carbamazepine in a non-sterile fungal bioreactor. *Bioresour. Technol.* **2012**, *112*, 221–227. [CrossRef]
23. Pylypchuk, I.V.; Daniel, G.; Kessler, V.G.; Seisenbaeva, G.A. Removal of diclofenac, paracetamol, and carbamazepine from model aqueous solutions by magnetic sol–gel encapsulated horseradish peroxidase and lignin peroxidase composites. *Nanomaterials* **2020**, *10*, 282. [CrossRef]
24. Bellona, C.; Drewes, J.E.; Oelker, G.; Luna, J.; Filteau, G.; Amy, G. Comparing nanofiltration and reverse osmosis for drinking water augmentation. *J. Am. Water Works Assoc.* **2008**, *100*, 102–116. [CrossRef]
25. Radjenović, J.; Petrović, M.; Ventura, F.; Barceló, D. Rejection of pharmaceuticals in nanofiltration and reverse osmosis membrane drinking water treatment. *Water Res.* **2008**, *42*, 3601–3610. [CrossRef]
26. Li, Y.; Yang, Y.; Lei, J.; Liu, W.; Tong, M.; Liang, J. The degradation pathways of carbamazepine in advanced oxidation process: A mini review coupled with DFT calculation. *Sci. Total Environ.* **2021**, *779*, 146498. [CrossRef]
27. Alharbi, S.K.; Price, W.E. Degradation and Fate of Pharmaceutically Active Contaminants by Advanced Oxidation Processes. *Curr. Pollut. Reports* **2017**, *3*, 268–280. [CrossRef]
28. Simon, A.; Price, W.E.; Nghiem, L.D. Effects of chemical cleaning on the nanofiltration of pharmaceutically active compounds (PhACs). *Sep. Purif. Technol.* **2012**, *88*, 208–215. [CrossRef]
29. Comerton, A.M.; Andrews, R.C.; Bagley, D.M. The influence of natural organic matter and cations on the rejection of endocrine disrupting and pharmaceutically active compounds by nanofiltration. *Water Res.* **2009**, *43*, 613–622. [CrossRef]
30. Hajibabania, S.; Verliefde, A.; Drewes, J.E.; Nghiem, L.D.; McDonald, J.; Khan, S.; Le-Clech, P. Effect of fouling on removal of trace organic compounds by nanofiltration. *Drink. Water Eng. Sci.* **2011**, *4*, 71–82. [CrossRef]
31. Alharbi, S.K.; Kang, J.; Nghiem, L.D.; van de Merwe, J.P.; Leusch, F.D.L.; Price, W.E. Photolysis and UV/H2O2 of diclofenac, sulfamethoxazole, carbamazepine, and trimethoprim: Identification of their major degradation products by ESI–LC–MS and assessment of the toxicity of reaction mixtures. *Process. Saf. Environ. Prot.* **2017**, *112*, 222–234. [CrossRef]
32. Décima, M.A.; Marzeddu, S.; Barchiesi, M.; Di Marcantonio, C.; Chiavola, A.; Boni, M.R. A review on the removal of carbamazepine from aqueous solution by using activated carbon and biochar. *Sustainability* **2021**, *13*, 11760. [CrossRef]
33. Baghdadi, M.; Ghaffari, E.; Aminzadeh, B. Removal of carbamazepine from municipal wastewater effluent using optimally synthesized magnetic activated carbon: Adsorption and sedimentation kinetic studies. *J. Environ. Chem. Eng.* **2016**, *4*, 3309–3321. [CrossRef]
34. To, M.H.; Hadi, P.; Hui, C.W.; Lin, C.S.K.; McKay, G. Mechanistic study of atenolol, acebutolol and carbamazepine adsorption on waste biomass derived activated carbon. *J. Mol. Liq.* **2017**, *241*, 386–398. [CrossRef]
35. Alaghmand, M.; Alizadeh-Saei, J.; Barakat, S. Adsorption and removal of a selected emerging contaminant, carbamazepine, using Humic acid, Humasorb and Montmorillonite. Equilibrium isotherms, kinetics and effect of the water matrix. *J. Environ. Sci. Health Part A Toxic/Hazard. Subst. Environ. Eng.* **2020**, *55*, 1534–1541. [CrossRef]
36. Kryuchkova, M.; Batasheva, S.; Akhatova, F.; Babaev, V.; Buzyurova, D.; Vikulina, A.; Volodkin, D.; Fakhrullin, R.; Rozhina, E. Pharmaceuticals removal by adsorption with montmorillonite nanoclay. *Int. J. Mol. Sci.* **2021**, *22*, 9670. [CrossRef]
37. Thiebault, T.; Boussafir, M.; Fougère, L.; Destandau, E.; Monnin, L.; Le Milbeau, C. Clay minerals for the removal of pharmaceuticals: Initial investigations of their adsorption properties in real wastewater effluents. *Environ. Nanotechnol. Monit. Manag.* **2019**, *12*, 100266. [CrossRef]
38. Özçelik, G.; Bilgin, M.; Şahin, S. Carbamazepine sorption characteristics onto bentonite clay: Box-Behnken process design. *Sustain. Chem. Pharm.* **2020**, *18*, 100323. [CrossRef]
39. Mahouachi, L.; Rastogi, T.; Palm, W.U.; Ghorbel-Abid, I.; Ben Hassen Chehimi, D.; Kümmerer, K. Natural clay as a sorbent to remove pharmaceutical micropollutants from wastewater. *Chemosphere* **2020**, *258*, 127213. [CrossRef]
40. Khazri, H.; Ghorbel-Abid, I.; Kalfat, R.; Trabelsi-Ayadi, M. Removal of ibuprofen, naproxen and carbamazepine in aqueous solution onto natural clay: Equilibrium, kinetics, and thermodynamic study. *Appl. Water Sci.* **2017**, *7*, 3031–3040. [CrossRef]
41. Rytwo, G. Securing The Future: Clay-Based Solutions For a Comprehensive and Sustainable Potable-Water Supply System. *Clays Clay Miner.* **2018**, *66*, 315–328. [CrossRef]
42. Haroune, L.; Salaun, M.; Ménard, A.; Legault, C.Y.; Bellenger, J.P. Photocatalytic degradation of carbamazepine and three derivatives using TiO2 and ZnO: Effect of pH, ionic strength, and natural organic matter. *Sci. Total Environ.* **2014**, *475*, 16–22. [CrossRef] [PubMed]
43. Bo, L.; Liu, H.; Han, H. Photocatalytic degradation of trace carbamazepine in river water under solar irradiation. *J. Environ. Manag.* **2019**, *241*, 131–137. [CrossRef] [PubMed]
44. Li, C.; Zhu, N.; Yang, S.; He, X.; Zheng, S.; Sun, Z.; Dionysiou, D.D. A review of clay based photocatalysts: Role of phyllosilicate mineral in interfacial assembly, microstructure control and performance regulation. *Chemosphere* **2021**, *273*, 129723. [CrossRef] [PubMed]
45. Ong, C.B.; Ng, L.Y.; Mohammad, A.W. A review of ZnO nanoparticles as solar photocatalysts: Synthesis, mechanisms and applications. *Renew. Sustain. Energy Rev.* **2018**, *81*, 536–551. [CrossRef]
46. Ning, F.; Shao, M.; Xu, S.; Fu, Y.; Zhang, R.; Wei, M.; Evans, D.G.; Duan, X. TiO$_2$/graphene/NiFe-layered double hydroxide nanorod array photoanodes for efficient photoelectrochemical water splitting. *Energy Environ. Sci.* **2016**, *9*, 2633–2643. [CrossRef]

47. Chong, M.N.; Jin, B.; Laera, G.; Saint, C.P. Evaluating the photodegradation of Carbamazepine in a sequential batch photoreactor system: Impacts of effluent organic matter and inorganic ions. *Chem. Eng. J.* **2011**, *174*, 595–602. [CrossRef]
48. Vimonses, V.; Jin, B.; Chow, C.W.K.; Saint, C. An adsorption-photocatalysis hybrid process using multi-functional-nanoporous materials for wastewater reclamation. *Water Res.* **2010**, *44*, 5385–5397. [CrossRef]
49. Zou, Y.; Hu, Y.; Shen, Z.; Yao, L.; Tang, D.; Zhang, S.; Wang, S.; Hu, B.; Zhao, G.; Wang, X. Application of aluminosilicate clay mineral-based composites in photocatalysis. *J. Environ. Sci.* **2022**, *115*, 190–214. [CrossRef]
50. Pérez-Carvajal, J.; Aranda, P.; Obregón, S.; Colón, G.; Ruiz-Hitzky, E. TiO2-clay based nanoarchitectures for enhanced photocatalytic hydrogen production. *Microporous Mesoporous Mater.* **2016**, *222*, 120–127. [CrossRef]
51. Rytwo, G.; Levy, S.; Shahar, Y.; Lotan, I.; Zelkind, A.L.; Klein, T.; Barak, C. Health protection using clay minerals: A case study based on the removal of BPA and BPS from water. *Clays Clay Miner.* **2022**, *69*, 641–653. [CrossRef]
52. Ben Moshe, S.; Rytwo, G. Thiamine-based organoclay for phenol removal from water. *Appl. Clay Sci.* **2018**, *155*, 50–56. [CrossRef]
53. Gonen, Y.; Rytwo, G. Using the dual-mode model to describe adsorption of organic pollutants onto an organoclay. *J. Colloid Interface Sci.* **2006**, *299*, 95–101. [CrossRef] [PubMed]
54. Ohtani, B.; Prieto-Mahaney, O.O.; Li, D.; Abe, R. What is Degussa (Evonik) P25? Crystalline composition analysis, reconstruction from isolated pure particles and photocatalytic activity test. *J. Photochem. Photobiol. A Chem.* **2010**, *216*, 179–182. [CrossRef]
55. Schneider, J.; Matsuoka, M.; Takeuchi, M.; Zhang, J.; Horiuchi, Y.; Anpo, M.; Bahnemann, D.W. Understanding TiO$_2$ Photocatalysis Mechanisms and Materials. *Chem. Rev.* **2014**, *114*, 9919–9986. [CrossRef] [PubMed]
56. Alonso-Tellez, A.; Masson, R.; Robert, D.; Keller, N.; Keller, V. Comparison of Hombikat UV100 and P25 TiO$_2$ performance in gas-phase photocatalytic oxidation reactions. *J. Photochem. Photobiol. A Chem.* **2012**, *250*, 58–65. [CrossRef]
57. Rytwo, G.; Serban, C.; Nir, S.; Margulies, L. Use of methylene blue and crystal violet for determination of exchangeable cations in montmorillonite. *Clays Clay Miner.* **1991**, *39*, 551–555. [CrossRef]
58. Rytwo, G.; Zakai, R.; Wicklein, B. The Use of ATR-FTIR Spectroscopy for Quantification of Adsorbed Compounds. *J. Spectrosc.* **2015**, *2015*, 727595. [CrossRef]
59. Dultz, S.; Rytwo, G. Effects of different organic cations on the electrokinetic surface charge from organo-montmorillonites—consequences for the adsorption properties. In Proceedings of the Clays of Geotechnical and Economical Interest. Swiss, Austrian and German Clay Group—DTTG Annual Meeting, Celle, Germany, 5 October 2005; Volume 11, pp. 6–14.
60. Rytwo, G.; Chorsheed, L.L.; Avidan, L.; Lavi, R. Three Unusual Techniques for the Analysis of Surface Modification of Clays and Nanocomposites. In *Surface Modification of Clays and Nanocomposites*; Beall, G., Ed.; Clay Minerals Society: Boulder, CO, USA, 2016; Volume 20, pp. 73–86. ISBN 9781881208457.
61. Vieth, W.R.; Sladek, K.J. A model for diffusion in a glassy polymer. *J. Colloid Sci.* **1965**, *20*, 1014–1033. [CrossRef]
62. Buttersack, C. Modeling of type IV and V sigmoidal adsorption isotherms. *Phys. Chem. Chem. Phys.* **2019**, *21*, 5614–5626. [CrossRef]
63. Wang, J.; Guo, X. Adsorption isotherm models: Classification, physical meaning, application and solving method. *Chemosphere* **2020**, *258*, 127279. [CrossRef] [PubMed]
64. Al-Ghouti, M.A.; Da'ana, D.A. Guidelines for the use and interpretation of adsorption isotherm models: A review. *J. Hazard. Mater.* **2020**, *393*, 122383. [CrossRef] [PubMed]
65. Rendel, P.; Rytwo, G. Degradation kinetics of caffeine in water by UV/H$_2$O$_2$ and UV/TiO$_2$. *Desalin. Water Treat.* **2020**, *173*, 231–242. [CrossRef]
66. Rendel, P.M.; Rytwo, G. The Effect of Electrolytes on the Photodegradation Kinetics of Caffeine. *Catalysts* **2020**, *10*, 644. [CrossRef]
67. Rytwo, G.; Klein, T.; Margalit, S.; Mor, O.; Naftaly, A.; Daskal, G. A continuous-flow device for photocatalytic degradation and full mineralization of priority pollutants in water. *Desalin. Water Treat.* **2015**, *57*, 16424–16434. [CrossRef]
68. Rytwo, G.; Zelkind, A.L. Evaluation of Kinetic Pseudo-Order in the Photocatalytic Degradation of Ofloxacin. *Catalysts* **2022**, *12*, 24. [CrossRef]
69. Atkins, P.; de Paula, J. *Physical Chemistry*; W.H.Freemand and Co.: New York, NY, USA, 2006; Volume 8, ISBN 0-7167-8759-8.
70. Efron, B. Bootstrap Methods: Another Look at the Jackknife. *Ann. Stat.* **1979**, *7*, 569–593. [CrossRef]
71. Mishra, D.K.; Dolan, K.D.; Yang, L. Bootstrap confidence intervals for the kinetic parameters of degradation of anthocyanins in grape pomace. *J. Food Process Eng.* **2011**, *34*, 1220–1233. [CrossRef]
72. Zadaka-Amir, D.; Bleimann, N.; Mishael, Y.G. Sepiolite as an effective natural porous adsorbent for surface oil-spill. *Microporous Mesoporous Mater.* **2013**, *169*, 153–159. [CrossRef]
73. Gutman, R.; Rauch, M.; Neuman, A.; Khamaisi, H.; Jonas-Levi, A.; Konovalova, Y.; Rytwo, G. Sepiolite Clay Attenuates the Development of Hypercholesterolemia and Obesity in Mice Fed a High-Fat High-Cholesterol Diet. *J. Med. Food* **2019**, *23*, 289–296. [CrossRef]
74. Sparks, D.L. *Environmental Soil Chemistry*; Academic Press: Cambridge, MA, USA, 1995; ISBN 9780126564457.
75. De Oliveira, T.; Boussafir, M.; Fougère, L.; Destandau, E.; Sugahara, Y.; Guégan, R. Use of a clay mineral and its nonionic and cationic organoclay derivatives for the removal of pharmaceuticals from rural wastewater effluents. *Chemosphere* **2020**, *259*, 127480. [CrossRef] [PubMed]
76. Guégan, R.; De Oliveira, T.; Le Gleuher, J.; Sugahara, Y. Tuning down the environmental interests of organoclays for emerging pollutants: Pharmaceuticals in presence of electrolytes. *Chemosphere* **2020**, *239*, 124730. [CrossRef] [PubMed]

77. Cabrera-Lafaurie, W.A.; Román, F.R.; Hernández-Maldonado, A.J. Transition metal modified and partially calcined inorganic-organic pillared clays for the adsorption of salicylic acid, clofibric acid, carbamazepine, and caffeine from water. *J. Colloid Interface Sci.* **2012**, *386*, 381–391. [CrossRef] [PubMed]
78. Rivera-Jimenez, S.M.; Lehner, M.M.; Cabrera-Lafaurie, W.A.; Hernández-Maldonado, A.J. Removal of naproxen, salicylic acid, clofibric acid, and carbamazepine by water phase adsorption onto inorganic-organic-intercalated bentonites modified with transition metal cations. *Environ. Eng. Sci.* **2011**, *28*, 171–182. [CrossRef]
79. Sunarić, S.; Pavlović, D.; Stanković, M.; Živković, J.; Arsić, I. Riboflavin and thiamine content in extracts of wild-grown plants for medicinal and cosmetic use. *Chem. Pap.* **2020**, *74*, 1729–1738. [CrossRef]
80. Gonen, Y.; Rytwo, G. Using a Matlab Implemented Algorithm for UV-vis Spectral Resolution for pKa Determination and Multicomponent Analysis. *Anal. Chem. Insights* **2009**, *2009*, 21. [CrossRef]
81. König, T.N.; Shulami, S.; Rytwo, G. Brine wastewater pretreatment using clay minerals and organoclays as flocculants. *Appl. Clay Sci.* **2012**, *67–68*, 119–124. [CrossRef]
82. Rytwo, G.; Kohavi, Y.; Botnick, I.; Gonen, Y. Use of CV- and TPP-montmorillonite for the removal of priority pollutants from water. *Appl. Clay Sci.* **2007**, *36*, 182–190. [CrossRef]
83. Steudel, A.; Emmerich, K. Strategies for the successful preparation of homoionic smectites. *Appl. Clay Sci.* **2013**, *75–76*, 13–21. [CrossRef]
84. Rytwo, G.; Gonen, Y.; Huterer-Shveky, R. Evidence of degradation of triarylmethine dyes on Texas vermiculite. *Clays Clay Miner.* **2009**, *57*, 555–565. [CrossRef]
85. Thilak Kumar, R.; Umamaheswari, S. FTIR, FTR and UV-Vis analysis of carbamazepine. *Res. J. Pharm. Biol. Chem. Sci.* **2011**, *2*, 685–693.
86. Madejová, J.; Komadel, P. Baseline studies of the clay minerals society source clays: Infrared methods. *Clays Clay Miner.* **2001**, *49*, 410–432. [CrossRef]
87. Vogna, D.; Marotta, R.; Andreozzi, R.; Napolitano, A.; d'Ischia, M. Kinetic and chemical assessment of the UV/H_2O_2 treatment of antiepileptic drug carbamazepine. *Chemosphere* **2004**, *54*, 497–505. [CrossRef]
88. Liu, N.; Lei, Z.-D.; Wang, T.; Wang, J.-J.; Zhang, X.-D.; Xu, G.; Tang, L. Radiolysis of carbamazepine aqueous solution using electron beam irradiation combining with hydrogen peroxide: Efficiency and mechanism. *Chem. Eng. J.* **2016**, *295*, 484–493. [CrossRef]
89. Bouyarmane, H.; El Bekkali, C.; Labrag, J.; Es-saidi, I.; Bouhnik, O.; Abdelmoumen, H.; Laghzizil, A.; Nunzi, J.-M.; Robert, D. Photocatalytic degradation of emerging antibiotic pollutants in waters by TiO_2/Hydroxyapatite nanocomposite materials. *Surf. Interfaces* **2021**, *24*, 101155. [CrossRef]
90. Al-Mamun, M.R.; Kader, S.; Islam, M.S.; Khan, M.Z.H. Photocatalytic activity improvement and application of UV-TiO_2 photocatalysis in textile wastewater treatment: A review. *J. Environ. Chem. Eng.* **2019**, *7*, 103248. [CrossRef]
91. Carabin, A.; Drogui, P.; Robert, D. Photo-degradation of carbamazepine using TiO_2 suspended photocatalysts. *J. Taiwan Inst. Chem. Eng.* **2015**, *54*, 109–117. [CrossRef]
92. Im, J.-K.; Son, H.-S.; Kang, Y.-M.; Zoh, K.-D. Carbamazepine Degradation by Photolysis and Titanium Dioxide Photocatalysis. *Water Environ. Res.* **2012**, *84*, 554–561. [CrossRef]
93. Kostich, M.S.; Batt, A.L.; Lazorchak, J.M. Concentrations of Prioritized Pharmaceuticals in Effluents from 50 Large Wastewater Treatment Plants in the US and Implications for Risk Estimation I US EPA. Available online: https://www.epa.gov/water-research/concentrations-prioritized-pharmaceuticals-effluents-50-large-wastewater-treatment (accessed on 11 May 2022).
94. Dudziak, S.; Bielan, Z.; Kubica, P.; Zielinska-Jurek, A. Optimization of carbamazepine photodegradation on defective TiO2-based magnetic photocatalyst. *J. Environ. Chem. Eng.* **2021**, *9*, 105782. [CrossRef]
95. Mohapatra, D.P.; Brar, S.K.; Daghrir, R.; Tyagi, R.D.; Picard, P.; Surampalli, R.Y.; Drogui, P. Photocatalytic degradation of carbamazepine in wastewater by using a new class of whey-stabilized nanocrystalline TiO_2 and ZnO. *Sci. Total Environ.* **2014**, *485–486*, 263–269. [CrossRef]
96. Cates, E.L. Photocatalytic water treatment: So where are we going with this? *Environ. Sci. Technol.* **2017**, *51*, 757–758. [CrossRef] [PubMed]

Article

Ecological Synthesis of CuO Nanoparticles Using *Punica granatum* L. Peel Extract for the Retention of Methyl Green

Mongi ben Mosbah [1,2], Abdulmohsen Khalaf Dhahi Alsukaibi [3], Lassaad Mechi [3], Fathi Alimi [3] and Younes Moussaoui [2,4,*]

[1] Laboratory for the Application of Materials to the Environment, Water and Energy (LR21ES15), Faculty of Sciences of Gafsa, University of Gafsa, Gafsa 2112, Tunisia; mbenmosbah@yahoo.fr
[2] Faculty of Sciences of Gafsa, University of Gafsa, Gafsa 2112, Tunisia
[3] Department of Chemistry, College of Science, University of Ha'il, Ha'il 81451, Saudi Arabia; a.alsukaibi17@gmail.com (A.K.D.A.); mechilassaad@yahoo.fr (L.M.); alimi.fathi@gmail.com (F.A.)
[4] Organic Chemistry Laboratory (LR17ES08), Faculty of Sciences of Sfax, University of Sfax, Sfax 3029, Tunisia
* Correspondence: y.moussaoui2@gmx.fr

Abstract: The aqueous extract from the bark of *Punica granatum* L. was invested to generate CuO nanoparticles from $CuSO_4$ using a green, economical, ecological, and clean method. The synthesized nanoparticles were characterized and were successfully used as adsorbents for methyl green retention of an absorptive capacity amounting to 28.7 mg g^{-1}. Methyl green equilibrium adsorption data were correlated to the Langmuir model following the pseudo-second order kinetics model. This study clearly corroborates that copper nanoparticles exhibit a high potential for use in wastewater treatment.

Keywords: CuO nanoparticles; green synthesis; *Punica granatum* L.; biosorption; methyl green

1. Introduction

Currently, nanotechnology is a very interesting field of research due to the successful application of nanomaterials in numerous areas, namely medicine, electronics, environment, biology, and optics [1–4]. Nanoparticles are known for their small size, their large area surface, and their dimensions ranging from 1 to 100 nm [5,6]. The physico-chemical parameters of these materials make them good candidates for a variety of applications such as the removal of pollutants from contaminated waters, biological activities, energy storage systems, etc., [1,7–13]. Numerous techniques have been used to synthesize nanomaterials in the form of particles, colloids, powders, rods, aggregates, tubes, wires, and thin films. These techniques are classified into physical, biological, and chemical methods. Physical methods involve evaporation and mechanical treatments [14,15], whereas chemical methods involve sonochemistry, microemulsion, and electrochemistry [4,14,16]. The latter have the merit of synthesis at a temperature below 350 °C [14,17]. Biological methods rely on the preparation of nanoparticles by such microorganisms as plant extracts, bacteria, yeasts, and fungi [18–20]. These methods have the benefits of easy scaling, non-toxicity, respect for the environment, and reproducibility of manufacturing [19,21]. Thus, the orientation towards natural products gained the widest attention owing to worldwide development, favoring good health, and mitigating the risk of disease. In this respect, the valorization of plant biomass is of paramount importance relating to the protection of the environment [22,23]. Renewable natural resources perform a key role in economic activities [24,25]. In this context, multiple natural resources have been investigated as sources of bioproducts. Components of biomass extracts, including phenols, flavonoids, tannins and other bioproducts, proved to be responsible for reducing metal salts to nanoparticles [13,23,26,27].

CuO nanoparticles are known as a p-type semiconductor [28,29] used as materials for Li batteries, sensors, solar energy, magnetic storage media, semiconductors, catalysis, and adsorption [29–31]. CuO nanoparticles can be synthesized by chemical reduction-based processes including sol-gel, precipitation, bio-template, microwave heating, electrospinning, and ion exchange [28]. For example, Kumar et al. [32] synthesized CuO nanoparticles from copper nitrate through a hydrothermal reaction and used them for the adsorption of methyl orange from aqueous solution. They found that this synthetic method provided CuO nanoparticles with good adsorption performance. In another study, Darwish et al. [33] prepared CuO nanoparticles through the microwave heating method and the obtained materials were used for the retention of methyl orange. The obtained results indicate that experimental kinetics data correlated with the second-order model and the adsorption was endothermic and spontaneous and governed by a physical process [33]. Moreover, CuO nanoparticles biosynthesized using *Tamarindus indica* extracts exhibit excellent photostability and significant antibacterial activity against bacterial strains [34]. Likewise, Anand et al. [35] showed that CuO nanoparticles synthesized by the microwave combustion method can be used to cure urinary tract infection.

Within this framework, the basic objective of this work lies in using an ecological method for the elaboration of a nanomaterial for the adsorption of aqueous pollutants. In this case, it is a question of examining the possibility of using *Punica granatum* L. bark extract for the preparation of CuO nanoparticles. Indeed, Tunisia is a country with a Mediterranean climate that has several types of plants, among which *Punica granatum* is a drought tolerant species, and it benefits from great capacities to adapt to environmental conditions characterized by marked climatic aridity. The fruits of this plant are rich in secondary metabolites [36–38] which can be used as binding agents to generate nanoscale materials.

2. Materials and Methods

2.1. Chemicals

Copper sulfate ($CuSO_4 \geq 99\%$), hydrochloric acid (HCl, 37%), sodium hydroxide ($NaOH \geq 98\%$), and methyl green ($C_{23}H_{25}N_3Cl_2 \cdot ZnCl_2$) were purchased from Sigma-Aldrich (Tunis; Tunisia). All chemicals were invested as received without any additional purification.

2.2. Extract Preparation

The bark extract of *Punica granatum* L. was prepared according to the steps illustrated in Figure 1.

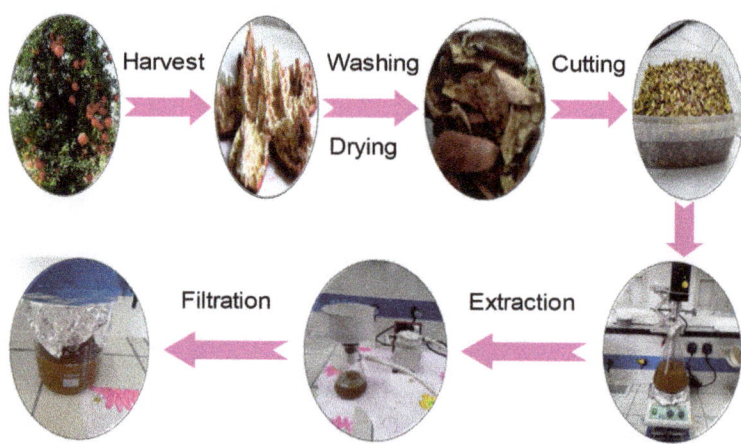

Figure 1. Schematic representation of the extract preparation.

Punica granatum bark was washed and dried in the shade in a dry and aerated environment at room temperature (25 ± 2 °C). After drying, the material was cut into small pieces. To prepare the extracts, 200 g of obtained material were mixed with 800 mL of distilled water. The obtained mixture underwent mechanical stirring at a temperature between 50 and 60 °C, until we obtained a pasty solution. Subsequently, the mixture was filtered using Whatman filter paper No.01, before it was recovered and stored at the temperature of 4 °C.

2.3. Chemical Composition of Punica granatum Bark

Chemical composition of *Punica granatum* bark was specified according to standardized methods. The ash content was estimated according to the T211 om-07 method. A given mass (m_0) of the material was calcined in an oven at 600 °C for 6 h to provide a mass (m_1). The ash content is provided by Equation (1):

$$\text{Ash content} = \frac{m_1}{m_0} \times 100 \quad (1)$$

TAPPI methods, namely, T212 om-07, T207 cm-08, and T 204-cm-07 were invested to determine the extractives in 1% NaOH, for cold and hot water, and for ethanol-toluene, respectively. Then, the α-cellulose, Klason lignin and holocellulose were quantified using T203 cm-99, T222 om-06, and Wise et al. [39] methods, respectively. All the experiments were duplicated and the difference between the values was within and experimental error of 5%.

2.4. CuO Nanoparticles Biosynthesis

A total of 100 mL of 0.1 mol L^{-1} copper sulphate solution (24.96 g of $CuSO_4 \cdot 5H_2O$ in 100 mL distilled water) was incorporated with 100 mL of the freshly resulting extract of the *Punica granatum* bark. The solution was agitated at 70 ± 2 °C for 3 h. The solution turned to a very dark green color. Then, after filtration, the residue was recovered and dried for 24 h at 50 °C. The resulting solid was eventually crushed (Figure 2).

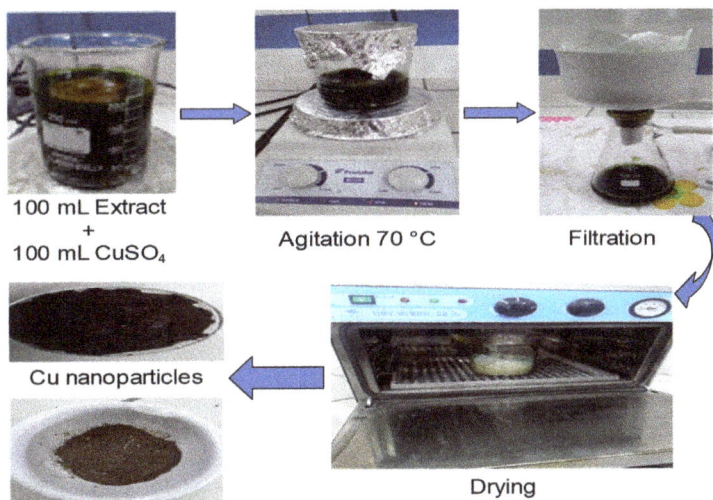

Figure 2. Graphical representation for CuO nanoparticles formation using extract of *Punica granatum* bark.

2.5. Methyl Green Adsorption on CuO Nanoparticles

The adsorption tests were carried out with a solution of methyl green. To improve the adsorbent capacity of the prepared CuO nanoparticles, we opted for the optimization of the following conditions: pH of the solution, mass of the adsorbent, initial concentration, and contact time.

The pH effect was explored through adjusting the pH in the range 2–10 by integrating a solution of sodium hydroxide or hydrochloric acid (0.1 mol L^{-1}). A total of 20 mg of CuO nanoparticles was added to 10 mL of methyl green solutions (50 mg L^{-1}). The solutions were agitated at 120 rpm for 12 h. The specimens were then centrifuged and the methyl green concentration was specified by UV-Visible spectroscopy at a wavelength of 630 nm using a Beckman DU 800 spectrophotometer. The pH study was indicated in the form of a graphical representation of the adsorbed amount versus pH.

The impact of the adsorbent dose was traced through varying the adsorbent mass from 2 to 500 mg in 10 mL of 50 mg L^{-1} methyl green solution at natural pH solution (pH = 6.5). Next, the solutions were agitated for 12 h, then centrifuged, filtered and the residual concentration was estimated. The study of the adsorbent dose corresponds to a graphical representation of the adsorbed amount versus the adsorbent dose.

To ensure good adsorption, we examined the impact of contact time on the adsorption of methyl green on CuO nanoparticles. Likewise, the previously reported steps were followed. Adsorbent (20 mg) was added to 10 mL of methyl green solution (50 mg L^{-1}, pH = 6.5). The solutions were stirred for a time of 20 to 420 min. The impact of contact time was obtained through a graphical representation of the quantity adsorbed versus time.

The kinetics study was carried out by following the same steps as those described previously with a solution of methyl green (50 mg L^{-1}). The mixtures were stirred for a time ranging from 20 to 420 min.

For adsorption isotherm, experiments were conducted by varying the initial concentration of methyl green from 5 to 120 mg g^{-1} (pH = 6.5, amount of adsorbent = 20 mg, time = 60 min). The models of Langmuir [40], Freundlich [41], Temkin [42], and Dubinin–Radushkevich [43] (Table 1) were invested to model the experimental results.

Table 1. Theoretical models.

Isotherm	Non-Linear Form	Linear Form
Langmuir	$q_{ads} = \frac{C_e q_e K_L}{1+(K_L C_e)}$	$\frac{1}{q_{ads}} = \frac{1}{K_L q_e} \frac{1}{C_e} + \frac{1}{q_e}$
Freundlich	$q_{ads} = K_F C_e^{\frac{1}{n}}$	$\ln(q_{ads}) = \ln(K_F) + \frac{1}{n}\ln(C_e)$
Temkin	$q_{ads} = \frac{RT}{B} \ln(A_T C_e)$	$q_{ads} = \frac{RT}{B} \ln(A_T) + \frac{RT}{B} \ln(C_e)$
Dubinin–Radushkevich	$q_{ads} = q_e e^{-\beta \varepsilon^2}$	$\ln(q_{ads}) = \ln(q_e) - \beta \varepsilon^2$

q_{ads} and q_e correspond to the adsorbed amount and the adsorbed amount at equilibrium (mg g^{-1}), respectively; C_e refers to the concentration of the solution at equilibrium (mg L^{-1}); K_L and K_F indicate Langmuir and Freundlich constants, respectively; (1/n) denotes Freundlich coefficient; B represents Temkin constant relating to heat of adsorption; A_T signifies Temkin constant relating to adsorption potential; R expresses universal gas constant (kJ kg^{-1} mol^{-1} K^{-1}); T refers to temperature (K); β corresponds to constant related to the mean free energy of adsorption; ε stands for Polanyi potential.

For desorption experiments, the recuperated adsorbent after adsorption was recovered in 50 mL of a solution of 1.0 mol L^{-1} hydrochloric acid, for 3 h. Then the sample was shaken and centrifuged and washed with distilled water. Finally, the material was dried and reused for another adsorption cycle.

2.6. Characterization Techniques

Scanning electron microscopy was performed out by a HITACHI S-3400N (Paris, France) device (voltage: 0.5–30 KW, resolution: 4.5 nm).

Energy dispersive X-ray spectrometry (EDS) was carried out using a Perkin Elmer DIFFRAX (Paris, France).

The thermograms were obtained by testing the samples under nitrogen flow for the temperature ranges from 30 to 950 °C for the raw material and from 30 to 600 °C for the synthesized nanoparticles (heating rate: 10 °C min^{-1}). The used device was the Analyzer Pyris 6 Perkin Elmer thermogravimetric analyzer (Paris, France).

The infrared analysis was conducted with a Nicolet iS 50 FT-IR Spectrometer Thermo Scientific (Paris, France). The pellets were obtained with anhydrous KBr 1% (wt/wt) of product.

3. Results and Discussion

3.1. Characterization of Punica granatum Bark

By applying standard methods, we determined the chemical composition of the bark of *Punica granatum* (Table 2).

Table 2. Chemical composition (content in%) of the bark of *Punica granatum* and some plant biomass collected from literature.

	C. water	H. water	ET	1% NaOH	Ash	Holocell.	Lignin	Cellulose
Punica granatum bark	28	37.6	46.5	32.4	4	16.2	22	7.9
Wood								
Palmier Dattier [44]	n.a.	n.a.	3	n.a.	6.5	59.5	27	33.5
Olivier [45]	15.5	17	10.4	30	1.4	65.8	15.6	41.5
Tamarisk aphylla [46]	12.8	19.0	4.5	18.5	3.5	50.0	30.0	39.0
Prunus amygdalus [47]	11.3	12.3	5	28.7	3.6	60.7	19.2	40.7
Non-wood								
Rachis de Palmier Dattier [45]	5	8.1	6.3	20.8	5	74.8	27.2	45
Tiges de vignes [48]	8.2	13.9	11.3	37.8	3.9	65.4	28.1	35
FiguierBarbarie [49]	24	36.3	8.8	29.6	5.5	64.5	4.8	53.6
Ziziphus lotus [50]	n.a.	n.a.	11	n.a.	1.49	n.a.	19.6	30.8
Annual plants								
Alfa [51]	9.1	11.1	7.9	19.4	5.1	69.7	17.71	47.6
Stipagrostispungens [52]	19.3	20.5	4.8	42.9	4.6	71	12	44
Astragalus armatus [53]	26.2	33	13	32.7	3	54	16.8	35
Retama raetam [54]	32	31.5	10	47	3.5	58.7	20.5	36
Nitraria retusa [54]	23	25.5	3	40	6.2	52	26.3	41
Pithuranthoschloranthus [54]	25	26.7	9.5	49	5	62	17.6	46.5

n.a.—not available; Holocell.—holocellulose content; 1% NaOH—content in 1% sodium hydroxide; ET—content in ethanol/toluene; H. water—hot water content; C. water—cold water content.

The bark of *Punica granatum* has an approximately high rate of extractives in cold and hot water (28%, 37.6%). These rates were higher than those of annual and perennial plants, non-wood and wood. However, extractives in a 1% NaOH solution (32.4%) were lower than those of some annual and perennial plants as well as vine stems but was higher than those of non-wood depicted in Table 2. Indeed, the bark of *Punica granatum* is rich in ester of hydrolysable ellagitannins, mainly punicalagin. Ellagitannins are high molecular weight, water-soluble polyphenols which contain several hydroxyl functions [55,56].

The extractables in the ethanol/toluene mixture displayed a rate of 46.5% which was higher than that of the materials reported in Table 2. Furthermore, the lignin content (22%) proved to be lower than that of wood plants such as date palm (27%) [44], and vine stems (28.1%) [48]. Nevertheless, it remained higher than that of olive wood (15.6%) [45], and higher than some non-wood and annual plants such as prickly pear (4.8%) [44], *Astragalus armatus* (16.8%) [53], and *Stipagrostis pungens* (12%) [52]. The holocellulose content (16.2%) and α-cellulose (7.2%) were lower than those of non-wood, annual, and wood plants exhibited in Table 2. On the other side, the bark of *Punica granatum* presented ash content comparable to that encountered for the sources outlined in Table 2.

The infrared spectrum displayed broad stretching vibrations around 3200–3600 cm^{-1} of OH groups and a band around 1360 cm^{-1} reflecting alcoholic, phenolic, and acidic groups. In addition, bands of punicalagin alkyl groups appeared at 2800–2970 cm^{-1}. The band at 1740 cm^{-1} was assigned to carbonyl groups. Moreover, the emergence of two bands at 1620 cm^{-1} and 1430 cm^{-1} was attributed to the stretching of the C=C aromatic rings and the band at 1052 cm^{-1} was ascribed to the vibration d stretch CO. Moreover, the bands at 1229 and 637 cm^{-1} were associated respectively with the aromatic C-H bending and the C-O bond.

The atomic composition of *Punica granatum* bark powder was specified using EDX analysis. The spectrum (Figure 3) suggested the presence of carbon, oxygen, iron, sodium, potassium, and calcium.

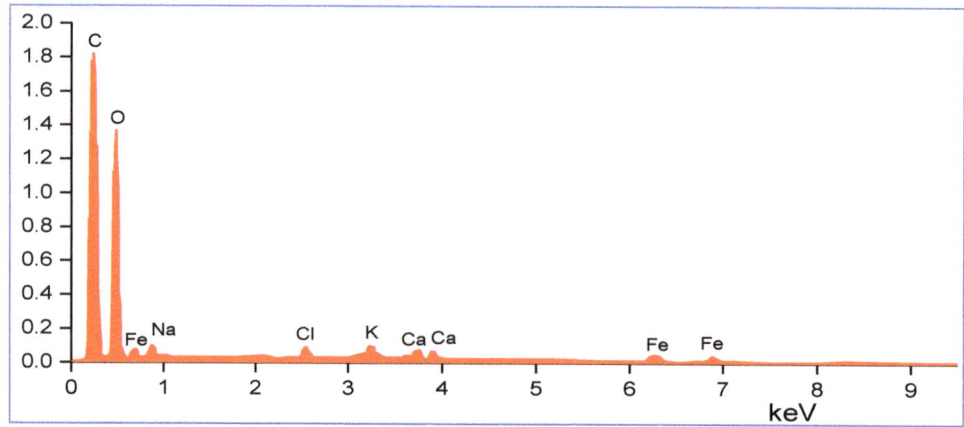

Figure 3. EDX analysis of *Punica granatum* bark.

The thermogram of the *Punica granatum* bark (Figure 4) revealed four stages. The first stage presented a mass loss of 5% from 30 °C to 200 °C. This loss was due to material dehydration. The second stage presented the largest mass loss (90%) occurring between 200 and 450 °C, during which the components of the lignocellulosic fibers and part of the lignin chain were degraded [57]. The third stage ranged between 450 and 600 °C. This region characterizes the thermal degradation of the remaining lignin and the complete decomposition of the carbon chains of the *Punica granatum* bark. Above 600 °C (the fourth stage) ash appeared and a constant mass was obtained.

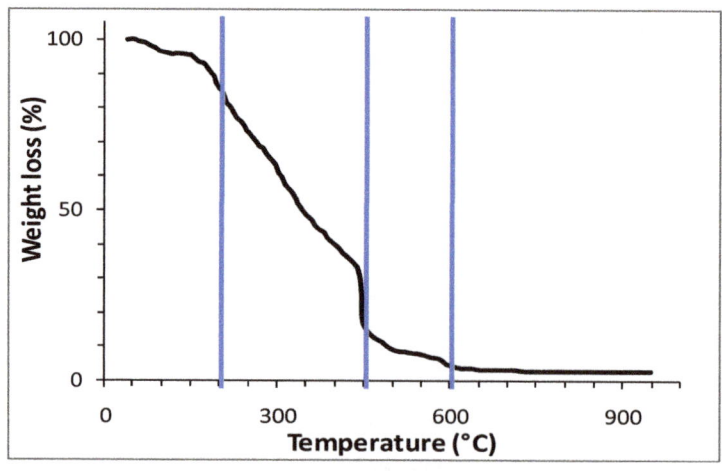

Figure 4. TGA curve of *Punica granatum* bark. (Blue lines indicate the three stages of mass loss).

3.2. Preparation and Characterization of CuO Nanoparticles

The *Punica granatum* bark is an intrinsic source of biomolecules, involving condensed tannins, ellagic acids, catechins, punicalin, and punicalagin [55,56]. These biomolecules act as ligating agents that are responsible for stabilizing and styling the copper nanoparticles. Indeed, the hydroxyl groups of the biomolecules perform the role of complexing agents that fix the copper ions of the copper sulphate solution. Although the structure of the complex formed was not determined, it canbe the product of the coordination of copper ions with the functional groups of the molecules present in the *Punica granatum* bark extract. The formed complexes underwent decomposition and led to the release of CuO nanoparticles.

The SEM images (Figure 5) of the synthesized CuO nanoparticles display a smooth surface with the arrangement of the layers on top of each other forming a multilayer structure connected with strong molecular interactions with a large variation in the distribution of the particle size and shape. This accounts for the grafting of copper ions on the organic chains of the *Punica granatum* extract. Indeed, we noticed a beam of white light reflected on the material surface referring to the response of the copper.

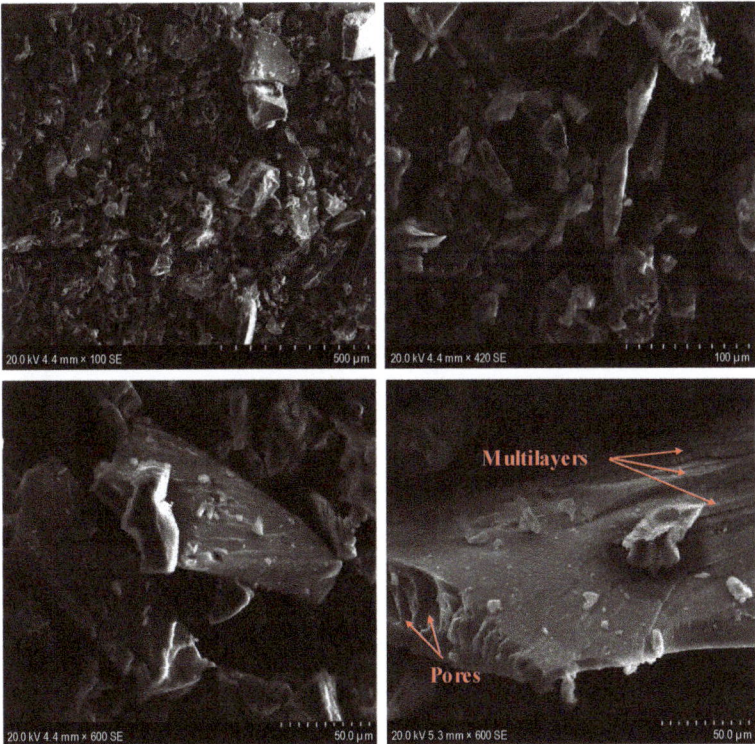

Figure 5. SEM microphotographs of synthesized CuO nanoparticles.

The spectrum obtained through X-ray microanalysis (Figure 6) is indicative of peaks with strong energy for the copper found between 1 and 8 KeV, which confirms the obtaining of CuO nanoparticles. Additionally, about 54.58% of copper was quantified in the synthesized material. In addition, it is noteworthy that oxygen (19.49%) and carbon (25.93%) were noticed as constituents of the capping molecules surrounding the copper nanoparticles.

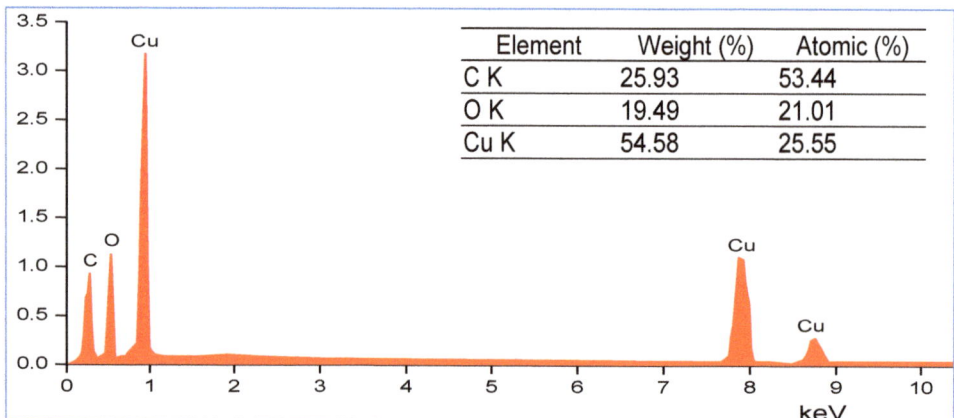

Figure 6. EDX analysis of CuO nanoparticles.

The thermogravimetric analysis of the synthesized CuO nanoparticles (Figure 7) revealed three main areas of mass loss: The first zone representing about 10% located between 30 and 120 °C corresponded to the release of adsorbed water. The second zone of mass loss (approximately 40%) lies between 120 and 500 °C, subdivided into two sub-regions of degradation: The first sub-region of mass loss (approximately 25%) observed between 120 and 300 °C, was attributed to the degradation and opening of the organic molecules constituting bio-capping [58], which stands for the organic support of the extract present in the prepared nanoparticles. The second sub-region (about 15%) ranging between 300 and 500 °C, was ascribed to the thermal resistance of copper which is evidenced through the slowing down of degradation processes followed by thermal decomposition of organic chain residues. The third zone located above 500 °C presented a mass stability that was recorded with a total mass loss of 51.45%. This proves that a percentage of 48.55% metallic copper was present in the prepared nanoparticles. This behavior is in good agreement with facts reported by EDX analysis.

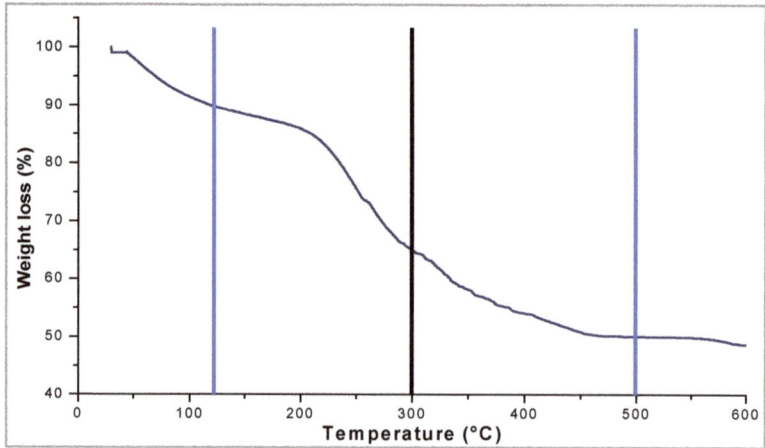

Figure 7. TGA curve of CuO nanoparticles. (Blue lines for main areas of mass loss and black line for sub-region of mass loss).

3.3. Study of the Adsorption of Methyl Green on CuO Nanoparticles

The synthesized CuO nanoparticles were used as an adsorbent for the retention of methyl green. Thus, various parameters were investigated such as methyl green concentration, contact time, adsorbent dose, and pH.

3.3.1. Influence of pH, Adsorbent Dose and Contact Time on the Methyl Green Adsorption onto CuO Nanoparticles

The influence of pH is illustrated in Figure 8a. The adsorbed amount (q_{ads} (mg g^{-1})) increased with the pH, increasing from 2 to 6. The maximum adsorbed amount was reached for a pH ranging from 6 to 8 such as those reported using loofah fiber [59], titanium dioxide [60], CoFe$_2$O$_4$/rGO nanocomposites [61], Graphene oxide [62], activated bentonite [63], and activated carbon [64]. Indeed, when the pH value increased, many of the anionic sites were available, contributing to the increased retention of methyl green. For low pH values, H$^+$ ions were present on the surface of CuO nanoparticles, which entailed an electrostatic repulsion between the surface of the CuO nanoparticles and the molecules of methyl green, hampering the process of adsorption. Therefore, the natural pH (pH = 6.5) of the methyl green solution was used for the rest of the work.

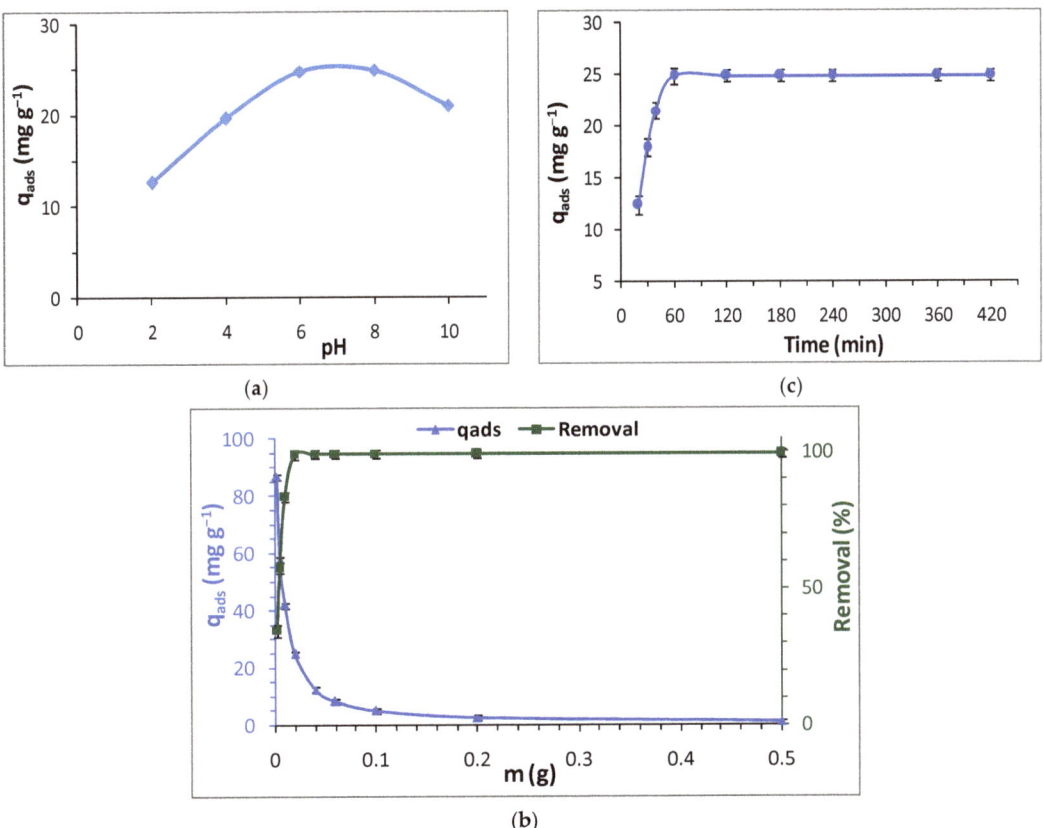

Figure 8. Influence of pH (**a**), adsorbent amount (**b**), and contact time (**c**) on the methyl green adsorption onto CuO nanoparticles. (Blue line for adsorbed quantity and green line for removal).

The adsorbed amount of methyl green decreased with an increase in adsorbent dose (Figure 8b). This is accounted for in terms of adsorption sites, which remained unsaturated throughout the adsorption process. Thus, a mass of 20 mg of adsorbent was considered sufficient to ensure good adsorption of the methyl green on the CuO nanoparticles with complete elimination of the methyl green (elimination rate = 100%). This adsorbent mass value was selected for the current work.

To determine the time required to ensure good elimination of the methyl green, a study of the influence of the contact time on the adsorbed amount was conducted (Figure 8c). The profile of the adsorption process was characterized by two different stages. The first stage was described throughout the first 60 min of which several adsorption sites were available on the CuO nanoparticles surface. In the second stage, the elimination of methyl green remained unchanged, reflecting the saturation of the active adsorption sites. Therefore, a contact time of 60 min was considered sufficient to ensure adsorption equilibrium.

3.3.2. Kinetic Study

Both the pseudo-first order Equation (2) [65–67] and pseudo-second order Equation (3) [66–68] models were invested to characterize the kinetics of adsorption of methyl green on CuO nanoparticles.

$$\ln(q_e - q_t) = \ln q_e - k_1 t \quad (2)$$

$$\frac{t}{q_t} = \frac{1}{k_2 q_e^2} + \frac{t}{q_t} \quad (3)$$

where q_e and q_t (mg g^{-1}) refer, respectively, to the adsorbed quantities at equilibrium and at time "t"; k_1 and k_2 represent the constants for pseudo-first and pseudo-second order model, respectively.

The value of the adsorbed amount of methyl green calculated (q_{cal} = 25.6 mg g^{-1}) using the pseudo-second order model was comparable to the experimental value (q_{exp} = 24.9 mg g^{-1}) (Table 3). Additionally, the regression coefficient value in the case of the pseudo-second order model (R^2 = 0.999) was high in comparison with that recorded for the pseudo-first order model. This suggests that the adsorption of methyl green on CuO nanoparticles adheres to the pseudo-second order kinetics. Likewise, according to Δq, RMSE and χ^2, the pseudo-second order model displays lower errors and is in good consistency with the experimental data. Similar phenomena have been reported for the adsorption of methyl green onto various adsorbent such as loofah fiber [59], titanium dioxide [60], zeolite H-ZSM-5 [69], carbon nanotubes [70], and graphene sheets [71].

Table 3. Kinetics parameters of methyl green adsorption on CuO nanoparticles.

	q_{exp} (mg g^{-1})	24.9
Pseudo-first order	q_{cal} (mg g^{-1})	7.0
	k_1 (min^{-1})	0.0287
	R^2	0.7911
	χ^2	232.297
	Δq	0.5199
	RMSE	15.2870
Pseudo-second order	q_{cal} (mg g^{-1})	25.6
	k_2(g mg^{-1} min^{-1})	0.0047
	R^2	0.9990
	χ^2	8.238
	Δq	0.1125
	RMSE	4.8492

3.3.3. Adsorption Isotherms

The Temkin, Langmuir, Freundlich, and Dubinin–Radushkevich isotherms were used to characterize the adsorption of methyl green on the prepared nanoparticles (Table 4). The goodness of fit of the theoretical models at the experimental isotherm was confirmed by comparing the values of the regression coefficients R^2.

Table 4. Isothermal parameters.

Langmuir	q_{max} (mg g^{-1})	28.7
	K_L (L mg^{-1})	6.96
	R_L	0.0011–0.0279
	R^2	0.9892
	χ^2	1.9458
	Δq	0.0144
Freundlich	$1/n$	0.2294
	K_F	16.683
	R^2	0.7084
	χ^2	12.5882
	Δq	0.2419
Temkin	R^2	0.8699
	B (J mol^{-1})	3.3198
	A_T (L g^{-1})	424.766
	χ^2	5.2396
	Δq	0.1960
Dubinin–Radushkevich	R^2	0.9649
	B	0.0215
	q_{max} (mg g^{-1})	27.709
	χ^2	6.7088
	Δq	0.0217

It seems that the Langmuir fit ($R^2 = 0.9892$) fits the methyl green adsorption data better than the other three models. It turns out that the Langmuir model is the most adequate to characterize the adsorption of methyl green on CuO nanoparticles. This result was checked by the low value of the chi-square test χ^2, and the standard deviation Δq, reflecting monolayer adsorption. Moreover, the values of dimensionless equilibrium constant R_L, (Equation (4)) were lower than 1, reflecting the favorable adsorption of methyl green on the CuO nanoparticles. This fact was confirmed by the value of the Freundlich coefficient ($1/n = 0.2294$), which was lower than 1 [67,72–74].

$$R_L = \frac{1}{1 + K_L C_0} \quad (4)$$

The maximum adsorbed amount of methyl green on CuO nanoparticles calculated from the Langmuir model, was 28.7 mg g^{-1}.

A comparison of the adsorption capacity of methyl green onto CuO nanoparticles with other adsorbents reported by other researchers was shown in Table 5. It can be seen that CuO nanoparticles showed comparable adsorption capacity with respect to zeolite H-ZSM-5 [69] and grapheme oxide [62]. The adsorption capacity of CuO nanoparticles was higher than those of loofah fiber (18.2 mg g^{-1}) [59] and bamboo [75], and was lower than that of titanium dioxide (384.6 mg g^{-1}) [60], activated carbon from *Brachychiton Populneus* fruit shell [64], activated bentonite [63], and graphene sheets [71]. Therefore, the CuO nanoparticles were suitable and promising for the removal of methyl green from aqueous solutions since they have a relatively higher adsorption capacity.

Table 5. Reported maximum adsorption capacities in the literature for methyl green adsorption.

Adsorbent	Adsorption Conditions	q_{max} (mg g^{-1})
Loofah fiber [59]	pH = 6.5; dose = 0.4 g/L; T = 25 °C	18.2
Bamboo [75]	pH of solution; dose = 1 g/L; T = 25 °C	20.4
Graphene oxide [62]	pH = 7; dose = 0.1 g/L; T = 25 °C	28.5
Zeolite H-ZSM-5 [69]	pH = 4; dose = 1 g/L; T = 25 °C	31.3
Graphene sheets-CoFe$_2$O$_4$ nanparticles [71]	pH of solution; dose = 1 g/L; T = 25 °C	47.2
Activated carbon from *Brachychiton Populneus* fruit shell [64]	pH of solution; dose = 1 g/L; T = 20 °C	67.9
CoFe$_2$O$_4$/rGO nanocomposites [61]	pH of solution; dose = 0.25 g/L	88.3
(CNTs-NiFe$_2$O$_4$) [70]	pH of solution; dose = 1 g/L; T = 25 °C	88.5
CNTs [70]	pH of solution; dose = 1 g/L; T = 25 °C	146
Graphene sheets [71]	pH of solution; dose = 1 g/L; T = 25 °C	203.5
Activated bentonite [63]	pH of solution; dose = 0.1 g/L; T = 25 °C	335.3
Titanium dioxide [60]	pH = 6.6; dose = 1 g/L; T = 25 °C	384.6
CuO nanoparticles (this work)	pH of solution; dose = 2 g/L; T = 25 °C	28.7

3.3.4. Desorption and Adsorbent Reuse

To assess the feasibility and applicability of adsorbent reuse and therefore to increase the use effectiveness of the prepared nanoparticles, desorption of the adsorbed methyl green was performed. Departing from the study of the pH effect, it was demonstrated that adsorption of methyl green was defective in acidic medium. Thus, desorption was conducted by washing the recuperated adsorbent after adsorption with a hydrochloric acid and afterwards with distilled water. Next, the adsorbent was dried and reused for the adsorption of methyl green. The adsorption/desorption procedure was repeated five times and results were summarized in Table 6.

Table 6. Adsorption-desorption of methyl green.

Cycle	q_{ads} (mg g^{-1})	Adsorption (%)	Desorption (%)
1	24.9	99.8	99.2
2	24.3	97.3	98.6
3	23.1	92.4	94.1
4	20.8	83.3	93.2
5	17.1	68.4	90.4

The maximum adsorption capacity of CuO nanoparticles decreased from 24.9 mg g^{-1} to 17.1 mg g^{-1} after five cycles. The regeneration results revealed that the adsorption efficiency of CuO nanoparticles dropped of only about 8% after three cycles. The adsorption/desorption tests corroborated appropriate reuse characteristics of the adsorbent after three cycles with a better performance for methyl green sorption.

4. Conclusions

The green synthesis of metallic nanoparticles has accumulated interest over the past decade due to their distinctive properties that make them applicable in various fields. Metal nanoparticles synthesized using plants have been shown to be non-toxic and environmentally friendly. In this study, a very cheap and simple conventional method was used to obtain the CuO nanoparticles using the extract from the bark of *Punica granatum*. The nanoparticles' purity was confirmed through EDX, yielding the presence of elemental copper oxide containing 25.93, 19.49, and 54.58%, of carbon, oxygen, and copper, respectively. The synthesized CuO nanoparticles demonstrated a great performance in the adsorption of methyl green with an adsorption capacity of 28.7 mg g^{-1} under optimal conditions (60 min, contact time; 2 g L^{-1}, adsorbent dose; pH, 6.5; 25 ± 2 °C). The adsorption of methyl green on the CuO nanoparticles follows the pseudo-second order model and the adsorption isotherm is in good conformity with the Langmuir model. CuO nanoparticles can be reused with good recycle properties for three cycles. Regarding the good sorption performance of the prepared CuO nanoparticles, the use of environmentally friendly *Punica granatum* bark can constitute an ecological, simple, and economical method for the elaboration of materials in order not only to purify contaminated waters but also to enact additional environmental applications. The nanoparticles obtained through plant extracts have the potential to be widely used in the medical field as dressings or therapeutic drugs. Thus, green chemistry provides a new innovative technique that does not use hazardous substances for the design and development of variable activity materials.

Author Contributions: Literature review, data collection, field work, M.b.M.; methodology and writing—original draft preparation, M.b.M. and Y.M.; designed and revised the manuscript, L.M. and F.A.; writing—review and editing, A.K.D.A.; visualization and supervision, Y.M. All authors have read and agreed to the published version of the manuscript.

Funding: This research was funded by scientific research deanship at University of Ha'il—Saudi Arabia through project number RG-191251. The APC was funded by scientific research deanship at University of Ha'il—Saudi Arabia through project number RG-191251.

Institutional Review Board Statement: Not applicable.

Informed Consent Statement: Not applicable.

Data Availability Statement: Not applicable.

Acknowledgments: This research has been funded by scientific research deanship at University of Ha'il—Saudi Arabia through project number RG-191251.

Conflicts of Interest: The authors declare no conflict of interest.

References

1. Kazemzadeh, Y.; Shojaei, S.; Riazi, M.; Sharifi, M. Review on application of nanoparticles for EOR purposes: A critical review of the opportunities and challenges. *Chin. J. Chem. Eng.* **2019**, *27*, 237–246. [CrossRef]
2. Yasmin, M. Compressive Strength Prediction for Concrete Modified with Nanomaterials. *Case Stud. Constr. Mater.* **2021**, *15*, e00660.
3. Azhdarzadeh, M.; Saei, A.; Sharifi, A.S.; Hajipour, M.J.; Alkilany, A.M.; Sharifzadeh, M.; Ramazani, F.; Laurent, S.; Mashaghi, A.; Mahmoudi, M. Nanotoxicology: Advances and pitfalls in research methodology. *Nanomedicine* **2015**, *10*, 2931–2952. [CrossRef] [PubMed]
4. Nakamura, S.; Sato, M.; Sato, Y.; Ando, N.; Takayama, T.; Fujita, M.; Ishihara, M. Synthesis and Application of Silver Nanoparticles (Ag NPs) for the Prevention of Infection in Healthcare Workers. *Int. J. Mol. Sci.* **2019**, *20*, 3620. [CrossRef]
5. Garnett, M.C.; Kallinteri, P. Nanomedicines and nanotoxicology: Some physiological principles. *Occup. Med.* **2006**, *565*, 307–311. [CrossRef]
6. Ben Mosbah, M.; Lassaad, M.; Khiari, R.; Moussaoui, Y. Current State of Porous Carbon for Wastewater Treatment. *Processes* **2020**, *8*, 1651. [CrossRef]
7. Hamad, H.T.; Al-Sharify, Z.T.; Al-Najjar, S.Z.; Gadooa, Z.A. A review on nanotechnology and its applications on Fluid Flow in agriculture and water recourses. *IOP Conf. Ser. Mater. Sci. Eng.* **2020**, *870*, 012038. [CrossRef]
8. Xavier, M.; Parente, I.; Rodrigus, P.; Cerqueira, M.A.; Pastrana, L.; Gonçalves, C. Safety and fate of nanomaterials in food: The role of in vitro tests. *Trends Food Sci. Technol.* **2021**, *109*, 593–607. [CrossRef]
9. Huang, Z.; Wang, X.; Sun, F.; Fan, C.; Sun, Y.; Jia, F.; Yin, G.; Zhou, T.; Liu, B. Super response and selectivity to H_2S at room temperature based on CuO nanomaterials prepared by seed-induced hydrothermal growth. *Mater. Des.* **2021**, *201*, 109507. [CrossRef]
10. Pathakoti, K.; Manubolu, M.; Hwang, H.M. Nanostructures: Current uses and future applications in food science. *J. Food Drug Anal.* **2017**, *25*, 245–253. [CrossRef]
11. Tian, X.; Fan, T.; Zhao, W.; Abbas, G.; Han, B.; Zhang, K.; Li, N.; Liu, N.; Liang, W.; Huang, H.; et al. Recent advances in the development of nanomedicines for the treatment of ischemic stroke. *Bioact. Mater.* **2021**, *6*, 2854–2869. [CrossRef] [PubMed]
12. Hafiz, M.; Talhami, M.; Ba-Abbad, M.M.; Hawari, A.H. Optimization of Magnetic Nanoparticles Draw Solution for High Water Flux in Forward Osmosis. *Water* **2021**, *13*, 3653. [CrossRef]
13. Marouzi, S.; Sabouri, Z.; Darroudi, M. Greener synthesis and medical applications of metal oxide nanoparticles. *Ceram. Int.* **2021**, *47*, 19632–19650. [CrossRef]
14. Satyanarayana, T.; Reddy, S.S. A Review on Chemical and Physical Synthesis Methods of Nanomaterials. *Int. J. Res. Appl. Sci. Eng. Technol.* **2018**, *6*, 2885–2889. [CrossRef]
15. Ubaidullah, M.; Al-Enizi, A.M.; Shaikh, S.; Ghanem, M.A.; Mane, R.S. Waste PET plastic derived ZnO@NMC nanocomposite via MOF-5 construction for hydrogen and oxygen evolution reactions. *J. King Saud Uni. Sci.* **2020**, *32*, 2397–2405. [CrossRef]

16. Yakoot, S.M.; Salem, N.A. A Sonochemical-assisted Simple and Green Synthesis of Silver Nanoparticles and its Use in Cosmetics. *Int. J. Pharmacol.* **2016**, *12*, 572–575. [CrossRef]
17. Sahoo, M.; Vishwakarma, S.; Panigrahi, C.; Kumar, J. Nanotechnology: Current applications and future scope in food. *Food Front.* **2021**, *2*, 3–22. [CrossRef]
18. Singh, H. Nanotechnology applications in functional foods; opportunities and challenges. *Prev. Nutr. Food Sci.* **2016**, *21*, 1–8. [CrossRef]
19. Naseem, T.; Farrukh, M.A. Antibacterial Activity of Green Synthesis of Iron Nanoparticles Using Lawsoniainermis and Gardenia jasminoides Leaves Extract. *J. Chem.* **2015**, *2015*, 912342. [CrossRef]
20. Manjare, S.B.; Pendhari, P.D.; Badade, S.M.; Thopate, S.R. Palladium Nanoparticles: Plant Aided Biosynthesis, Characterization, Applications. *Chem. Afr.* **2021**, *4*, 715–730. [CrossRef]
21. Siddiqui, S.; Alrumman, S.A. Influence of nanoparticles on food: An analytical assessment. *J. King Saud Uni. Sci.* **2021**, *33*, 101530. [CrossRef]
22. Suwalsky, M.; Vargas, P.; Avello, M.; Villena, F.; Sotomayor, C.P. Human erythrocytes are affected in vitro by flavonoids of Aristotelia chilensis (Maqui) leaves. *Int. J. Pharm.* **2008**, *363*, 85–90. [CrossRef] [PubMed]
23. Naikoo, G.A.; Mustaqeem, M.; Hassan, I.U.; Awan, T.; Arshad, F.; Salim, H.; Qurashi, A. Bioinspired and green synthesis of nanoparticles from plant extracts with antiviral and antimicrobial properties: A critical review. *J. Saudi Chem. Soc.* **2021**, *25*, 101304. [CrossRef]
24. Makul, N.; Fediuk, R.; Amran, M.; Al-Akwaa, M.S.; Pralat, K.; Nemova, D.; Petropavlovskii, K.; Novichenkova, T.; Petropavlovskaya, V.; Sulman, M. Utilization of Biomass to Ash: An Overview of the Potential Resources for Alternative Energy. *Materials* **2021**, *14*, 6482. [CrossRef]
25. Mora-Sandí, A.; Ramírez-González, A.; Castillo-Henríquez, L.; Lopretti-Correa, M.; Vega-Baudrit, J.R. Persea Americana Agro-Industrial Waste Biorefinery for Sustainable High-Value-Added Products. *Polymers* **2021**, *13*, 1727. [CrossRef]
26. Shahrashoub, M.; Bakhtiari, S.; Afroosheh, F.; Googheri, M.S. Recovery of iron from direct reduction iron sludge and biosynthesis of magnetite nanoparticles using green tea extract. *Colloids Surf. A Physicochem. Eng. Asp.* **2021**, *622*, 126675. [CrossRef]
27. Ajayi, A.; Larayetan, R.; Yahaya, A.; Falola, O.O.; Ude, N.A.; Adamu, H.; Oguche, S.M.; Abraham, K.; Egbagba, A.O.; Egwumah, C.; et al. Biogenic Synthesis of Silver Nanoparticles with Bitter Leaf (*Vernonia amygdalina*) Aqueous Extract and Its Effects on Testosterone-InducedBenign Prostatic Hyperplasia (BPH) in Wistar Rat. *Chem. Afr.* **2021**, *4*, 791–807. [CrossRef]
28. Ighalo, J.O.; Sagboye, P.A.; Umenweke, G.; Ajala, O.J.; Omoarukhe, F.O.; Adeyanju, C.A.; Ogunniyi, S.; Adeniyi, A.G. CuO nanoparticles (CuO NPs) for water treatment: A review of recent advances. *Environ. Nanotechnol. Monit. Manag.* **2021**, *15*, 100443. [CrossRef]
29. Nayak, R.; Ali, F.A.; Mishra, D.K.; Ray, D.; Aswal, V.K.; Sahoo, S.K.; Nanda, B. Fabrication of CuO nanoparticle: An efficient catalyst utilized for sensing and degradation of phenol. *J. Mater. Res. Technol.* **2020**, *9*, 11045–11059. [CrossRef]
30. Baylan, N.; Ilalan, I.; Inci, I. Copper oxide nanoparticles as a novel adsorbent for separation of acrylic acid from aqueous solution: Synthesis, characterization, and application. *Water Air Soil Pollut.* **2020**, *231*, 1–15. [CrossRef]
31. Sharifpour, E.; AlipanahpourDil, E.; Asfaram, A.; Ghaedi, M.; Goudarzi, A. Optimizing adsorptive removal of malachite green and methyl orange dyes from simulated wastewater by Mn-doped CuO-Nanoparticles loaded on activated carbon using CCD-RSM: Mechanism, regeneration, isotherm, kinetic, and thermodynamic studies. *Appl. Organomet. Chem.* **2019**, *33*, e4768.
32. Kumar, K.Y.; Archana, S.; Vinuth Raj, T.N.; Prasana, B.P.; Raghu, M.S.; Muralidhara, H.B. Superb adsorption capacity of hydrothermally synthesized copper oxide and nickel oxide nanoflakes towards anionic and cationic dyes. *J. Sci. Adv. Mater. Devices* **2017**, *2*, 183–191. [CrossRef]
33. Darwish, A.; Rashad, M.; AL-Aoh, H.A. Methyl orange adsorption comparison on nanoparticles: Isotherm, kinetics, and thermodynamic studies. *Dye. Pigm.* **2019**, *160*, 563–571. [CrossRef]
34. Zaman, M.B.; Poolla, R.; Singh, P.; Gudipati, T. Biogenic synthesis of CuO Nanoparticles using *Tamarindus indica* L. and a study of their photocatalytic and antibacterial activity. *Environ. Nanotechnol. Monit. Manage.* **2020**, *14*, 100346. [CrossRef]
35. Anand, G.T.; Sundaram, S.J.; Kanimozhi, K.; Nithiyavathi, R.; Kaviyarasu, K. Microwave assisted green synthesis of CuO nanoparticles for environmental applications. *Mater. Today Proc.* **2021**, *36*, 427–434. [CrossRef]
36. Xi, J.; He, L.; Yan, L.-g. Continuous extraction of phenolic compounds from pomegranate peel using high voltage electrical discharge. *Food Chem.* **2017**, *230*, 354–361. [CrossRef]
37. Hasnaoui, N.; Wathelet, B.; Jiménez-Araujo, A. Valorization of pomegranate peel from 12 cultivars: Dietary fiber composition, antioxidant capacity and functional properties. *Food Chem.* **2014**, *160*, 196–203. [CrossRef]
38. El-Hadary, A.E.; Taha, M. Pomegranate peel methanolic-extract improves the shelf-life of edible-oils under accelerated oxidation conditions. *Food Sci. Nutr.* **2020**, *8*, 1798–1811. [CrossRef]
39. Wise, L.E.; Murphy, M. Chlorite holocellulose, its fractionation and bearing on summative wood analysis and studies on the hemicelluloses. *Pap. Trade J.* **1946**, *122*, 35–43.
40. Langmuir, I. The adsorption of gases on plan surfaces of glass, mica and platinum. *J. Am. Chem. Soc.* **1918**, *40*, 1361–1403. [CrossRef]
41. Freundlich, H.M.F. Uber die adsorption in losungen. *Z. Phys. Chem.* **1906**, *57*, 385–470. [CrossRef]
42. Temkin, M.J.; Pyzhev, V. Recent Modifications to Langmuir Isotherms. *ActaPhysiochimca URSS* **1940**, *12*, 217–225.

43. Dubinin, M.M.; Radushkevich, L.V. Equation of the Characteristic Curve of Activated Charcoal. *Proc. USSR Phys. Chem.* **1947**, *55*, 331–333.
44. Bendahou, A.; Dufrense, A.; Kaddamia, H.; Habibi, Y. Isolation and structural characterization of hemicelluloses from palm of *Phoenix dactylifera* L. *Carbohydr. Polym.* **2007**, *68*, 601–608. [CrossRef]
45. Khiari, R.; Mhini, M.F.; Belgacem, M.N.; Mauret, E. Chemical composition and pulping of date palm rachis and Posidonia oceanica—A comparison with other wood and non-wood fibre sources. *Bioresour. Technol.* **2010**, *101*, 775–780. [CrossRef]
46. M'barek, I.; Slimi, H.; AlSukaibi, A.K.D.; Alimi, F.; Lajimi, R.H.; Mechi, L.; Bensalem, R.; Moussaoui, Y. Cellulose from Tamarixaphylla's stem via acetocell for cadmium adsorption. *Arab. J. Chem.* **2022**, *15*, 103679. [CrossRef]
47. Mechi, N.; Khiari, R.; Elaloui, E.; Belgacem, M.N. Preparation of paper sheets from cellulosic fibres obtained from Prunus Amygdalus and Tamarisk SP. *Cellul. Chem. Technol.* **2016**, *50*, 863–872.
48. Samar, M.; Khiari, R.; Bendouissa, N.; Saadallah, S.; Mhenni, F.; Mauret, E. Chemical composition and pulp characterization of Tunisian vine stems. *Ind. Crops Prod.* **2012**, *36*, 22–27.
49. Mannai, F.; Ammar, M.; Yanez, J.G.; Elaloui, E. Alkaline Delignification of Cactus Fibres for Pulp and Papermaking Applications. *J. Polym. Environ.* **2018**, *26*, 798–806. [CrossRef]
50. Saad, S.; Dávila, I.; Mannai, F.; Labidi, J.; Moussaoui, Y. Effect of the autohydrolysis treatment on the integral revalorisation of Ziziphus lotus. *Biomass Conv. Bioref.* **2022**, 1–13. [CrossRef]
51. Jimenez, L.; Sanchez, I.; Lopez, F. Olive wood as a raw material for paper manufacture. *Tappi J.* **1992**, *75*, 89–91.
52. Ferhi, F.; Das, S.; Moussaoui, Y.; Elaloui, E.; Yanez, J.G. Paper from *Stipagrostis pungens*. *Ind. Crops Prod.* **2014**, *59*, 109–114. [CrossRef]
53. Moussaoui, Y.; Ferhi, F.; Elaloui, E.; Bensalem, R.; Belgacem, M.N. Utilisation of *Astragalus Armatus* roots in papermaking. *Bioresources* **2011**, *6*, 4969–4978.
54. Ferhi, F.; Dass, S.; Elaloui, E.; Moussaoui, Y.; Yanz, J.G. Chemical characterisation and suitability for papermaking applications studied on four species naturally growing in Tunisia. *Ind. Crops Prod.* **2014**, *61*, 180–185. [CrossRef]
55. Benchagra, L.; Berrougui, H.; Islam, M.O.; Ramchoun, M.; Boulbaroud, S.; Hajjaji, A.; Fulop, T.; Ferretti, G.; Khalil, A. Antioxidant Effect of Moroccan Pomegranate (*Punica granatum* L. Sefri Variety) Extracts Rich in Punicalag in against the Oxidative Stress. *Process. Foods* **2021**, *10*, 2219.
56. Sanhueza, L.; García, P.; Giménez, B.; Benito, J.M.; Matos, M.; Gutiérrez, G. Encapsulation of Pomegranate Peel Extract (*Punica granatum* L.) by Double Emulsions: Effect of the Encapsulation Method and Oil Phase. *Foods* **2022**, *11*, 310. [CrossRef]
57. Mannai, F.; Elhleli, H.; Ammar, M.; Passas, R.; Elaloui, E.; Moussaoui, Y. Green process for fibrous networks extraction from Opuntia (*Cactaceae*): Morphological design, thermal and mechanical studies. *Industrial. Ind. Crops Prod.* **2018**, *126*, 347–356. [CrossRef]
58. Kasthuri, J.; Kathiravan, K.; Rajendiran, N. hyllanthin-Assisted Biosynthesis of Silver and Gold Nanoparticles: A Novel Biological Approach. *J.Nanopart. Res.* **2009**, *11*, 1075–1085. [CrossRef]
59. Tang, X.; Li, Y.; Chen, R.; Min, F.; Yang, J.; Dong, Y. Evaluation and modeling of methyl green adsorption from aqueous solutions using loofah fibers. *Korean J. Chem. Eng.* **2015**, *32*, 125–131. [CrossRef]
60. Abbas, M. Adsorption of methyl green (MG) in aqueous solution by titanium dioxide (TiO_2): Kinetics and thermodynamic study. *Nanotechnol. Environ. Eng.* **2021**, *7*, 1–12. [CrossRef]
61. Yin, W.; Hao, S.; Cao, H. Solvothermal synthesis of magnetic $CoFe_2O_4$/rGO nanocomposites for highly efficient dye removal in wastewater. *RSC Adv.* **2017**, *7*, 4062–4069. [CrossRef]
62. Sharma, P.; Saikia, B.K.; Das, M.R. Removal of methyl green dye molecule from aqueous system using reduced graphene oxide as an efficient adsorbent: Kinetics, isotherm and thermodynamic parameters. *Colloids Surf. A* **2014**, *457*, 125–133. [CrossRef]
63. Maghni, A.; Ghelamallah, M.; Benghalem, A. Sorptive removal of methyl green from aqueous solutions using activated bentonite. *Acta Phys. Pol. A* **2017**, *132*, 448–450. [CrossRef]
64. Rida, K.; Chaibeddra, K.; Cheraitia, K. Adsorption of cationic dye methyl green from aqueous solution onto activated carbon prepared from *Brachychiton populneus* fruit shell. *Indian J. Chem. Technol.* **2020**, *27*, 51–59.
65. Lagergren, S. Zurtheorie der sogenannten adsorption geloesterstoffe. *K. Sven. Vetensk. Handl.* **1898**, *24*, 1–39.
66. Saad, M.K.; Mnasri, N.; Mhamdi, M.; Chafik, T.; Elaloui, E.; Moussaoui, Y. Removal of methylene blue onto mineral matrices. *Desalin. Water Treat.* **2015**, *56*, 2773–2780. [CrossRef]
67. Saad, M.K.; Khiari, R.; Elaloui, E.; Moussaoui, Y. Adsorption of anthracene using activated carbon and Posidonia oceanica. *Arab. J. Chem.* **2014**, *7*, 109–113. [CrossRef]
68. Ho, Y.S.; McKay, G. Pseudo-second-order model for sorption processes. *Process. Biochem.* **1999**, *34*, 451–465. [CrossRef]
69. Maaza, L.; Djafri, F.; Djafri, A. Adsorption of Methyl Green onto Zeolite ZSM-5(pyrr.) in Aqueous Solution. *Orient J. Chem.* **2016**, *32*, 171–180.
70. Bahgat, M.; Farghali, A.A.; El Rouby, W.; Khedr, M.; Mohassab-Ahmed, Y. Adsorption of methyl green dye onto multi-walled carbon nanotubes decorated with Ni nanoferrite. *Appl. Nanosc.* **2013**, *3*, 251–261. [CrossRef]
71. Farghali, A.A.; Bahgat, M.; El Rouby, W.; Khedr, M. Preparation, decoration and characterization of graphene sheets for methyl green adsorption. *J. Alloys Compd.* **2013**, *555*, 193–200. [CrossRef]
72. Araissi, M.; Ayed, I.; Elaloui, E.; Moussaoui, Y. Removal of barium and strontium from aqueous solution using zeolite 4A. *Water Sci. Technol.* **2016**, *77*, 1628–1636. [CrossRef] [PubMed]

73. Taleb, F.; Ben Mosbah, M.; Elaloui, E.; Moussaoui, Y. Adsorption of ibuprofen sodium salt onto Amberlite resin IRN-78: Kinetics, isotherm and thermodynamic investigations. *Korean J. Chem. Eng.* **2017**, *34*, 1141–1148. [CrossRef]
74. Khadhri, N.; Saad, M.K.; Ben Mosbah, M.; Moussaoui, Y. Batch and continuous column adsorption of indigo carmine onto activated carbon derived from date palm petiole. *J. Environ. Chem. Eng.* **2019**, *7*, 102775. [CrossRef]
75. Atsahan, A.A. Adsorption of methyl green dye onto bamboo in batch and continuous system. *Iraqi J. Chem. Pet. Eng.* **2014**, *15*, 65–72.

Article

Introducing a Calculator for the Environmental and Financial Potential of Drain Water Heat Recovery in Commercial Kitchens

Isabel Schestak [1,*], Jan Spriet [2], David Styles [3] and A. Prysor Williams [1]

1. School of Natural Sciences, Bangor University, Bangor LL57 2UW, UK; prysor.williams@bangor.ac.uk
2. Department of Civil, Structural and Environmental Engineering, Trinity College, The University of Dublin, College Green, D02 PN40 Dublin, Ireland; sprietj@tcd.ie
3. Bernal Institute, School of Engineering, University of Limerick, V94 T9PX Limerick, Ireland; david.Styles@ul.ie
* Correspondence: isabel.schestak@bangor.ac.uk

Abstract: Food service providers like restaurants, cafes, or canteens are of economic importance worldwide, but also contribute to environmental impacts through water and energy consumption. Drain water heat recovery from commercial kitchens, using a heat exchanger, has shown large potential to decarbonise hot water use across food services, but is rarely deployed. This work translates previous findings on the technical feasibility and heat recovery potential for commercial kitchens into a publicly available calculator. It facilitates decision-making towards recovery and reuse of the freely available heat in kitchen drains by estimating both financial costs and payback time, as well as environmental burdens associated with the installation and environmental savings from avoided energy consumption. Environmental burdens and savings include, but are not limited to, carbon emissions. Further, the tool highlights key aspects of the technical implementation to understand installation requirements. The tool is freely available and could contribute to the uptake of heat recovery in the food service sector, ideally in conjunction with policy support through financial incentives or subsidies.

Keywords: carbon footprint; catering; climate change mitigation; cooking; energy efficiency; greenhouse gas emissions; hospitality; life cycle assessment; meal preparation; wastewater reuse

Citation: Schestak, I.; Spriet, J.; Styles, D.; Williams, A.P. Introducing a Calculator for the Environmental and Financial Potential of Drain Water Heat Recovery in Commercial Kitchens. *Water* 2021, *13*, 3486. https://doi.org/10.3390/w13243486

Academic Editors: Martin Wagner and Sonja Bauer

Received: 30 September 2021
Accepted: 2 December 2021
Published: 7 December 2021

Publisher's Note: MDPI stays neutral with regard to jurisdictional claims in published maps and institutional affiliations.

Copyright: © 2021 by the authors. Licensee MDPI, Basel, Switzerland. This article is an open access article distributed under the terms and conditions of the Creative Commons Attribution (CC BY) license (https://creativecommons.org/licenses/by/4.0/).

1. Introduction

The food service sector is an important economic part of the global tourism and hospitality industry. After transport and accommodation, tourist expenditure on food services generates the third highest revenue in tourism [1,2]. Globally, an estimated 75 billion meals are served in international tourism [3]. Additionally, food services are enjoyed by local visitors as well as at their places of work or education. They include restaurants (including those in hotels), pubs, cafes, catering services, canteens (e.g., in schools, businesses, or hospitals), and quick services (e.g., takeaways, fast food) [4]. In the US, food services are the largest private-sector employer, and 42% of the food expenditure in 2008 by US consumers was in food service establishments [5,6]. In 2019, consumers in the Republic of Ireland spent approximately 1300 Euro per person on food and non-alcoholic drinks consumed outside their home [7,8]. These statements relate to pre-pandemic times, but it can be expected that the tourism and food service sector will continue its pre-pandemic growth, once recovered [9,10].

1.1. Water, Energy, and Greenhouse Gas Emissions in Food Services

The hospitality and food service sectors not only contribute to economic revenue but are also major consumers of water and energy. In the United Kingdom (UK), for instance, approximately 143 M m³ of water was used in hospitality and food services in 2010 to prepare 8.2 billion meals, which compares to 190 M m^3 in the same year in the total food and drink manufacturing subsectors [4].

Energy consumption for the preparation of meals in food services in the UK stands at 36 TWh of primary energy per year, equalling 46% of primary energy consumption in hospitality and 10% of the primary energy of the whole service sector, i.e., more primary energy is used in food services than for domestic cooking, with 24 TWh [11]. Pubs and restaurants alone are estimated to have an energy consumption of approximately 7.5 TWh [12]. In Austria, 450 million meals were consumed by foreign and domestic tourists in 2017 and energy consumption ranges between 5–10 kWh/meal, resulting in 2.3–4.4 TWh for all tourist meals [10].

Energy consumption in food services is associated with considerable amounts of greenhouse gas (GHG) emissions. In 2017, energy consumption in UK food services was met 70% by fossil fuels, 21% by electricity, and just 9% by bio- and waste energy [11]. In light of the urgency to reduce global carbon emissions and mitigate climate change, the UK hospitality association has signed an agreement, the "Courtauld Commitment", targeting a 50% reduction of GHG emissions between 2015 and 2030 [13,14].

1.2. Heat Recovery for GHG Mitigation

Due to the use of hot water, e.g., for dishwashing and boiling food, wastewater from commercial kitchens constitutes a considerable source of heat, which is available independently of the season or weather as can be the case with renewable energy such as wind or solar energy. Drain water heat recovery therefore offers the opportunity to decarbonise thermal energy consumption in food services. This work is based on previous research from [15,16] on the drain water heat recovery potential and its technical, environmental, and financial aspects in the widely unexplored field of commercial kitchens. A study by [15] found that approximately 1.4 TWh of heat could be recovered every year in the UK food service sector, 90% of which could be achieved in a financially feasible way. In a case study restaurant, recovered heat from drain water was estimated to be capable of replacing 30% of the thermal energy needed for water heating [16]. GHG emissions of approximately 490 kt CO_2 eq. could potentially be avoided if appliances for direct heat recovery from drain water were installed in the over 250,000 commercial food outlets in the UK [16]. This would equal 52 kt CO_2 eq. in Ireland's 27,000 outlets [7]. Further environmental benefits beyond climate change mitigation were found to apply for drain water heat recovery from kitchens, such as the reduction of eutrophication, acidification, human and eco-toxicity, or resource depletion, even under consideration of the life cycle environmental impacts of the required heat recovery equipment. Environmental savings could be shown for the avoidance of both fossil fuels, such as natural gas, and renewable energy, such as geothermal or solar energy (Supplementary Materials S1). Despite promising energy and emission savings, decentralised drain water heat recovery is rarely applied to commercial kitchens outside of case study restaurants and thus represents a novel field of application.

Instead, research and application of decentralised drain water heat recovery has to date mostly focused on shower drains. Energy savings from shower drain heat recovery for hot water heating have been reported in the range of 4–15% annually in a high-rise residential building in Hong Kong [17] and an estimated 9–27% depending on the size of the heat exchanger and water use pattern, amongst others [18]. McNabola and Shields estimated the national savings potential for shower drain water heat recovery for Ireland with 808 GWh of energy or 400 kt CO_2 eq. of emissions per year, using a horizontally installed heat exchanger [19]. The quantity of energy recovered and hence, financial and environmental viability, depends on the design and orientation (vertical or horizontal) of the heat exchanger, and increases with the volume of wastewater and its temperature flowing through the heat exchanger [20–23]. Intensive use of a heat exchanger is therefore preferable and some studies have been conducted on drain water heat recovery in hospitals, hotels, or sport facilities [21–24]. In the UK, the average domestic daily water consumption is about 350 L/day [25], while commercial kitchens reach an average daily water flow of up to 12,500 L/day [15], presenting a benefit of kitchen drain water heat recovery.

As a lack of awareness of the possibility of drain water heat recovery from commercial kitchens and its potential environmental and financial gains exists amongst food outlets, there is a need for a tool to support the uptake of heat recovery in this novel and promising area of application.

1.3. Facilitating the Implementation of Heat Recovery in Food Businesses

This work translates the findings from studies by Spriet et al. [15] and Schestak et al. [16] (see Section 2) into a freely available decision-support tool for commercial kitchens in order to facilitate the implementation of drain water heat recovery as a measure to reduce climate change and other environmental impacts from food services. The tool intends to help overcome barriers which have been identified for the uptake of efficiency measures in the food service industry. Becken and Dolinicar [26] conducted a survey amongst over 600 small and medium-sized (SME) accommodation and food service businesses in the US and Europe. They found the most important barriers to be costs or administrative hurdles, followed by the difficulty of choosing the right measure. On the other hand, cost savings and financial or fiscal incentives as well as environmental protection were said to be the top motivators. Strategies suggested to foster the implementation of efficiency measures were grants and subsidies, but also consultancy, the demonstration of new technologies, and support through reporting and self-learning tools. We pick up on these findings by providing an easy-to-use tool which informs about (a) the technology of drain water heat recovery, including the required equipment and implementation conditions; (b) the kitchen-specific/user-specific potential financial costs and savings; and (c) the individual potential for environmental benefits, including but not limited to, carbon emission savings. The latter, apart from addressing food businesses' own "green conscience", increases the attractiveness of the business given a growing consumer interest in sustainability [6]. According to the US National Restaurant Association, 60% of consumers claim to be "likely to choose a restaurant based on its environmental efforts" and 44% based on a restaurant's energy and water conservation efforts [6]. The results generated through the tool enable the users to judge if drain water heat recovery can be technically feasible, as well as financially and environmentally viable, in their own food establishment.

2. Methodology

The tool is based on the research and findings from [15,16], which estimated the heat recovery potential as well as the financial and environmental performance of kitchen drain heat recovery based on monitoring campaigns in the UK and Ireland. The following paragraphs summarise the most important background information on the development of the toolkit taken from [15,16], explaining the technical assumptions and data underlying the calculations of the toolkit, as well as exploring the user interface with data entry and generated results. For further methodological background, refer to [15,16].

Due to data availability from the monitoring campaigns, the toolkit has been developed to represent commercial kitchens in Ireland and the UK, using country-specific background data on costs and environmental impacts. However, the toolkit can be adapted to other countries worldwide.

2.1. Drain Water Heat Recovery Technology

The heat recovery system considered here uses a tube-in-tube heat exchanger and hence represents a kind of direct and passive heat recovery without the need for a heat pump. This and similar kinds of heat exchangers are typically used for shower water heat recovery, for which they have been studied and installed in domestic and commercial settings across the UK and Europe [20,21,27–30]. In 2020, the installation of a heat recovery system was completed at a case study restaurant at the Penrhyn Castle tourist attraction in Wales, UK, initiated and overseen by the authors as part of the Dŵr Uisce

INTERREG project, and was proven to be effective at recovering heat from kitchen drain water [15,16,31,32].

The heat exchanger is made from copper in the form of a double walled pipe. The inner pipe leads the warm kitchen drain water, while the outer pipe leads the incoming cold water which is pre-warmed in the heat exchanger and then enters the boiler or other conventional water heating system (Figure 1). For this calculator, the heat exchanger is assumed to replace a part of the kitchen drain of approximately 2 m in height; however, shorter heat exchangers are available on the market as well. In order to reach its maximum heat exchange effectiveness of about 60%, it needs to be installed vertically [33]. With currently available designs, vertical heat exchangers have been proven to reach higher heat exchange effectiveness and are less prone to blockage than horizontal ones, which is important for the use with kitchen wastewater carrying higher organic contaminations than shower wastewater [19,30]. In order to connect the heat exchanger to the cold-water supply and boiler, fittings, and further pipework are required, the length of which depends on site-specific conditions, and which are assumed to be insulated.

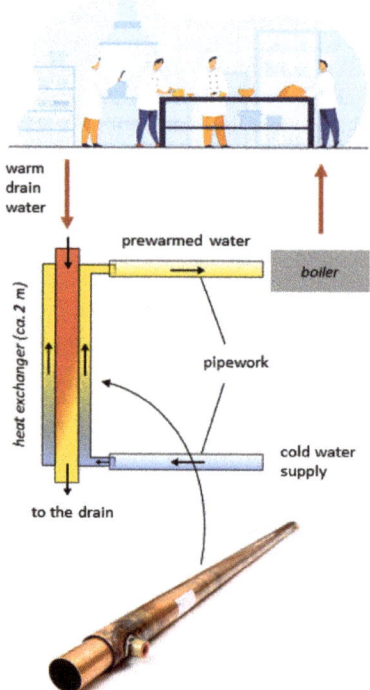

Figure 1. Overview of heat recovery from kitchen drain water with a pipe-in-pipe heat exchanger. With graphical elements modified from [27,34].

Although previous studies revealed that financial and environmental savings are likely for the large majority of commercial kitchens in the UK—90% when considering an average kitchen in regard to the heating energy sources used [15,16]—the suitability of a kitchen for drain water heat recovery ultimately depends on three main factors. These are (1) the amount of recoverable heat, depending on the water consumption or kitchen size; (2) the type of energy source replaced through heat recovery; and (3) the equipment needed and its material (e.g., copper, steel, or polyethylene pipework), which differs from kitchen to kitchen. For pipework, the user of the toolkit can therefore choose between three materials: copper, steel, and polyethylene (PE).

2.2. Heat Recovery Potential

The calculation of the heat recovery potential is based on data from a monitoring campaign at several kitchens in the UK and Ireland, during which drain water temperatures were measured, amongst other parameters [15]. Spriet and McNabola [15] determined the average heat recovery potential of a commercial kitchen taking into account varying drain water flows over the time of day during kitchen operation, the temporal mismatch between drain water flow and hot water consumption, a retention time of water through kitchen appliances of no longer than 1 h, a 90% return rate of the consumed water into the drain, and the use of several heat exchangers in parallel for high flow rates. The remaining 10% of the water consumption was considered to leave the kitchen via different means such as evaporation or incorporation into food. Here, we convert the information on the amount of heat recovered, expressed as kWh/L of water consumed at a specific water flow rate (derived from [15], Table 1), into the amount of heat recovered based on user data entry (see Section 3.1). The amount of recovered heat equals the amount of energy saved which would otherwise have been provided through the conventional water heating energy source of the respective kitchen. For further methodological details about the calculation of the heat recovery potential, refer to [15].

Table 1. Background data on (A) flow-specific heat recovery potential and (B) water consumption per meal.

(A) Flow, Heat Recovery, and Number of Heat Exchanger Pipes Derived from [15,16]		
Flow (L/h)	Heat Recovery (kWh/L)	# Of Heat Exchanger Pipes
45	0.0083	1
69	0.0083	1
75	0.0083	1
94	0.0083	1
120	0.0083	1
185	0.0083	1
200	0.0083	1
250	0.0082	1
390	0.0078	1
601	0.0081	2
650	0.0080	2
750	0.0078	2
813	0.0077	2
1156	0.0078	3
1250	0.0077	3
1563	0.0078	4
(B) Water Consumption from [4]		
(L/meal)	Type of Food Outlet	
12	Quick service category: pub, fast food, cafe, takeaway, mobile catering	
18.5	Canteen category: staff catering, schools, universities	
20	Hotel category: restaurant in a hotel	
25	Restaurant category: restaurant with table service, food services in hospitals and care/nursing homes	

As the heat recovery potential considers an average water consumption and wastewater generation in a kitchen, the toolkit only provides an estimate of the heat recovery potential of a kitchen. This, however, has the advantage of not requiring data on water use through different appliances or at a specific time of the day and hence simplifies the use of the toolkit (see Section 3.1).

2.3. Financial Assessment

Financial cost-effectiveness is considered a key driver for the uptake of kitchen drain heat recovery. The calculator considers both operational financial savings through avoided energy use, and the investment costs (capital costs) for the heat exchanger, pipework, fittings, and insulation. Purchase costs are taken from manufacturers or suppliers of the respective parts in Ireland and the UK. A heat exchanger such as the one suggested by the toolkit costs around £500, or €600 [15]. Savings depend on the heat recovery potential and the energy price of the conventional heating source provided by the user. Labour costs have not been included in the financial assessment due to high variability: they depend not only on specific hourly rates charged by the installing company, but also on site-specific conditions, such as distance and obstacles between the drainpipe and boiler. Nevertheless, the total operational savings for a chosen service life provide the user with a financial frame for labour costs while ensuring payback.

2.4. Environmental Assessment

Similar to the financial assessment, the environmental assessment considers operational environmental savings, such as GHG savings, but also environmental burdens connected to the equipment. Burdens related to the environmental footprint from the life cycle (manufacture, use, end of life) of the heat recovery system, including all parts (heat exchanger, pipework, insulation, fittings), have been determined via a complete cradle-to-grave life cycle assessment (LCA) [16] (Supplementary Materials S2). Environmental burdens across seven impact categories have been shown to derive mainly from the manufacture of the copper heat exchanger, linked to emissions from mining and energy use during processing of the material, such as forming and finishing [16]. The environmental impacts considered in the toolkit are climate change (GHG emissions), acidification, freshwater eutrophication, freshwater ecotoxicity, and resource depletion (of mineral, fossil, and renewable resources), as recommended by the International Reference Life Cycle Data System (ILCD) handbook [35].

The environmental profile greatly depends on the amount and material of the pipework used, which can exceed burdens from the heat exchanger, especially when copper or steel is used. Further, environmental savings are influenced by the amount and the type of water heating energy source replaced through heat recovery. The user can choose between the following energy sources: natural gas, grid electricity, green electricity, light fuel oil, geothermal energy, solar thermal energy, and wood chips, modelled as in [16]. Grid electricity is modelled as the marginal grid electricity from a natural gas power plant, while green electricity represents the current mix of renewable energies in the UK [36].

3. User Interface

The calculator is based on MS Excel to offer accessibility to a wide range of users in line with similar energy efficiency and carbon emission calculators [37,38], and has been developed to be used by those without a sciences or technical background. Functionality and user-friendliness have been validated with kitchen managers, and the calculator was adapted upon feedback provided during a public webinar where the calculator was presented. The tool is divided into three main sections: (1) an introduction which describes the purpose of the tool, intended users, considered technology, technical requirements, and expected outcomes in layperson language; (2) data entry; and (3) individual results and conclusions.

3.1. Data Entry

The calculator requires the user to enter the following data:
- Water consumption: This can be entered as yearly water consumption in cubic metres or, alternatively—depending on the data availability of the user—as the number of meals served per year and specification of the type of food outlet. To facilitate data entry, sample values are provided. If served meals are used, yearly water consumption is derived from benchmark values for water consumption per meal for specific food outlet types from [4] (Table 1). This is available for restaurants with table service, hospital and nursing home kitchens, hotel restaurants, canteens for staff catering,

schools, or universities, and quick service restaurants which include pubs, fast food, cafes, takeaways, and mobile catering.
- Opening times: This information is required to determine the hourly water flow rate which links to the heat recovery potential.
- Currently used energy source for hot water: This determines the environmental savings through avoided energy consumption.
- Country: Either UK or Ireland in the current version of the calculator, to derive country-specific installation costs.
- Energy price for water heating per kWh: To determine financial operational savings.
- Approximate distance between the kitchen drainpipe and boiler: This information serves to determine the amount of pipework required for the installation.

The calculator assumes a service life of 10 years and polyethylene pipework as default values; however, the user can change these values to understand how they affect the results. As no long-term study has been performed using the proposed heat recovery system with kitchen drain water, the calculator assumes a conservative maximum service life of 20 years compared to a service life of up to 50 years with shower water [21].

3.2. Results

The following results are generated by the tool:
- Heat recovery potential: The calculator provides the amount of heat which can potentially be recovered per year, and during the whole service life, expressed in kWh (Figure 2).
- Financial assessment: The financial results include the operational savings per year and over the service life, total investment costs (capital costs), and simple payback time in years (Figure 3).
- Environmental assessment: This section is divided into savings and impacts related to carbon emissions and other environmental categories. The carbon section contains—equivalent to the financial results—operational carbon savings per year and per service life, carbon costs (footprint), and carbon payback time (Figure 4). The carbon savings are also translated into the amount of vehicle kilometres saved, considering the emissions of an average gasoline car in Europe [39], for context.
- Further to carbon savings, the results for four other environmental impact categories are shown. In a compromise between comprehensiveness and consideration of a user without an environmental sciences background, the following four categories have been chosen for display: acidification of soils and water bodies, freshwater eutrophication, freshwater ecotoxicity, and resource depletion (mineral, fossil, and renewable resources) (Figure 5).
- Conclusion: A conclusion field contains a summary of the overall results, concluding if heat recovery could be applied in a financially and environmentally viable way, based on the user-specific data entry.

Results are displayed in parallel for (a) baseline and (b) custom assumptions regarding the material of the pipework and the service lifetime. This enables the direct comparison of results and the evaluation of individual choices which is relevant to understand the higher environmental burden and financial costs of copper and steel compared to polyethylene. Figures 2–5 exemplify the results for a kitchen with a water consumption of 2000 m^3/year, which can be considered a medium-sized kitchen compared to average rates of different food outlet categories in the UK, which range from 360 to 12,500 L/day or 130 to 4600 m^3/year [15]. The results in the example show that, when the recovered heat replaces natural gas at a rate of 3.1 Eurocent/kWh [40], and 10 m of pipework (polyethylene) is needed in addition to the heat exchanger, the capital costs would be paid back within just over 4 years. The carbon footprint would be offset in under one month. In case of a heat recovery system with steel pipework (example for customised data entry), environmental payback times would range from under one month (carbon footprint) to over three years (resource depletion), i.e., the heat recovery system in the example would offer environmental viability, as environmental burdens are paid back within its service life.

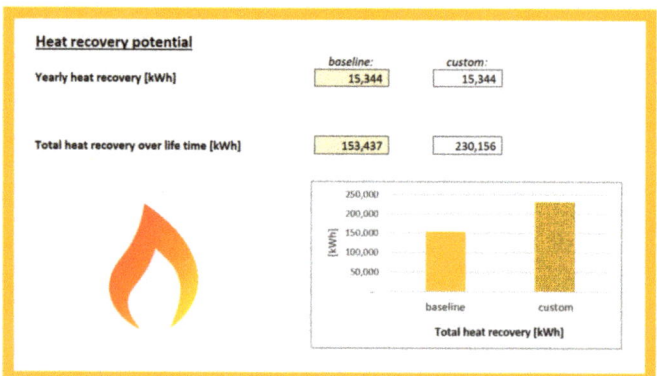

Figure 2. Example for results of the heat recovery potential from the toolkit. Assumptions: water consumption of 2000 m^3/year and a service life of 10 years (baseline) and 15 years (custom).

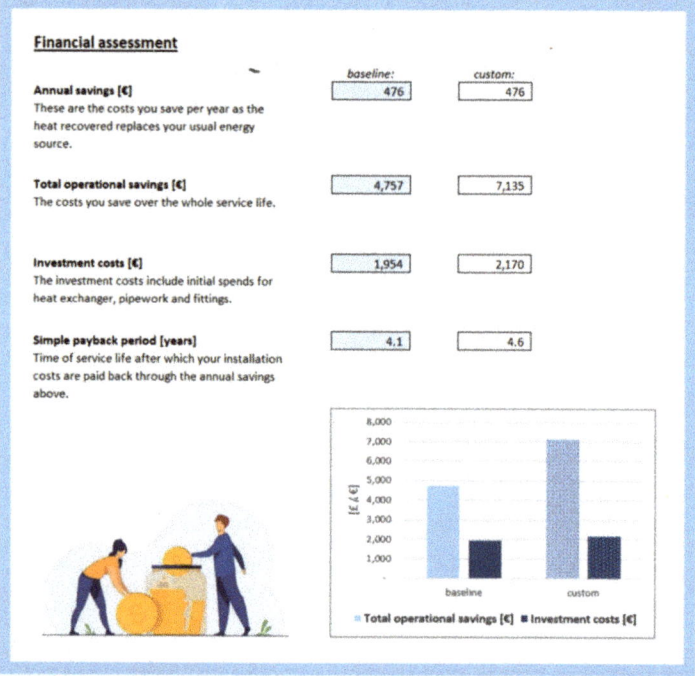

Figure 3. Example for results of the financial assessment of the heat recovery toolkit. Assumptions: water consumption of 2000 m^3/year; natural gas as heating fuel at a rate of 3.1 Eurocent/kWh [40]; country: Ireland; distance between kitchen drain and boiler of 10 m. Baseline: 10-year service life and pipework from polyethylene. Custom: 15-year service life and pipework from steel. With graphical element modified from [41].

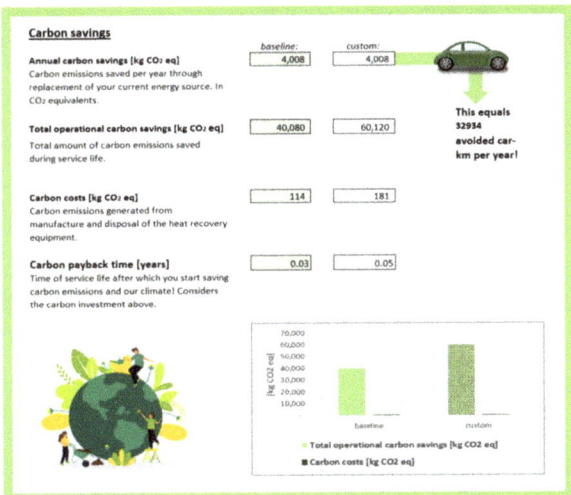

Figure 4. Example for results on carbon costs (footprint) and savings from the heat recovery toolkit. Assumptions: see Figure 3. With graphical element modified from [42].

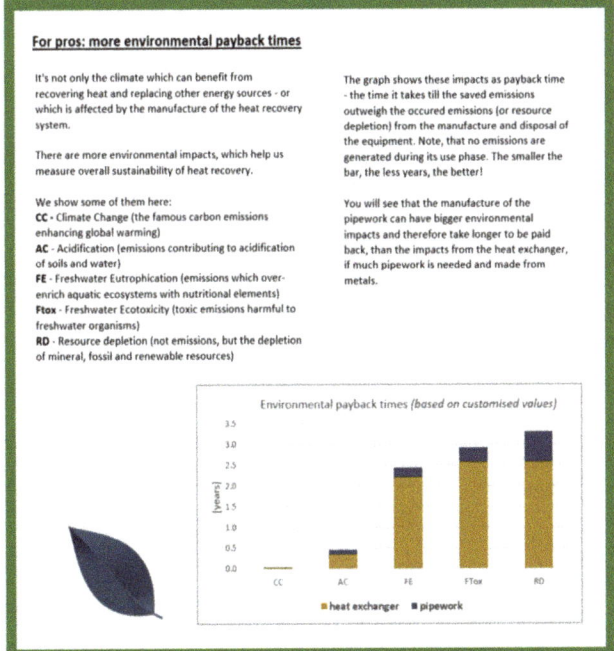

Figure 5. Example for results for payback time (in years) for carbon emissions and other environmental impacts of the heat recovery system, with a water consumption of 2000 m^3/year, natural gas as replaced heating fuel, and 10 m pipework from steel. With graphical element modified from [43].

4. Expected Learnings

The toolkit delivers an estimate for the heat recovery potential of commercial kitchens, as a basis for encouraging heat recovery in appropriate contexts. It facilitates decision-making for kitchen owners interested in reducing the carbon footprint of their business (whether in favour of or against heat recovery depending on, e.g., kitchen size and drain water flow). Following the use of the toolkit, the user will be able to answer the following questions:

- Is it technically possible to install this heat recovery system within my kitchen?
- What are the capital costs?
- Does the financial payback fit the business plan?
- Is it environmentally beneficial to recover heat? Are burdens of the system paid back within the intended service life?
- Can pipe material choice be adapted to reduce the footprint of the equipment, and reduce environmental payback time?
- Is it worth undertaking further planning into heat recovery?

5. Discussion of Heat Recovery Application in Kitchens

The achievable benefits of recovering heat from commercial kitchens regarding environmental and financial aspects have been outlined in the Introduction. Some challenges, however, need to be addressed for its application. This article focused on overcoming potential issues such as a lack of awareness by kitchen owners of the availability of heat recovery and its benefits. A lack of awareness and access to information has also been identified as an issue for the implementation of heat recovery systems for showers by domestic users [30]. From a technological perspective, the necessity to access a part of the drainpipe with an approximately 2 m vertical drop can constrain the installation of a heat exchanger, as can the existence of obstacles between the drainpipe and the boiler. The number of potentially affected kitchens is unknown. In shower heat recovery systems, this issue can be overcome by the installation of a horizontal heat exchanger [19,20,30], which could, however, lead to increased blockage when used with kitchen drain water. Heat recovery from a kitchen's grease trap could pose a more accessible heat recovery option worthwhile to study in future research. Currently, a heat recovery system for kitchen drains—mainly the connecting pipes—must be customised to fit site-specific conditions, which could increase investment costs. With a broad uptake of (kitchen) drain water heat recovery, increased installations experience, and specialisation of engineering companies into this type of heat recovery application, a decrease in these costs can be expected.

Rising energy prices such as the recently observed rise in the natural gas price in Europe [44] will further increase the financial profitability of a heat recovery installation. Nevertheless, as financial advantages play a vital role in the realisation of energy efficiency measures, public support and state aid through subsidies could further accelerate heat recovery implementation. This would be specifically beneficial for smaller food outlets where heat recovery proves to be environmentally, but not financially viable. Another policy measure could be the promotion of kitchen heat recovery through its inclusion into the reference document on best environmental management practice of the EU Eco-Management and Audit Scheme (EMAS) for the tourism sector [45,46]. The sectoral reference document for the tourism sector contains best environmental practices for restaurant and hotel kitchens and proposes measures for the reduction of water or energy consumption, for instance [45]. Heat recovery from drain water could be included as a measure for energy conservation, and the energy consumption of kitchens applying heat recovery could serve as an energy benchmark.

6. Conclusion and Outlook

A toolkit has been developed which provides technology guidance and estimates thermal energy, financial, and environmental savings potential associated with heat recovery from the drain water of commercial kitchens. It includes on the one hand environmental burdens from the life cycle of the required equipment as well as investment costs, and on the other hand operational environmental and financial savings. Based on empirically monitored data from several kitchens, the toolkit delivers a customised estimate for heat recovery potential and a first decision pro/contra installation of a heat recovery system.

The calculator is a first step towards enhancing the implementation of heat recovery in practice, which has already shown its theoretical potential for the cost-effective reduction of GHG emissions in the food service sector across the UK. The toolkit is adapted in language and content to be accessible to users lacking an engineering or environmental sciences background and requires no extensive data collection. If the toolkit generates results in favour of heat recovery, a kitchen owner is encouraged to seek professional advice for an individual assessment, which then enables detailed site-specific information to be considered. Although the toolkit cannot replace professional planning, it can raise awareness and interest by showcasing financial and environmental gains achievable from the "free heat" embedded in drain water.

Passive drain water heat recovery with a heat exchanger can be regarded as a low-cost decarbonisation measure. Nevertheless, further support of this measure in the form of subsidies could augment its uptake, especially in cases where there is a gap between a short environmental, and a long financial, payback time, or where investment costs are higher due to site-specific conditions.

The tool is available for download free of charge at the Dŵr Uisce project website: https://www.dwr-uisce.eu/heat-recovery-tool.

Supplementary Materials: The following are available online at https://www.mdpi.com/article/10.3390/w13243486/s1, Figure S1: Comparison of environmental impacts from heating water through heat recovery or conventional energy sources, Figure S2: Life Cycle steps included in the environmental footprint of the heat recovery system.

Author Contributions: Conceptualization, I.S., D.S. and A.P.W.; formal analysis, I.S.; funding acquisition, A.P.W.; investigation, I.S.; methodology, I.S.; project administration, A.P.W.; supervision, D.S. and A.P.W.; validation, I.S. and J.S.; visualization, I.S.; writing—original draft preparation, I.S.; writing—review and editing, J.S., D.S. and A.P.W.; All authors have read and agreed to the published version of the manuscript.

Funding: This research is part of the Dŵr Uisce project, which aims at improving the long-term sustainability of water supply, treatment, and end-use in Ireland and Wales. The project has been funded by the European Regional Development Fund (ERDF) Interreg Ireland-Wales Programme 2014–2023 (grant number 14122).

Informed Consent Statement: Not applicable.

Data Availability Statement: Data sharing is not applicable to this article.

Acknowledgments: Thank you to Roberta Bellini for the professional support in setting up the webinar and for making the toolkit available online.

Conflicts of Interest: The authors declare no conflict of interest.

References

1. Özgen, I.; Binboğa, G.; Güneş, S.T. An assessment of the carbon footprint of restaurants based on energy consumption: A case study of a local pizza chain in Turkey. *J. Foodserv. Bus. Res.* **2021**, *24*, 711–731. [CrossRef]
2. World Trade Organization. *Tourism Services: Background Note by the Secretariat*; World Trade Organization, Council for Trade in Services: Geneva, Switzerland, 1998.
3. Gössling, S.; Garrod, B.; Aall, C.; Hille, J.; Peeters, P. Food management in tourism: Reducing tourism's carbon 'foodprint,'. *Tour. Manag.* **2011**, *32*, 534–543. [CrossRef]
4. Bromley-Challenor, K.; Kowalski, M.; Barnard, R.; Lynn, S. *Water Use in the UK Food and Drink Industry*; Technical Report; Waste & Resources Action Programme (WRAP): Banbury, UK, 2013.
5. U.S. Department of Labor – Bureau of Labor Statistics. *Consumer Expenditures in 2008*; U.S. Department of Labor – Bureau of Labor Statistics: Washington, DC, USA, 2010.
6. Baldwin, C.; Wilberforce, N.; Kapur, A. Restaurant and food service life cycle assessment and development of a sustainability standard. *Int. J. Life Cycle Assess.* **2011**, *16*, 40–49. [CrossRef]
7. Bord Bia. *Irish Foodservice Market and Consumer Insights*; Bord Bia Irish Food Board: Dublin, Ireland, 2019.
8. Central Statistics Office of the Republic of Ireland. *Ireland's Facts and Figures 2019*; Central Statistics Office of the Republic of Ireland: Cork, Ireland, 2019.

9. United Nations World Tourism Organization. World Tourism Barometer and Statistical Annex, January 2020. Available online: https://www.e-unwto.org/doi/epdf/10.18111/wtobarometereng.2020.18.1.1 (accessed on 23 September 2021).
10. Lund-Durlacher, D.; Gössling, S. An analysis of Austria's food service sector in the context of climate change. *J. Outdoor Recreat. Tour.* **2021**, *34*, 100342. [CrossRef]
11. Department for Business, Energy & Industrial Strategy. *Energy Consumption in the UK (ECUK)*; Department for Business, Energy & Industrial Strategy: London, UK, 2018.
12. Mudie, S. Energy benchmarking in UK commercial kitchens. *Build. Serv. Eng. Res. Technol.* **2016**, *37*, 205–219. [CrossRef]
13. UK Hospitality. Sustainability and the Hospitality Industry. 2021. Available online: https://www.ukhospitality.org.uk/page/sustainability (accessed on 30 August 2021).
14. Waste and Resources Action Programme. The Courtauld Commitment 2030. 2021. Available online: https://wrap.org.uk/taking-action/food-drink/initiatives/courtauld-commitment (accessed on 30 August 2021).
15. Spriet, J.; McNabola, A. Decentralized drain water heat recovery from commercial kitchens in the hospitality sector. *Energy Build.* **2019**, *194*, 247–259. [CrossRef]
16. Schestak, I.; Spriet, J.; Styles, D.; Williams, A.P. Emissions down the drain: Balancing life cycle energy and greenhouse gas savings with resource use for heat recovery from kitchen drains. *J. Environ. Manage.* **2020**, *271*, 110988. [CrossRef] [PubMed]
17. Wong, L.T.; Mui, K.W.; Guan, Y. Shower water heat recovery in high-rise residential buildings of Hong Kong. *Appl. Energy* **2010**, *87*, 703–709. [CrossRef]
18. Zaloum, C.; Gusdorf, J.; Parekh, A. *Performance Evaluation of Drain Water Heat Recovery Technology at the Canadian Centre for Housing Technology—Final Report*; Sustainable Buildings and Communities, Natural Resources Canada: Ottawa, ON, Canada, 2007.
19. McNabola, A.; Shields, K. Efficient drain water heat recovery in horizontal domestic shower drains. *Energy Build.* **2013**, *59*, 44–49. [CrossRef]
20. Pochwat, K.; Kordana-Obuch, S.; Starzec, M.; Piotrowska, B. Financial analysis of the use of two horizontal drain water heat recovery units. *Energies* **2020**, *13*, 15. [CrossRef]
21. Ip, K.; She, K.; Adeyeye, K. Life-cycle impacts of shower water waste heat recovery: Case study of an installation at a university sport facility in the UK. *Environ. Sci. Pollut. Res.* **2018**, *25*, 19247–19258. [CrossRef] [PubMed]
22. Kordana, S.; Słyś, D.; Dziopak, J. Rationalization of water and energy consumption in shower systems of single-family dwelling houses. *J. Clean. Prod.* **2014**, *82*, 58–69. [CrossRef]
23. Słyś, D.; Kordana, S. Financial analysis of the implementation of a Drain Water Heat Recovery unit in residential housing. *Energy Build.* **2014**, *71*, 1–11. [CrossRef]
24. Nagpal, H.; Spriet, J.; Murali, M.K.; McNabola, A. Heat recovery from wastewater—A review of available resource. *Water* **2021**, *13*, 9. [CrossRef]
25. EST. *At Home with Water*; Energy Saving Trust: London, UK, 2013.
26. Becken, S.; Dolnicar, S. Uptake of resource efficiency measures among European small and medium-sized accommodation and food service providers. *J. Hosp. Tour. Manag.* **2016**, *26*, 45–49. [CrossRef]
27. BPD Ltd. Showersave Vertical System. 2021. Available online: https://showersave.com/vertical-wwhrs/ (accessed on 25 March 2021).
28. Recoup. Passively Recovering Waste Heat Energy with Every Shower—Case Studies. 2021. Available online: https://recoupwwhrs.co.uk/case-studies/ (accessed on 25 September 2021).
29. Joulia. Joulia Switzerland—Auszug Referenzen. 2021. Available online: https://joulia.com/en/ (accessed on 25 September 2021).
30. Kordana-Obuch, S.; Starzec, M.; Słyś, D. Assessment of the feasibility of implementing shower heat exchangers in residential buildings based on users' energy saving preferences. *Energies* **2021**, *14*, 17. [CrossRef]
31. Spriet, J.; McNabola, A. Drain Water Heat Recovery in Commercial Kitchens: Case of Tourist Attraction. *Dubrovnik.* 2019. Available online: https://www.dwr-uisce.eu/conference-proceedings-and-posters (accessed on 6 July 2021).
32. Dŵr Uisce. Dŵr Uisce—Distributing our Water Resources: Utilising Integrated, Smart and Low-Carbon Energy. *Project Funded by the European Regional Development Fund (ERDF) through the Ireland Wales Co-Operation Programme 2014–2023*. 2021. Available online: https://www.dwr-uisce.eu/ (accessed on 24 September 2021).
33. Q-Blue b.v. Installation Manual Showersave QB1-12-16-21. 2018. Available online: https://www.q-blue.nl/en/products/q-blue-showersave-en (accessed on 2 February 2021).
34. Freepik. Chefs Cooks and Waiters Working at Restaurant Kitchen. *Food Vector Created by PCH.Vector.* 2021. Available online: https://www.freepik.com/vectors/food (accessed on 12 June 2021).
35. E. JRC-IES. *ILCD Handbook—Recommendations for Life Cycle Impact Assessment in the European Context*; European Commission—Joint Research Centre—Institute for Environment and Sustainability (EC JRC-IES) Publication Office of the European Union: Luxembourg, 2011.
36. BEIS. Digest of UK Energy Statistics (DUKES): Renewable Sources of Energy, Chapter 6. 2020. Available online: https://assets.publishing.service.gov.uk/government/uploads/system/uploads/attachment_data/file/904823/DUKES_2020_Chapter_6.pdf (accessed on 29 March 2021).
37. Carbon Trust. Catering Cut Costs & Carbon Calculator. 2013. Available online: https://www.carbontrust.com/resources/catering-cut-costs-and-carbon-calculator (accessed on 24 September 2021).
38. Energy Star, Cash Flow Opportunity Calculator. 2018. Available online: https://www.energystar.gov/CFOcalculator (accessed on 24 September 2021).

39. Transport & Environment. *CO$_2$ Emissions from Cars: The Facts*; European Federation for Transport & Environment: Brussels, Belgium, 2018.
40. SEAI—Sustainable Energy Authority of Ireland. Average Gas Price to Business, ex-VAT. 2021. Available online: https://www.seai.ie/data-and-insights/seai-statistics/key-statistics/prices/ (accessed on 24 September 2021).
41. Freepik. Family Couple Saving Money. People Vector Created by PCH.Vector. 2021. Available online: https://www.freepik.com/vectors/people (accessed on 12 June 2021).
42. Freepik. Save Planet Concept with People Taking Care Earth. World Vector Created by Freepik. 2021. Available online: https://www.freepik.com/vectors/world (accessed on 12 June 2021).
43. Clipart Library. Leaf with Transparent Background. 2021. Available online: http://clipart-library.com/clip-art/leaf-with-transparent-background-8.htm (accessed on 12 June 2021).
44. ACER. *High Energy Prices October 2021*; ACER—European Union Agency for the Cooperation of Energy Regulators: Ljubljana, Slovenia, 2021.
45. Styles, D.; Schönberger, H.; Martos, J.L.G. *JRC Scientific and Policy Report on Best Environmental Management Practice in the Retail Trade Sector*; Learning from Frontrunners; Publications Office of the European Union: Luxembourg, 2013.
46. European Commission. Eco-Management and Audit Scheme. 2021. Available online: https://ec.europa.eu/environment/emas/index_en.htm (accessed on 14 November 2021).

MDPI
St. Alban-Anlage 66
4052 Basel
Switzerland
Tel. +41 61 683 77 34
Fax +41 61 302 89 18
www.mdpi.com

Water Editorial Office
E-mail: water@mdpi.com
www.mdpi.com/journal/water

www.ingramcontent.com/pod-product-compliance
Lightning Source LLC
LaVergne TN
LVHW070446100526
838202LV00014B/1677